GW01034982

Groundwater Lowering in Construction

Groundwater Lowering in Construction

A practical guide

P. M. Cashman and M. Preene

London and New York

First published 2001 by Spon Press
11 New Fetter Lane, London EC4P 4EE

Simultaneously published in the USA and Canada
by Spon Press
29 West 35th Street, New York, NY 10001

Spon Press is an imprint of the Taylor & Francis Group

Typeset in 10/12 Sabon by Newgen Imaging Systems (P) Ltd.
Printed and bound in Great Britain by Biddles Ltd, Guildford and King's Lynn.

British Library Cataloguing in Publication Data
A catalogue record for this book is available from the British Library

Library of Congress Cataloging in Publication Data

Cashman, P.M. (Pat M.)
Groundwater lowering in construction: a practical guide/P.M. Cashman and M. Preene
p. cm.
Includes bibliographical references and index.
ISBN 0-419-21110-1 (alk. paper)
1. Earthwork. 2. Drainage. 3. Building sites. 4. Groundwater flow.
I. Preene, M. (Martin) II. Title.

TA715.C35 2001
624.1'5–dc21 2001032252

ISBN 0-419-21110-1

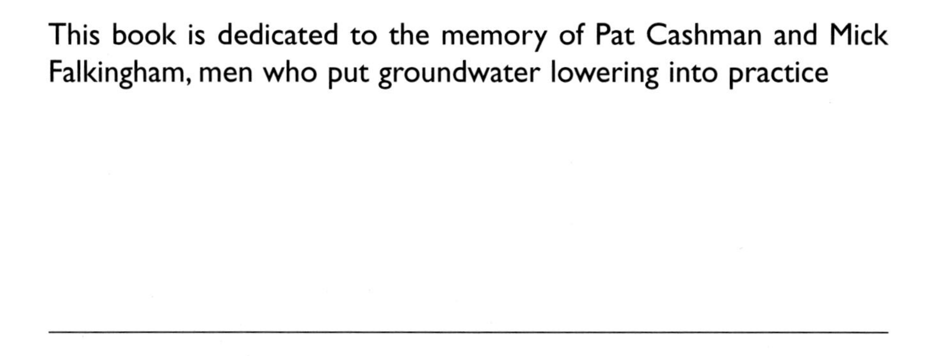
This book is dedicated to the memory of Pat Cashman and Mick Falkingham, men who put groundwater lowering into practice

Patrick Michael Cashman

Pat Cashman, the leading British exponent of groundwater control of his generation, died on 25th June 1996. For more than forty years, during the growth of soil mechanics into the practice of geotechnical engineering, Pat was responsible, through the organizations he ran and later as a consultant, for maintaining a practical and straightforward approach to the art of groundwater control. This book, the manuscript of which was well advanced at the time of his death, sets out that approach.

Following war service with the Royal Engineers, Pat Cashman graduated from the University of Birmingham, and joined Soil Mechanics Limited, soon transferring to the Groundwater Lowering Department, so beginning his lifelong interest in this field. He became head of the department in 1961 and later became responsible for the joint venture with Soletanche which introduced French techniques into the United Kingdom (UK). In 1969 he became contracts director for Soletanche (UK) Limited.

In 1972 Pat joined Groundwater Services Limited (later Sykes Construction Services Limited) as Managing Director. Over a ten year period he designed and managed a huge number and range of groundwater lowering projects. Commercial and financial success was achieved alongside technical innovation and practical advancements. In the 1980s he joined Stang Wimpey Dewatering Limited as Managing Director. Again he achieved commercial success as well as introducing American ideas into British practice. During this period Pat made a major contribution to the production of CIRIA Report 113 *Control of Groundwater for Temporary Works* – the first comprehensive dewatering guide produced in the UK in the modern era.

In 1986 he 'retired' and commenced an active role as a consultant, often working closely with Ground Water Control Limited, a contracting company formed by men who had worked for Pat during the Sykes years. His practical approach to problems meant that he was always in demand, particularly by contractors when they were in trouble. Although a practical man with a healthy suspicion of arcane theory and in particular computer modelling of problems, he took to the computer to document his own experience. This book is record of a singular approach to a challenging business.

Martin Preene

Contents

Figures

Tables

Acknowledgements

It is difficult to make fulsome and comprehensive acknowledgements to the many who have encouraged me and thereby helped me to persist with the task of writing this book. My wife has been a steadfast 'encourager' throughout its very lengthy gestation period. It was Daniel Smith who first persuaded me to compile this work. I have lost track of his present whereabouts but I hope that he will approve of the finished work. I am grateful for his initial stimulation.

I owe many thanks to my neighbour, Andy Belton. Without his patient sorting out of my computer usage problems, this text would never have been fit to send to a publisher. Andy will be the first to admit that this is not a subject that he has knowledge of, but having read some of the draft text and many of my faxes, on occasions he asked some very pertinent questions. These made me think back to first principles and so have – I am sure – resulted in improved content of this work.

I thank the many organizations that have so kindly allowed use of their material, photographs and diagrams. Where appropriate, due acknowledgement is made in the text beneath each photo or diagram. I hope that this will be acceptable to all who have helped me and given their permissions for their material to be reproduced; a complete list of each and every one would be formidable indeed: I do ask to be forgiven for not so doing. Following this philosophy it would be invidious to single out and name a few of the major assistors for fear of upsetting others.

P. M. Cashman, Henfield, West Sussex, 1996

At the time of Pat Cashman's death in 1996 the manuscript for this book was well advanced. I am grateful to William Powrie and the editorial team at Spon Press for encouraging me to complete his work. I would also like to acknowledge the support and contribution of David Hartwell, who contributed closely with Pat on an earlier draft of the text, and who has provided information and assistance to me during the preparation of the manuscript.

Valuable comments on parts of the text were provided by Lesley Benton, Rick Brassington, Steve Macklin, Duncan Nicholson, Jim Usherwood and Gordon Williams. I am also grateful to Toby Roberts for kindly writing the afterword.

Illustrations and photographs provided by others, or from published works, are acknowledged in the captions, and I would like to thank all those who provided such material and permissions. The libraries of the Institution of Civil Engineers and Ove Arup & Partners have also been of great assistance in the gathering together many of the references quoted herein.

Finally, I would like to thank my wife, Pam, for her invaluable and unstinting support throughout this project.

M. Preene, Wakefield, 2001

Chapter 1

Groundwater lowering

A personal view and introduction by P. M. Cashman

Many engineering projects, especially major ones, entail excavations into water-bearing soils. For all such excavations, appropriate system(s) for the management and control of the groundwater and surface water run-off, should be planned before the start of each project. In practice this can only be done with knowledge of the ground and groundwater conditions likely to be encountered by reference to site investigation data. The control of groundwater (and also surface water run-off) is invariably categorized as 'temporary works' and so almost always is regarded by the client and his engineer or architect as the sole responsibility of the contractor and of little or no concern of theirs. In many instances this philosophy has been demonstrated to be shortsighted and ultimately costly to the client.

Sometimes as work proceeds, the actual soil and groundwater conditions encountered may differ from what was expected. Should this happen, all concerned should be willing to consider whether to modify operations and construction methods as the work progresses and as more detailed information is revealed. Based upon this philosophy, I advocate, particularly for large projects, frequent 'engineering oriented' reappraisal meetings between client/owner and contractor or both (as distinct from 'cost oriented' meetings). This will afford the best assurance that his project will be completed safely, economically, and within a realistic programme time and cost.

On a few occasions I have been privileged to be involved in the resolution of some difficult excavation and construction projects when the engineer succeeded in persuading the client to share the below ground risks with the contractor. During the progress of the contract there were frequent engineering oriented meetings with the contractor, to discuss and mutually agree how to proceed. I believe that the engineers concerned with these complex projects realized that it would not be in the best interests of their client to adhere rigidly to the traditional view that the contractor must take all of the risk. They were enlightened, had a wealth of practical experience and so had a realistic awareness that the soil and groundwater conditions likely to be encountered were complex. Also they realized that the measures for effecting stable soil conditions during construction may not be straightforward.

The few occasions when I have experienced this joint risk sharing approach have, without exception in my view, resulted in sound engineering solutions to problems that needed to be addressed: they were resolved sensibly and the projects were completed within realistic cost to the client. Furthermore, claims for additional payments for dealing with unforeseen conditions were not pressed by the contractor.

I found these experiences most interesting and enlightening and I learned much by having direct access to different points of view of the overall project as distinct from my own view as a specialist contractor. I find it encouraging that in recent years the target form of contract – the client and the contractor sharing the risks of unforeseen conditions – is being implemented more frequently. Thereby the contractor is confident of a modest but reasonable profit and the client is not eventually confronted with a multitude of claims for additional payments, some of which may be spurious, but all requiring costly time consuming analysis and investigation.

There are three groups of methods available for temporary works control of groundwater:

(a) Lowering of groundwater levels in the area of construction by means of water abstraction, in other words – groundwater lowering or dewatering.
(b) Exclusion of groundwater inflow to the area of construction by some form of very low permeability cut-off wall or barrier (e.g. sheet-piling, diaphragm walls, artificial ground freezing).
(c) Application of a fluid pressure in confined chambers such as tunnels, shafts and caissons to counter balance groundwater pressures (e.g. compressed air, earth pressure balance tunnel boring machines).

This book deals in detail only with group (a) methods, the groundwater lowering and dewatering methods. The various methods of all three groups, their uses, advantages and disadvantages are only presented in summary form in Chapter 5.

Rudolf Glossop (1950) stated

> The term drainage embraces all methods whereby water is removed from soil. It has two functions in engineering practice: permanent drainage is used to stabilize slopes and shallow excavations; whilst temporary drainage is necessary in excavating in water bearing ground.

This book addresses the subject of temporary drainage only, though many of the principles are common to both temporary and permanent requirements.

The book is intended for use by the practical engineer (either contractor or consultant or client); but it is intended particularly for the guidance of the specialist 'dewatering practitioner' or advisor. In addition it is commended to the final year graduate or masters student reading civil engineering or engineering geology as well as to the civil engineering oriented hydrogeologist.

It is deliberately addressed to the practitioner involved in the many day-to-day small to medium scale dewatering projects for which a simplistic empirical approach is usually adequate. It is anticipated that the typical reader of this work will be one quite comfortable with this philosophy but one who is aware of the existence of – though perhaps wishing to avoid – the purist hydrogeologist philosophy and the seemingly unavoidable high level mathematics that come with it.

We, the writers and the readers, are pragmatic temporary works engineers – or, in the case of some readers, aspiring to be so – seeking the successful and economic completion of construction projects. For the small and medium size projects (which are our 'bread and butter') there seems little practical justification for the use of sophisticated and time consuming techniques, when simpler methods can give serviceable results. The analytical methods described in this book are based on much field experience by many practitioners from diverse countries and have thereby been proven to be practicable and adequate for most temporary works assessment requirements. I consider that Powers (1992) stated a great dewatering truism 'The successful practitioner in dewatering will be the man who understands the theory and respects it, but who refuses to let the theory overrule his judgement'.

Extensively, use is made of the Dupuit–Forcheimer analytical approaches. I am conscious that purists will question this simplistic approach. My riposte – based on some thirty or more years of dealing with groundwater lowering problems – is that in my experience and that of many others, this empirical philosophy has resulted in acceptably adequate pumping installations. Always provided, of course, that due allowance is made for the often limited reliability of ground and groundwater information available. It requires acquired practical field experience to assess the quality of the site investigation data. Whenever possible, reference should be made to other excavations in adjacent areas or in similar soil conditions to verify one's proposals. In the text there are some brief descriptions of a number of relevant case histories that I have dealt with in the past.

I readily acknowledge that for a groundwater lowering system design pertinent to large-scale and/or long-term projects – for example construction of a dry dock; a nuclear power station or an open cast mining project – more sophisticated methods of analysis will be appropriate. These can provide reassurance that the pragmatic solution is about right, but do we ever know the 'permeability value' to a similar degree of accuracy?

The underlying philosophy of this publication is to address the pragmatic approach. It follows that three questions arise:

- How does water get into the ground and how does it behave whilst getting there and subsequently behave whilst there?
- What is the inter-relationship between the soil particles and the groundwater in the voids between them?

- How to control groundwater and surface water run-off and so prevent it or them or both causing problems during excavation and construction?

A thorough site investigation should go a long way to providing the answers to these questions. Unfortunately, experience indicates that many engineers responsible for specifying the requirements for project site investigations consider only the *designer's* requirements and do not address the other important considerations, namely – *how can this be built?* Often the site investigation is not tailored to the obtaining of data pertinent to temporary works design requirements, nor of problems that may occur during construction.

The contractor should not expect always to encounter conditions exactly as revealed by the site investigation. Soils, due to the very nature of their deposition and formation are variable and rarely, if ever, isotropic and homogeneous, as is assumed in many of the analytical methods. The contractor should carry out the works using his professional skills and abilities and should be prepared to adjust, if changed circumstances are revealed as the work proceeds.

Throughout the planning, excavation and construction phases of each project, safety considerations must be of paramount importance. Regrettably, the construction industry historically has a poor safety performance record.

Let us consider another professional discipline! Hopefully no surgeon would contemplate commencing an operation on a patient without carrying out a thorough physical examination and having the information from X-ray; ECG; urine; blood and other test results and any pertinent scans, available to him beforehand. The surgeon will realize that these may not indicate everything and that during the operation complications may occur but the possibilities of such 'surprise' occurrences will have been minimized by having reliable site investigation data concerning his patient. Likewise, if the client's engineer/designer provides comprehensive ground and groundwater information at tender stage, the 'surprise' occurrences during construction should be minimized. An experienced contractor, with the cooperation of an experienced client/engineer should be able to agree how to adjust working techniques to deal adequately with the changed circumstances as or if they are revealed, and this at realistic final cost.

Structure of the rest of the book

At the commencement of each chapter the introduction acts as a summary of the subject matter to be covered therein. I hope that this approach will enable the reader to decide speedily which chapters he wishes to read forthwith when seeking guidance on how to deal with his individual requirements, and which chapters may be deferred till later.

Chapter 2 contains a brief historical review of the principal theories concerning seepage towards wells and of the technologies used to apply them. Many readers will probably consider this as superfluous and omit it from their initial reading. As their interest in this subject becomes further stimulated, I hope that they will turn back to it. I derived great pleasure when researching this aspect some years ago.

Chapter 3 contains a very brief summary of the hydrological cycle (i.e. how does water get into the ground?), together with a similarly brief summary concerning soils, their properties and permeability. The theoretical concepts of Darcy's law and the nature of flow to wells are introduced, which is much used in a practical sense later. Groundwater chemistry is also briefly discussed.

Chapter 4 discusses the mechanisms of instability problems in excavations. An understanding of these issues can be vital when selecting a groundwater control method.

Chapter 5 presents in summary form the principal features of the methods available for control of groundwater by exclusion and by pumping.

Chapter 6 addresses site investigation requirements, but only that specific to groundwater lowering, and does not detail the intricacies of the available methods of site investigation. This chapter also covers permeability determination in the field and laboratory. Guidance is also given on the relative reliability of permeability estimated by the various methods.

Chapter 7 describes various empirical and simple design methods for assessing the discharge flow rates required for groundwater lowering installations, and for determining the number of wells, etc. Several simple design examples are included in an appendix.

Chapters 8–11 address the various groundwater lowering methods: sump pumping; wellpointing; deepwells; and less commonly used methods.

Chapter 12 describes the types of pumps suitable for the various systems.

Chapter 13 deals with some associated side effects of groundwater lowering, including ground settlements and describes the method of artificial recharge which can be used to mitigate some of these side effects.

Chapter 14 presents appropriate methods of monitoring and maintenance to ensure that groundwater lowering systems operate effectively when they are first installed, and after extended periods of operation.

Chapter 15 covers safety, contractual and environmental issues (using British convention, but based on principles of good practice that are applicable in other countries).

Chapter 16 is a short afterword, looking at the future of groundwater lowering.

The book ends with a Glossary of Terms pertinent to the subject matter and also a summary of the Symbols and Notations used. Various appendices are included, providing detailed background information on various subjects.

Finally I reiterate my earlier statement: this book is intended for the guidance of practical engineers and those – many I hope! – desirous of joining us.

I trust that there are several new aspirants who will realize after reading this book that it is a challenging scientific field, and to some extent an applied art form as well, wherein the requirement for practical experience is paramount. No two sites have the same requirements – this is one of its fascinations!

References

Glossop, R. (1950). Classification of geotechnical processes. *Géotechnique*, **2**(1), 3–12.

Powers, J. P. (1992). *Construction Dewatering: New Methods and Applications*, 2nd edition. Wiley, New York, p. 11.

Chapter 2

The history of groundwater theory and practice

2.0 Introduction

Man has been aware of groundwater since prehistory, long before Biblical times. Over the centuries the mysteries of groundwater have been solved, and man has developed an increasing capability to manipulate it to his will. This chapter describes some of the key stages in the development of the understanding and control of groundwater. The history of some of the technologies now used for groundwater lowering is also discussed, especially in relation to early applications in the United Kingdom. Detailed knowledge of the history of groundwater control might not be considered essential for a practical engineer working today. Nevertheless, study of the past can be illuminating, not least by showing that even when theories are incomplete, and technology untried, the application of scientific principles and engineering judgement can still allow groundwater to be controlled.

2.1 The earliest times to the sixteenth century

The digging of wells for the exploitation of water and primitive implementation of water management dates back to Babylonian times and even earlier.

The source of the water flowing from springs and in streams was a puzzling problem, and the subject of much controversy and speculation. It was generally held that the water discharged from springs could not be derived from rain. Ingenious hypotheses were formulated to account for the occurrence of springs. Some early writers suggested large inexhaustible reservoirs while others recognized that there must be some form of replenishment of the supplying reservoirs. The Greek hypotheses, with many incredible embellishments, were generally accepted until near the end of the seventeenth century. The theory of rainfall infiltration was propounded by only a very few writers.

Though the Romans and other early cultures indulged in quite sophisticated water management projects by building imposing aqueducts to channel water from spring and other sources to centres of population, they had no understanding of the sources of replenishment of groundwater. The aqueducts of the Romans were remarkable, and showed great appreciation of the value of water, but their methods of measuring or estimating water quantities were crude. Generally they appear to have lacked any semblance of knowledge of either surface or groundwater hydrology.

According to Tolman (1937)

> Centuries have been required to free scientists from superstitions and wild theories handed down from the dawn of history regarding the unseen sub-surface water, ... even in this century there is still much popular superstition concerning underground water.

An elemental principle, that gravity controls the motion of water underground as well as at the surface, is not appreciated by all. A popular belief exists that 'rivers' of underground water pass through solid rock devoid of interconnected interstices and flow under intervening mountain ranges.

Marcus Vitruvius, a Roman architect who lived about the time of Christ, produced a book describing the methods of finding water. He wrote of rain and snow falling on mountains and then percolating through the rock strata and reappearing as springs at the foot of the mountains. He gave a list of plants and of other conditions indicative of groundwater such as colour and dampness of the soil and mists rising from the ground early in the morning. Vitruvius and two contemporaries of his, Cassiodorus and Plinz, were the first to make serious efforts to list practical methods for locating water, and this when geology was yet unknown!

2.2 The renaissance period to the nineteenth century

At the beginning of the sixteenth century Leonardo da Vinci directed his attention to the occurrence and behaviour of water. He correctly described the infiltration of rain and the occurrence of springs, and concluded that water goes from rivers to the sea and from the sea to the rivers and thus is constantly circulating and returning. About the same time Pallissy, a French Huguenot, presented a clear and reasoned argument that all water from springs is derived from rain.

The latter part of the seventeenth century was a watershed in the beginning of an understanding of the replenishment of groundwater. Gradually there arose the concept of an 'hydrological cycle'. This presumed that water was returned from the oceans by way of underground channels below the mountains. The removal of salt was thought to be either by distillation or by percolation and there were some highly ingenious theories of how water was raised up to the springs.

2.3 Progress from a qualitative to a quantitative science

It was in the seventeenth century that the quantitative science of hydrology was founded by Palissy, Pérrault and Mariotté in France, Ramazzini in Italy and the astronomer Halley in Britain.

Palissy, a sixteenth century potter and palaeontologist, stated that rain and melt snow were the only source of spring and river waters; and that rain water percolates into the earth, following 'a downward course until they reach some solid and impervious rock surface over which they flow until they make some opening to the surface of the earth.'

Pérrault made rainfall run-off measurements and demonstrated the fallacy of the long held view that the rainfall was not sufficient to account for the discharge from springs. He also measured and investigated evaporation and capillarity. Mariotté verified Pérrault's results and showed that the flow from springs increased in rainy weather and decreased in times of drought. Halley made observations of the rate of evaporation from the Mediterranean ocean and showed that this was adequate to supply the quantity returned to that sea by its rivers. His measurements of evaporation were conducted with considerable care but his estimates of stream flow were very crude.

Towards the end of the eighteenth century La Metherie extended the researches of Mariotté and brought them to the attention of meteorologists. He also investigated permeability and explained that some rain flows off directly (surface water run-off), some infiltrates into the top soil layers only and evaporates or feeds plants, whilst some rain penetrates underground whence it can issue as springs (i.e. infiltration or groundwater recharge). This is the first recorded mention of 'permeability' and so is the first link between hydrology and seepage to wells.

2.3.1 Seepage towards wells

The robust Newcomen engine greatly influenced mining practice during the eighteenth century but it was far too cumbersome for construction works. Generally speaking, until the early nineteenth century civil engineers, by the use of timber caissons and other devices, avoided pumping whenever possible. However, where there was no alternative pumping was usually done by hand – a very onerous task – using a rag and chain pump (Fig. 2.1), known also as 'le chaplet' (the rosary).

Some idea of the magnitude of the problem is given by de Cessart in his book *Oeuvres Hydrauliques*. Speaking of the foundations for the abutment of a bridge at Saumur in 1757, he says that forty-five chain pumps were in use, operated by 350 soldiers and 145 peasants. Work on this type of pump was, of course, most exhausting, and the men could only work in short spells.

Figure 2.1 Rag and chain pump, manually operated. 1556 (from Bromehead 1956). The balls, which are stuffed with horsehair, are spaced at intervals along the chain and act as one-way pistons when the wheel revolves.

Pryce, in his *Mineralogea Cornubiensis* in 1778, said that work on pumps of this sort led to a great many premature deaths amongst Cornish miners.

For permanent installations such as graving docks large horse-driven chaplet pumps were used. Perronnet, the famous French bridge builder, made use of elaborate pumping installations in the cofferdams for the piers of his larger bridges; for example, under-shot water wheels were used to operate both chaplet pumps and Archimedean screws.

According to Crèsy, writing in 1847, the first engineer to use steam pumps on bridge foundations was Rennie, who employed them on Waterloo Bridge in 1811. In the same year, Telford, on the construction of a lock on the Caledonian Canal at Clachnacarry, at first used a chain pump worked by six horses, but replaced it by a 9 horsepower steam engined pump. From then on steam pumps were used during the construction of all the principal locks on that canal. In 1825 Marc Brunel used a 14 horse-power steam engine when sinking the shafts for the Thames Tunnel. By this date steam pumping seems to have been the common practice for dealing with groundwater and so below ground excavations for construction were less problematical.

2.3.2 *Kilsby tunnel, London to Birmingham railway*

There appears to have been no important advance in pumping from excavations until the construction in the 1830s by Robert Stephenson, of the Kilsby Tunnel south of Rugby, on the London to Birmingham Railway. He pumped from two lines of wells sited parallel to and either side of the line of the tunnel drive (Fig. 2.2).

It is clear from Stephenson's own *Second Report to the Directors of the London, Westminster and Metropolitan Water Company*, 1841, that he was the first to observe and explain the seepage or flow of water through sand to pumped wells. The wells were sited just outside the periphery of the construction so as to lower the groundwater level in the area of the work by pumping from these water abstraction points. This is most certainly the first

Figure 2.2 Pumps for draining the Kilsby tunnel (from Bourne 1839: courtesy of the Institution of Civil Engineers Library). A pumping well is shown in the foreground, with the steam pumphouse in the distance.

temporary works installation of a deep well groundwater lowering system in Britain, if not in the world. The following extract (from Boyd-Dawkins 1898, courtesy of the Institution of Civil Engineers Library) quotes from the report and shows that Stephenson had understood the mechanism of groundwater flow towards a pumping installation.

> The Kilsby Tunnel, near Rugby, completed in the year 1838, presented extreme difficulties because it had to be carried through the water-logged sands of the Inferior Oolites, so highly charged with water as to be a veritable quicksand. The difficulty was overcome in the following manner. Shafts were sunk and steam driven pumps erected in the line of the tunnel. As the pumping progressed the most careful measurements were taken of the level at which the water stood in the various shafts and boreholes; and I was soon much surprised to find how slightly the depression of the water-level in the one shaft, influenced that of the other, notwithstanding a free communication existed between them through the medium of the sand, which was very coarse and open. It then occurred to me that the resistance which the water encountered in its passage through the sands to the pumps would be accurately measured by the angle or inclination which the surface of the water assumed towards the pumps, and that it would be unnecessary to draw the whole of the water off from the quicksands, but to persevere in pumping only in the precise level of the tunnel, allowing the surface of the water flowing through the sand to assume that inclination which was due to its resistance.
>
> The simple result of all the pumping was to establish and maintain a channel of comparatively dry sand in the immediate line of the intended tunnel, leaving the water heaped up on each side by the resistance which the sand offered to its descent to that line on which the pumps and shafts were situated.

As Boyd-Dawkins then comments:

> The result of observations, carried on for two years, led to the conclusion that no extent of pumping would completely drain the sands. Borings, put down within 200 yards [185 m] of the line of the tunnel on either side, showed further, that the water-level had scarcely been reduced after twelve months continuous pumping and, for the latter six months, pumping was at the rate of 1,800 gallons per minute [490 m^3/h]. In other words, the cone of depression did not extend much beyond 200 yards [185 m] away from the line of pumps.
>
> In this account, … it is difficult to decide which is the more admirable, the scientific method by which Stephenson arrived at the conclusion that the cone of depression was small in range, or the practical application of the results in making a dry [the authors would have

used the word 'workable' rather than 'dry'] pathway for the railway between the waters heaped up [in the soil] ... on either side.

It is astonishing that neither Robert Stephenson, nor any of his contemporaries, realized the significance of this newly discovered principle. That by sinking water abstraction points, and more importantly placing them clear of the excavation so that the flow of water in the ground will be away from the excavation rather than towards it – stable ground conditions were created. For many decades this most important principle was ignored.

2.3.3 *Early theory – Darcy and Dupuit*

In the 1850s and early 1860s Henri Darcy made an extensive study of the problems of obtaining an adequate supply of potable water for the town of Dijon. He is famous for his Darcy's law (Darcy 1856) postulating how to determine the permeability of a column of sand of selected grading, knowing the rate of water flow through it but this formed only a small part of his treatise. He compiled a very comprehensive report (two thick volumes) in which he analysed the available sources of water from both river and wells – some of them artesian – and how economically to harness all these for optimum usage.

Darcy investigated the then current volume of supply of water per day per inhabitant for about ten municipalities in Britain – Glasgow, Nottingham and Chelsea amongst others – as well as Marseille and Paris, and many other French towns. He concluded that the average water provision in Britain was 80–85 l/inhabitant/day and more than 60 l/inhabitant/day for Paris. Darcy designed the water supply system for Dijon on the basis of 150 l/inhabitant/-day – no doubt his Victorian contemporaries this side of 'la manche' would have applauded this philosophy.

In the mid-1860s Dupuit (1863), using Darcy's law to express soil permeability, propounded his equations for determining flow to a single well positioned in the middle of an island. Dupuit made certain simplifying assumptions and having stated them (i.e. truly horizontal flow) then discounted their implications! For this Dupuit has been much castigated by some later purists but most accept that the Dupuit concept, latter slightly modified by Forcheimer, is acceptable and adequate in many practical situations.

Exchange of information was not as simple in Darcy's and Dupuit's time as it is now. Much of the fundamental work of these two French engineers was duplicated by independent developments shortly afterwards in Germany and Austria and a little later, in the United States.

About 1883 Reynolds demonstrated that for linear flow – that is, flow in orderly layers – commonly known as laminar flow, there is a proportionality existing between the hydraulic gradient and the velocity of flow. This is in keeping with Darcy's Law but as velocity increased the pattern of flow becomes irregular (i.e. turbulent) and the hydraulic gradient approached the square of the velocity. Reynolds endorsed the conclusion that Darcy's Law gives an acceptable representation of the flow within porous media – that is the flow through the pore spaces of soils will remain laminar, save for very rare and exceptional circumstances. However, this may not always be true of flow through jointed rocks (e.g. karstic limestone formations).

2.4 Later theoretical developments

In his Rankine Lecture Glossop (1968) suggested that 'classical soil mechanics' was founded on the work of Terzaghi which dates from his first book published in 1925 and that it is strongly influenced by geological thinking. In 1913 Terzaghi published a paper dealing with the hydrology of the Karst region after studying the geology for a hydro-electric scheme in Croatia. He soon realized that geology would be of far more use to engineers if the physical constants of rocks and soils were available for design (Terzaghi 1960). This was the first positive marrying of civil engineering and geology.

2.4.1 Verifications and modification of Darcy

In 1870, a civil engineer in Dresden, Adolph Theim reviewed Darcy's experiments and went on to derive the same equations as Dupuit for gravity and artesian wells. Theim was the first to collect extensive field data in support of his and Dupuit's equations. Thus, he was the first researcher to apply practical field experience to test the validity of the analytical pronouncements. This fundamentally practical philosophy was to be typical of Terzaghi and other later pioneers of soil mechanics, where reliable field measurements were essential to verify assumptions.

The next contributor to the advance of groundwater flow theory was an Austrian engineer, Forcheimer. In the late nineteenth century he published his first paper on flow towards wells (Forcheimer 1886). Over a period of half a century he published many contributions to this field of technology. Based on an analogy with heat flow he developed the use of flow nets in tracing the flow of water through sands. His results were published in 1917 and certainly influenced Terzaghi. Also, Forcheimer undertook the analysis of gravity flow towards a group of wells and introduced the concept of a hypothetical equivalent single well. This concept, later endorsed by Weber, is of great practical importance to this day in analysing well systems in connection with civil engineering projects for construction temporary works.

In the 1950s an Australian research worker, Chapman (1959), investigated the problem of analysing long lines of closely-spaced wells or wellpoints. Building on earlier theoretical work by Muskat (1935) who studied plane seepage through dams, Chapman's modelling and analytical work produced solutions for single and double lines of wells or wellpoints. These solutions are still widely used today for the analysis of wellpoint systems for trench excavations.

2.4.2 Non-steady state flow

An important aspect of the theory of flow towards wells upon which the various investigators in Europe were relatively silent, was the non-steady state. The first European investigator to produce anything of importance was Weber (1928). His work is still one of the most complete treatments of the subject. Weber, like Forcheimer before him, was very thorough in gathering field performance data and correlating it with his theoretical analyses.

In 1942 Meinzer, Chief of the Ground Water Division of the United States Geological Survey edited a comprehensive outline on the development of fieldwork and theoretical analysis in groundwater hydrology up to that time (Meinzer 1942). His contributions had extended already over more than a decade. It was he who took on engineers, physicists and mathematicians as well as geologists to undertake the challenging responsibilities delegated to his division. Meinzer is regarded by many as the first hydrogeologist, and in this specialist field is considered as its 'father' in like manner that Terzaghi is considered as the 'father' of soil mechanics. Theis is probably the best known of many of Meinzer's protégés. Theis approached the treatment of non-steady state flow from a different angle to that of Weber. His conceptual approach (Theis 1935) is now almost universally used as the basis for non-steady state flow analysis.

During the early 1930s Muskat, chief physicist for the Gulf Research and Development Company, was the leader of a team who compiled a comprehensive and scholarly volume treating all phases of flow of fluids through homogeneous porous media (Muskat 1937). The work was concerned primarily with the problems involved in flow of oil and oil–gas mixtures through rocks and sands. Muskat's approach was consistently that of a theoretical physicist rather than that of a field engineer. His appraisals of analytical methods of test procedures tended to be more of the scientific purist than of one concerned with the practical pragmatic needs in the field. However, the accomplishments of Muskat and his colleagues have been of immeasurable value to an understanding of seepage flow.

Muskat made much use of electric analogue models and of sand tank models. He investigated the effects of stratification and anisotropy and developed the transformed section. He made extensive studies of multiple confined wells. He seems to have given much serious thought to the limited validity of the Dupuit equations and pronounced 'Their accuracy, when valid, is a lucky accident'. No one before or since has been so intolerant of Dupuit. Boulton (1951) and many other investigators do not subscribe to Muskat's dismissiveness of the practical usefulness of the Dupuit–Forcheimer approach.

2.5 Groundwater modelling

In a pedantic context, groundwater modelling involves the use of models or analogues to investigate or simulate the nature of groundwater flow. In modern parlance, groundwater modelling invariably refers to numerical models run on computers. In fact these are not true models but iterative mathematical solutions to a model or mesh is defined by the operator; a solution is considered to be acceptable when the errors reach a user-defined level. These models have been developed by many organizations from original esoteric research tools into the current generation of easier to use models with excellent presentation of results which, when used appropriately, can clearly demonstrate what is happening to the groundwater flow. The use of numerical modelling is described in Section 7.7.

The origins of groundwater models and analogues are to be found, not in groundwater theory, but in other scientific fields such as electricity and heat flow. In the mid-nineteenth century Kirkoff studied the flow of electrical current in a thin disk of copper. Just who recognized that the mathematical expressions or laws governing the flow of electricity were analogous to those governing the flow of thermal energy and groundwater is not clear.

This dawning of the electrical analogy to groundwater flow was given a major impetus in North America where the growing interest in the theoretical aspects of oil reservoir development led to major developments. As noted earlier, the significant contributions of Muskat during the 1930s included electrical analogues and sand tank models, Wyckoff, who had worked with Muskat, published in 1935 the first conductive paper model study of groundwater flow through a dam. The use of conductive paper was a direct development from Kirchoff's original work, nearly a century earlier, and became the practical basis for much of the two-dimensional isotropic analogue modelling for the next fifty years before numerical models finally superseded it. The conductive paper technique was given a further boost with the development of teledeltos paper in 1948. This aluminium coated carbon based paper enabled rapid two-dimensional studies to be undertaken.

Karplus (1958) presents an extensive discussion on the use and limitations of conductive paper models, as well as introducing resistance networks where the uniform conducting layer is replaced by a grid of resistors. The advantage of this system was that it could be developed into three dimensions and so the problems of making measurements within a solid were overcome. Herbert and Rushton (1966) developed resistance networks to introduce switching techniques to determine a free water surface, transistors to simulate storage and also evolved time variant solutions. Case histories of the practical application of resistance networks have been published for pressure relief wells under Mangla dam in Pakistan (Starr *et al.* 1969) and for a deep well system at Sizewell B Power Station in England (Knight *et al.* 1996).

Other analogues include: electrolytic tanks where a thin layer of conducting fluid is used instead of the conducting paper; or a stretched rubber membrane which, when distorted at right angles to the surface, forms a shape analogous to the cone of depression formed by a well. Neither of these techniques or others appear to have been used extensively for the solution of practical groundwater problems.

A different class of models are the sand tank type (see Cedergren 1989). These physical models consist of two parallel plates of clear material such as glass, closely spaced and filled with sand to represent the aquifer. Physical impediments to flow, such as dams or wells can be inserted and the model saturated. By injecting a tracer or dye on the upstream side the flow path(s) can be observed. The viscous flow or Hele-Shaw technique – so named after the man who first used this technique in England – is a variation to the sand tank. These techniques have been used regularly to demonstrate the form of groundwater flow, particularly beneath dams but have limited application to groundwater lowering in construction.

2.6 Early dewatering technology in Britain

Much of the technology which forms the basis of modern dewatering methods was developed in the United States or Germany and was introduced to Britain in the early part of the twentieth century. Up to that time any groundwater lowering required was achieved by the crude (but often effective) methods of pumping from timbered shafts or from open-jointed subdrains laid ahead and below trench or tunnel works.

The wellpoint method is probably the oldest of the modern techniques. Originally, wellpoints were a simple form of driven tube, developed by Sir Robert Napier on his march to Magdala in 1896, during his Abyssinian campaign. Each Abyssinian tube, as they were known, was driven to depth using a sledgehammer. If water was found the tube was then equipped with a conventional village type hand pump. According to Powers (1992), wellpoints were used for dewatering in North America from 1901, but the modern form of the method probably derives from equipment developed by Thomas Moore in New Jersey in 1925. His equipment was an advance in that installation was by jetting to form a clean hole which was backfilled with filter sand. His system, known as 'Moretrench' equipment, is identical in principle to that used today – indeed the Moretrench American Corporation is still in the dewatering business more than seventy-five years after Thomas Moore's original innovation.

In Britain, the civil engineer H. J. B. Harding (later Sir Harold) was one of the leading practitioners in the new art of geotechnology, which included groundwater lowering. In the 1930s, working for John Mowlem & Company, Harding was contractor's agent on the Bow-Leyton extension of

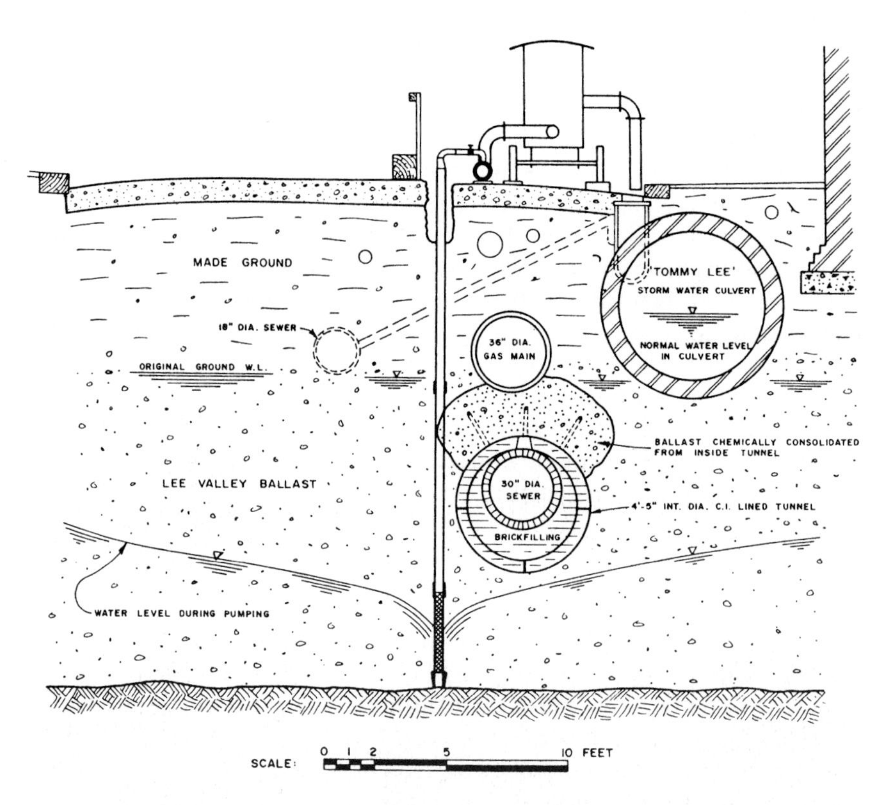

Figure 2.3 Sewer diversion under gas main using wellpoints and chemical injections (from Harding 1981: courtesy of Golder Associates). This shows an early British wellpoint application on the Bow-Leyton extension of the London Underground central line in the 1930s. The dewatering allowed tunnel works beneath an existing gas main.

the London Underground central line. Here he managed to acquire one of the first sets of Moretrench equipment to enter Britain and used it on sewer diversion work (Fig. 2.3). Harding and his colleagues became expert in the method and contributed to the development of British alternatives to the American equipment. One unexpected result of Harding's expertise was that, during the Second World War, Harding often assisted Royal Engineer bomb disposal units with dewatering for excavations in the search for unexploded bombs (Harding 1981).

In the early 1930s Harding was also instrumental in the introduction of the modern deep well method to Britain. Mowlem, with Edmund Nuttall Sons & Company, were awarded the contract to construct the King George V graving dock at Southampton to accommodate the liner Queen Mary which was being built on Clydeside. The dock was to be 100 ft (31 m) deep. The

Docks Engineer at Southampton was, according to Harding (1981), 'a wise and experienced man, he carried out his site investigation to unusual depths'. This revealed beds of Bracklesham sands containing water under artesian head which would reach above ground level.

At the time large deep wells using the recently developed submersible pump had been used for groundwater lowering in Germany from 1896 onwards, initially for the construction of the Berlin U-Bahn underground railway, but were not a method recognized in Britain. In 1932 Mowlem had obtained licencing agreements with Siemens Bau-Union to use their patents for, among other methods, groundwater lowering by deep wells, with Harding as the nominated British expert. This method was used at Southampton and

Figure 2.4 Early application of deep wells in Britain. A submersible pump is being prepared for installation into a well on site in the 1940s.

groundwater levels in the deep artesian aquifer were successfully lowered by ten deep wells, each equipped with a Siemens submersible pump, run from a central control (McHaffie 1938; Harding 1938). This was probably the first rational application of the deep well method in Britain since that by Robert Stephenson for the construction of the Kilsby tunnel a century earlier.

Further applications of deep well method with submersible pumps followed in the 1940s (Fig. 2.4), now unencumbered by licence agreements, and the method became an established technique for the control of groundwater (Harding 1946; Glossop and Collingridge 1948). The pioneering practical dewatering work on the Mowlem contracts was continued by Mowlem's subsidiary company, Soil Mechanics Limited, whose groundwater lowering department carried out numerous large-scale projects on power stations and docks in the 1950s and 1960s.

One of the most recent techniques to be introduced into the United Kingdom is the ejector dewatering method. Jet pumps, which are the basis of the ejector method, were first proposed in the 1850s by Thomson (1852) for the removal of water from water wheel sumps. Dewatering systems using ejectors were developed in the United States almost a century later based on jet pumps used in water supply wells (Prugh 1960; Werblin 1960). The ejector method does not seem to have been much employed in the United Kingdom until a few decades after its introduction in North America, although a small-scale ejector system was used in England in 1962 to dewater the Elm Park Colliery drift (Greenwood 1988). During the 1980s British engineers and contractors used ejectors on projects in Asia such as the HSBC headquarters in Hong Kong (Humpheson *et al.* 1986) and at the Benutan dam in Brunei (Cole *et al.* 1994). However, it was not until the late 1980s that a large-scale ejector system was used in the United Kingdom, for the casting basin and cut and cover sections of the river Conwy crossing project in North Wales (Powrie and Roberts 1990).

2.7 Practical publications

As described in Chapter 1, the control of groundwater is a practical problem, where theory is only part of the picture – how the theory is put into practice is vital. Historically, the best practical guidance came from in-house dewatering manuals produced by companies such as Geho Pumpen in Holland or the Moretrench American Corporation in North America. One of the first widely published more practical dewatering texts was Mansur and Kaufman (1962) which formed a chapter of the book *Foundation Engineering* edited by G. A. Leonards. This is a detailed statement-of-the-art of the time, with a strong bias towards the analytical but with some reference to practical considerations. Although it may seem a little dated, this book is essential reading for all who aspire to be specialist dewatering practitioners, and wish to understand some of the more accessible theory.

In the 1980s some useful publications became available. First, in 1981 J. P. Powers, then of the Moretrench American Corporation, produced his book *Construction Dewatering: A Guide to Theory and Practice* which is understandably oriented toward American practice. Updated as Powers (1992), this remains a thorough and readable book. In 1986 the Construction Industry Research and Information Association (CIRIA) produced Report 113 *Control of Groundwater for Temporary Works* (Somerville 1986), which was based on British practice. It was aimed at the non-specialist engineering designer and site staff, and drew on the expertise of experienced dewatering engineers. In the late 1990s CIRIA produced another report, covering the subject in more detail: Report C515 *Groundwater Control – Design and Practice* (Preene *et al.* 2000). Useful information and case histories can sometimes be found in groundwater conference proceedings. In the 1980s and 1990s relevant conferences were held on *Groundwater in Engineering Geology* (Cripps *et al.* 1986), *Groundwater Effects in Geotechnical Engineering* (Hanrahan *et al.* 1987), and *Groundwater Problems in Urban Areas* (Wilkinson 1994).

This book is intended to compliment and augment these texts, and will concentrate, in the main, on the practical requirements for groundwater lowering for temporary works.

References

Bourne, J. C. (1839). *Drawings of the London and Birmingham Railway*. Collection of the Institution of Civil Engineers library, London.

Boulton, N. S. (1951). The flow pattern near a gravity well in a uniform water bearing medium. *Journal of the Institution of Civil Engineers*, **36**, 534–550.

Boyd-Dawkins, W. (1898). On the relation of geology to engineering, James Forrest lecture. *Minutes of the Proceedings of the Institution of Civil Engineers*, **134**, Part 4, 2–26.

Bromehead, C. N. (1956). Mining and quarrying in the seventeenth century. *A History of Technology* (Singer, C., Holmyard, E. J., Hall, A. R. and Williams, T. L., eds), Volume 2. Oxford University Press, Oxford, pp 1–40.

Cedergren, H. R. (1989). *Seepage, Drainage and Flow Nets*, 3rd edition. Wiley, New York.

Chapman, T. G. (1959). Groundwater flow to trenches and wellpoints. *Journal of the Institution of Engineers, Australia*, 275–280.

Cole, R. G., Carter, I. C. and Schofield, R. J. (1994). Staged construction at Benutan Dam assisted by vacuum eductor wells. *Proceedings of the 18th International Conference on Large Dams*, Durban, South Africa, pp 625–640.

Cripps, J. C., Bell, F. G. and Culshaw, M. G. (eds) (1986). *Groundwater in Engineering Geology*. Geological Society Engineering Geology Special Publication No. 3, London.

Darcy, H. (1856). *Les Fontaines Publique de la Ville de Dijon*. Dalmont, Paris.

Dupuit, J. (1863). *Etudes Théoretiques et Practiques sur les Mouvement des Eaux dans les Canaux Decouverts et a Travers les Terrains Permeable*. Dunod, Paris.

Forcheimer, P. (1886). Uber die ergibigkeit von brunnen-anlagen und sickerschitzen. *Der Architekten-und Ingenieur-Verein*, **32**, No. 7.

Glossop, R. (1968). The rise of geotechnology and its influence on engineering practice. *Géotechnique*, **18**(2), 105–150.

Glossop, R. and Collingridge, V. H. (1948). Notes on groundwater lowering by means of filter wells. *Proceedings of the 2nd International Conference on Soil Mechanics and Foundation Engineering*, Rotterdam, **2**, pp 320–322.

Greenwood, D. A. (1988). Sub-structure techniques for excavation support. *Economic Construction Techniques*. Thomas Telford, London, pp 17–40.

Hanrahan, E. T., Orr, T. L. L. and Widdis, T. F. (eds) (1987). *Groundwater Effects in Geotechnical Engineering*. Balkema, Rotterdam.

Harding, H. J. B. (1938). Correspondence on Southampton docks extension. *Journal of the Institution of Civil Engineers*, **9**, 562–564.

Harding. H. J. B. (1946). The principles and practice of groundwater lowering. *Institution of Civil Engineers*, Southern Association.

Harding, H. J. B. (1981). *Tunnelling History and my Own Involvement*. Golder Associates, Toronto.

Herbert, R. and Rushton, K. R. (1966). Groundwater flow studies by resistance networks. *Géotechnique*, **16**(1), 53–75.

Humpheson, C., Fitzpatrick, A. J. and Anderson, J. M. D. (1986). The basements and substructure for the new headquarters of the Hongkong and Shanghai Banking Corporation, Hong Kong. *Proceedings of the Institution of Civil Engineers*, **80**, Part 1, 851–883.

Karplus, W. J. (1958). *Analog simulation*. McGraw-Hill. London.

Knight, D. J., Smith, G. L. and Sutton, J. S. (1996). Sizewell B foundation dewatering – system design, construction and performance monitoring. *Géotechnique*, **46**(3), 473–490.

Mansur, C. I. and Kaufman, R. I. (1962). Dewatering. *Foundation Engineering* (G. A. Leonards, ed.), McGraw-Hill, New York, pp 241–350.

McHaffie, M. G. J. (1938). Southampton docks extension. *Journal of the Institution of Civil Engineers*, **9**, 184–219.

Meinzer, O. E. (ed.) (1942). *Hydrology*. McGraw-Hill, New York.

Muskat, M. (1935). The seepage of water through dams with vertical faces. *Physics*, **6**, p 402.

Muskat, M. (1937). *The Flow of Homogeneous Fluids Through Porous Media*. McGraw-Hill, New York.

Powers, J. P. (1992). *Construction Dewatering: New Methods and Applications*, 2nd edition. Wiley, New York.

Powrie, W. and Roberts, T. O. L. (1990). Field trial of an ejector well dewatering system at Conwy, North Wales. *Quarterly Journal of Engineering Geology*, **23**, 169–185.

Prugh, B. J. (1960). New tools and techniques for dewatering. *Journal of the Construction Division, Proceedings of the American Society of Civil Engineers*, **86**, CO1, 11–25.

Preene, M., Roberts, T. O. L., Powrie, W. and Dyer, M. R. (2000). *Groundwater Control – Design and Practice*. Construction Industry Research and Information Association, CIRIA Report C515, London.

Somerville, S. H. (1986). *Control of Groundwater for Temporary Works*. Construction Industry Research and Information Association, CIRIA Report 113, London.

Starr, M. R., Skipp, B. O. and Clarke, D. A. (1969). Three-dimensional analogue used for relief well design in the Mangla Dam project. *Géotechnique*, **19**(1), 87–100.

Terzaghi, K. (1960). Land forms and subsurface drainage in the Gacka region in Yugoslavia. *From Theory to Practice in Soil Mechanics: Selections from the Writings of Karl Terzaghi*. Wiley, New York.

Theis, C. V. (1935). The relation between the lowering of the piezometric surface and the rate and duration of discharge of a well using groundwater storage. *Transactions of the American Geophysical Union*, **16**, 519–524.

Thomson, J. (1852). On a jet pump or apparatus for drawing water up by the power of a jet. *Report of the British Association*, 130–131.

Tolman, C. F. (1937). *Ground Water*. McGraw-Hill, New York.

Weber, H. (1928). *Die Reichweite von Grundwasserabsenkungen Mittels Rohrbunnen*. Springer, Berlin.

Werblin, D. A. (1960). Installation and operation of dewatering systems. *Journal of the Soil Mechanics and Foundations Division, Proceedings of the American Society of Civil Engineers*, **86**, SM1, 47–66.

Wilkinson, W. B. (ed.) (1994). *Groundwater Problems in Urban Areas*. Thomas Telford, London.

Chapter 3

Groundwater and permeability

3.0 Introduction

To allow even a basic approach to the control of groundwater, the practitioner should be familiar with some of the principles governing groundwater flow. Similarly, the specialist terms and language used must be understood. This chapter describes briefly the background to the flow of water through the ground. Chapter 4 will describe how groundwater flow affects the stability of construction excavations.

This chapter outlines the hydrological cycle and introduces the concepts of aquifers and permeability (a measure of how easily water can flow through a porous mass). Darcy's law – which is used to describe most groundwater flow regimes – is described, and typical values of permeability are presented (the problems of obtaining meaningful permeability values will be covered in Chapter 6). Since this is a practical text, our main interest is in abstracting water from the ground by means of wells or sumps of one kind or another. Accordingly, the principles of groundwater flow to wells are presented.

The importance of aquifer types and geological structure, both in the large and small scale, on groundwater flow is discussed. Two geological case histories are presented. Finally, this chapter discusses basic groundwater chemistry, and its relation to groundwater lowering problems.

3.1 Hydrology and hydrogeology

The study of water, and its occurrence in all its natural forms, is called 'hydrology'. This branch of science deals with water, its properties and all of its behavioural phases. It embraces geology, soil mechanics, meteorology and climatology, as well as hydraulics and the chemistry and bacteriology of water.

Terminology may vary from country to country, but in general those professionals who specialize in the hydrology of groundwater are known as 'hydrogeologists'. The study and practice of hydrogeology is of increasing importance both in developed and developing countries where groundwater is used as a resource, to supply water for the population and industries; for example, around one-third of the United Kingdom's drinking water is obtained from groundwater. The reader interested in hydrogeology is commended to an introductory text by Price (1996), and more theoretical treatments by Fetter (1994) or Todd (1980) among the copious literature on the subject.

For the dewatering practitioner detailed knowledge of some of the more arcane areas of hydrogeology is not necessary. However, if a rational approach is to be adopted, it is vital that the basic tenets of groundwater flow are understood in principle. Further study, beyond basic principles will require some understanding of higher mathematics and analysis – this can be very useful, but will not be for everyone. The references given at the end of this chapter should allow interested and motivated readers to pursue the subject as far as they wish.

It is worth mentioning that there is sometimes a little professional friction between dewatering practitioners (generally from a civil engineering background) and hydrogeologists (often with a background in earth sciences). For their part the dewatering engineers are often pragmatic. They have an unstable excavation and need to control groundwater accordingly – they see it very much as a local problem. Hydrogeologists, on the other hand, are trained to view groundwater as part of the wider environment. They see that a groundwater lowering system should not be considered in isolation – but sometimes they may not fully appreciate the practical limits of available dewatering methods. Like many professional rivalries, the wise observer can learn from both camps and gain a more rounded view of the problem in hand. This is the approach that the authors wish to encourage. Some of the misunderstandings between dewatering practitioners and hydrogeologists have resulted from slightly different terminology used by each group. This book unashamedly prefers the engineering terms used in dewatering practice, but introduces and explains alternative forms where appropriate.

3.1.1 The hydrological cycle

The hydrological cycle is now so widely accepted that it is difficult to conceive of any other concept. The hydrological cycle is based on the premise that the volume of water on earth is large, but finite. Most of this water is in continuous circulation. Water vapour is taken up into the atmosphere from surface water masses (principally the oceans) to form clouds; later, cold temperatures aloft cause the water to fall as precipitation (dew, fog, rain, hail or snow) on the earth's surface from whence it is eventually returned to the oceans (Fig. 3.1), from where the cycle is repeated.

Soil and rock are made up of mineral particles in contact with each other. The soil and rock masses contain voids, either widely distributed in the form

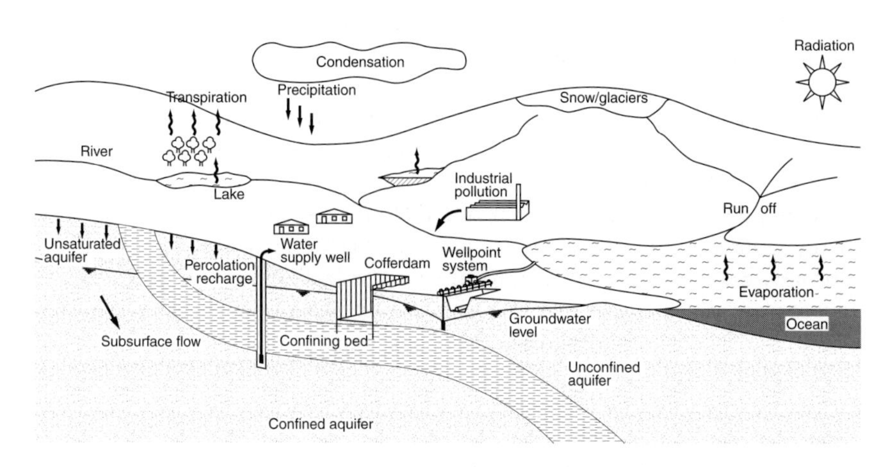

Figure 3.1 The hydrological cycle (from Preene *et al.* 2000: reproduced by kind permission of CIRIA).

of pores, or locally concentrated as fractures or fissures. In simple terms the water contained in the voids of the soil and rock is known as groundwater.

Generally a 'water table' will exist at some depth below ground surface; below the water table the soil and rock pores and fissures are full of water (and are said to be saturated). Above the water table the pores and fissures are unsaturated (i.e. they contain both water and air). The hydrological cycle illustrates clearly that precipitation replenishes the groundwater in the voids of the soils and rocks and that gravity flow causes movement of groundwater through the pores or fissures of soil and rock masses, towards rivers and lakes and eventually to the oceans.

It is unlikely that all of the precipitation that falls on landmasses will percolate downwards sufficiently far to replenish groundwater.

(a) Some precipitation may fall onto trees or plants and evaporate before ever reaching the ground.
(b) Some precipitation will contribute to surface run-off. This proportion will depend mainly upon the composition of the ground surface. For instance an urban paved surface area will result in almost total surface run-off, with little percolation into the ground.
(c) Of that precipitation which does enter the ground, not all of it will percolate sufficiently far to reach the water table. Some will be taken up by plant roots in the unsaturated zone just below the ground surface and will be lost as evapotranspiration from the plants.

One unavoidable conclusion from the hydrological cycle is that most groundwater is continually in motion, flowing from one area to another.

Under natural conditions (i.e. without interference by man in the form of pumping from wells), groundwater flow is generally relatively slow, with typical velocities being in the range of a few metres per day in high permeability soils and rocks down to a few millimetres per year in very low permeability deposits. Larger groundwater velocities generally only exist in the vicinity of pumped wells or sumps.

The stability of the hydrological cycle can be illustrated by considering groundwater temperature. Price (1996) states that in temperate climates the temperature of groundwater below a few metres depth tends not to vary with the seasons and remains remarkably constant at around the mean annual air temperature. In the United Kingdom this is around 10–12°C. This is obvious on site if you have to take a water sample from the pumped discharge. On a frosty winter morning the pumped groundwater will seem pleasantly warm on your hands, but take a similar sample on a warm summer day and the discharge water will seem icy cold. At greater depths in bedrock the temperature stays constant with time, but increases with depth. This is known as a geothermal temperature gradient, mainly the result of heat generated by the decay of radioactive materials in the rock.

3.1.2 Geology and soil mechanics

The successful design and application of groundwater lowering methods depends not just on the nature of the groundwater environment (such as where the site is within the hydrological cycle), but is also critically influenced by the geology or structure of the soils and rocks through which water flows. This is especially true when trying to assess the effect of groundwater conditions on the stability of engineering excavations (as will be discussed in Chapter 4). The eminent soil mechanics engineer, Karl Terzaghi wrote in 1945 'It is more than mere coincidence that most failures have been due to the unanticipated action of water, because the behaviour of water depends, more than anything else, on minor geological details that are unknown' (Peck 1969). This is still true today, but perhaps the last sentence should to be changed to 'minor geological details that are often overlooked during site investigation' – good practice for site investigation will be discussed in Chapter 6.

The reader wishing to apply groundwater lowering methods will need to be familiar with some of the aspects of hydrogeology, soil mechanics (including its applied form, geotechnical engineering) and engineering geology; texts by Powrie (1997) and Blyth and De Frietas (1984) respectively, are recommended as introductions to the latter two subjects. However, as has been stated before, study of theory is only part of the learning process. Many soils and rocks may not match the idealized homogeneous isotropic conditions assumed for some analytical methods. Judgement will be needed to determine how credible are the results of analyses based on such models.

An important geological distinction is made between uncemented 'drift' deposits (termed 'soil' by engineers) and cemented 'rock'.

Drift deposits are present near the ground surface and consist of sands, gravels, clays and silts, which may have resulted from weathering or from glacial or alluvial processes. Groundwater flow through drift deposits is predominantly intergranular – that is through the pores in the mass of the soil.

Rock may be exposed at the ground surface or may be covered by layers of drift and can consist of any of the many rock types that exist, such as sandstone, mudstone, limestone, basalt, gneiss, and so on. Groundwater flow through rock is often not through pores (which may be too small to allow significant passage of water) but along fissures, fractures and joints within the rock mass. This means that groundwater flow through rock may be concentrated locally where the deposition and subsequent solution or tectonic action has created fissure networks.

Because of their uncemented nature, drift deposits can cause severe stability problems for excavations below the groundwater level. Most groundwater lowering operations are carried out in drift, and this book will concentrate on those situations. Groundwater lowering is occasionally carried out in rock (see Section 4.5).

3.2 Permeability and groundwater flow

Permeability is a critical parameter for the assessment of how water flows through soil and rocks.

The precise meaning of the term 'permeability' is sometimes a cause of confusion between engineers and hydrogeologists. Civil and geotechnical engineers are interested almost exclusively in the flow of water through soils and rocks and use the term 'coefficient of permeability', given the symbol k. For convenience k is generally referred to simply as 'permeability', and throughout this book this terminology will be used. Therefore, for groundwater lowering purposes permeability, k, will be defined as 'a measure of the ease or otherwise with which groundwater can flow through the pores of a given soil mass'.

A slight complication is that the permeability of a porous mass is dependent not only on the nature of the porous media, but also on the properties of the permeating fluid. In other words, the permeability of a soil to water is different from the permeability of the soil to another fluid, such as air or oil. Hydrogeology references highlight this by calling the engineer's permeability 'hydraulic conductivity' to show that it is specific to water. Often, if the term 'permeability' appears in hydrogeology references it actually means the permeability of the porous media independent of the permeating fluid, sometimes also known as intrinsic permeability (see Price 1996: 52).

3.2.1 Darcy's law

The modern understanding of flow of groundwater through permeable ground originates with the researches of the French hydraulics engineer, Henri Darcy (1856), see Section 2.3. He investigated the purification of water by filtration through sand and developed an equation of flow through a granular medium based on the earlier work of Pouseuille concerning flow in capillary tubes. His conclusions can be expressed algebraically as equation (3.1), universally referred to as Darcy's law, which forms the basis for most methods of analysis of groundwater flow.

$$Q = kA\left(\frac{dh}{l}\right) \tag{3.1}$$

where (see Fig. 3.2 which schematically shows Darcy's experiment) Q is the volumetric flow of water per unit time (the 'flow rate'); A is the cross sectional area through which the water flows; l is the length of the flow path between the upstream and downstream ends; dh is the difference in total hydraulic head between the upstream and the downstream ends; and k is the permeability of the porous media through which the water flows.

The total hydraulic head at a given point is the sum of the 'pressure head' and the 'elevation head' at that point, as shown in Fig. 3.3. The elevation head is the height of the measuring point above an arbitrary datum, and the pressure head is the pore water pressure u, expressed as metres head of water. Total hydraulic head is important because it controls groundwater flow. Water will flow from high total hydraulic head to low total hydraulic head. It follows that water does not necessarily flow from high pressure to low pressure or from high elevations to low elevations – it will only flow in

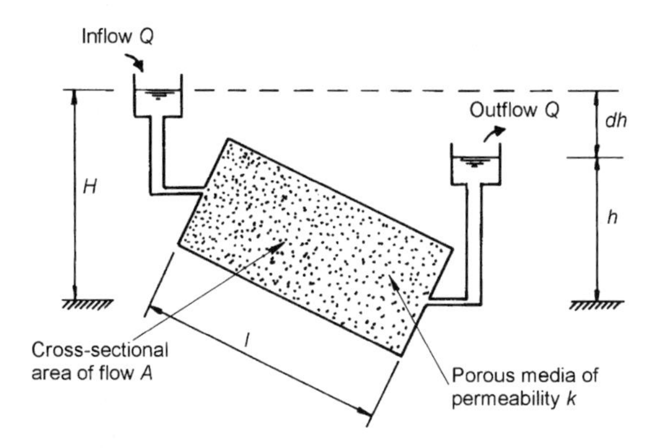

Figure 3.2 Darcy's experiment.

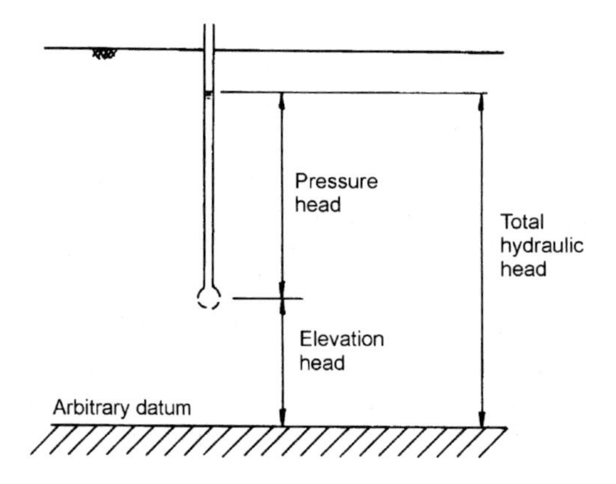

Figure 3.3 Definition of hydraulic head.

response to total hydraulic head, not pressure or elevation considered in isolation.

A term which is used frequently in any discussion of groundwater lowering is 'drawdown'. Drawdown is the amount of lowering (in response to pumping) of total hydraulic head, and is a key, measurable, performance target for any dewatering operation. Drawdown is equivalent to:

- i The amount of lowering of the 'water table' in an unconfined aquifer.
- ii The amount of lowering of the 'piezometric level' in a confined aquifer.
- iii The reduction (expressed as metres head of water) of pore water pressure observed in a piezometer (see Section 6.5). In this case the drawdown of total hydraulic head can be estimated directly from the change in pressure head, since the level of the piezometer tip does not change, so the change in elevation head is zero. If the reduction in pore water pressure is Δp, the drawdown is equal to $\Delta p/\gamma_w$, where γ_w is the unit weight of water.

Darcy's law is often written in terms of the hydraulic gradient i which is the change in hydraulic head divided by the length of the flow path ($i = dh/l$). Equation (3.1) then becomes

$$Q = kiA \tag{3.2}$$

In this form the key factors affecting groundwater flow are obvious.

- i If other factors are equal, an increase in permeability will increase the flow rate.
- ii If other factors are equal, an increase in the cross-sectional area of flow will increase the flow rate.
- iii If other factors are equal, an increase in hydraulic gradient will increase the flow rate.

These points are vital in beginning to understand how groundwater can be manipulated by groundwater lowering systems.

In the presentation of his equation Darcy left no doubt of its origin being empirical. His important contributions to scientific knowledge were based on careful observation in the field and in the laboratory, and on the conclusions that he drew from these. Permeability is in fact only a theoretical concept, but one vital to realistic assessments of groundwater pumping requirements and so an understanding of it is most desirable. In theory, permeability is the notional (or 'Darcy') velocity of flow of pore water through unit cross sectional area. In fact, the majority of the cross sectional area of a soil mass actually consists of soil particles, through which pore water cannot flow. The actual pore water flow velocity is greater than the 'Darcy velocity', and is related to it by the soil porosity n (porosity is the ratio of voids, or pore space, to total volume).

The main condition for Darcy's law to be valid is that groundwater flow should be 'laminar', a technical term meaning the flow is smooth. Flow will be laminar at low velocities but will become turbulent above some velocity, dependent on the porous media and the permeating fluid. Darcy's law is not valid for turbulent flow. In most groundwater lowering applications flow can safely be assumed to be laminar. The only locations where turbulent flow is likely to be generated is close to high flow rate wells pumping from coarse gravel aquifers. The implications of this flow to wells are discussed in Section 3.4.

For idealized, homogeneous, soils permeability depends primarily on the properties of the soil including the size and arrangement of the soil particles, and the resulting pore spaces formed when the particles are in contact. For example, consider an assemblage of billiard balls of similar size (Fig. 3.4(a)). This is analogous to the structure of a high permeability soil (such as a coarse gravel) where the voids (or pore water passages) are large, and the pore water can flow freely. Next, consider an assemblage of billiard balls with marbles placed in the spaces between the billiard balls (Fig. 3.4(b)); this is analogous to a soil of moderate permeability because the effective size of the pore water passages are reduced. Finally, consider a structure with lead shot particles placed in the voids between the billiard balls and the marbles – the passages for the flow of pore water are further reduced; this simulates a low permeability soil (Fig. 3.4(c)). It is logical to infer from this analogy that the 'finer' portion of a sample dominates permeability. The coarser particles are just the skeleton of the soil and may have little bearing on the permeability of a sample or of a soil mass *in situ*.

The proportion of different particle sizes in a sample of soil (as might be recovered from a borehole) can be determined by carrying out a particle size distribution (PSD) test, see Head (1982). The results of the test are normally presented as PSD curves like in Fig. 3.5, which shows typical PSD curves for a range of soils. PSD curves are sometimes known as grading curves or sieve analyses, after the sieving methods used to categorize the coarser particle

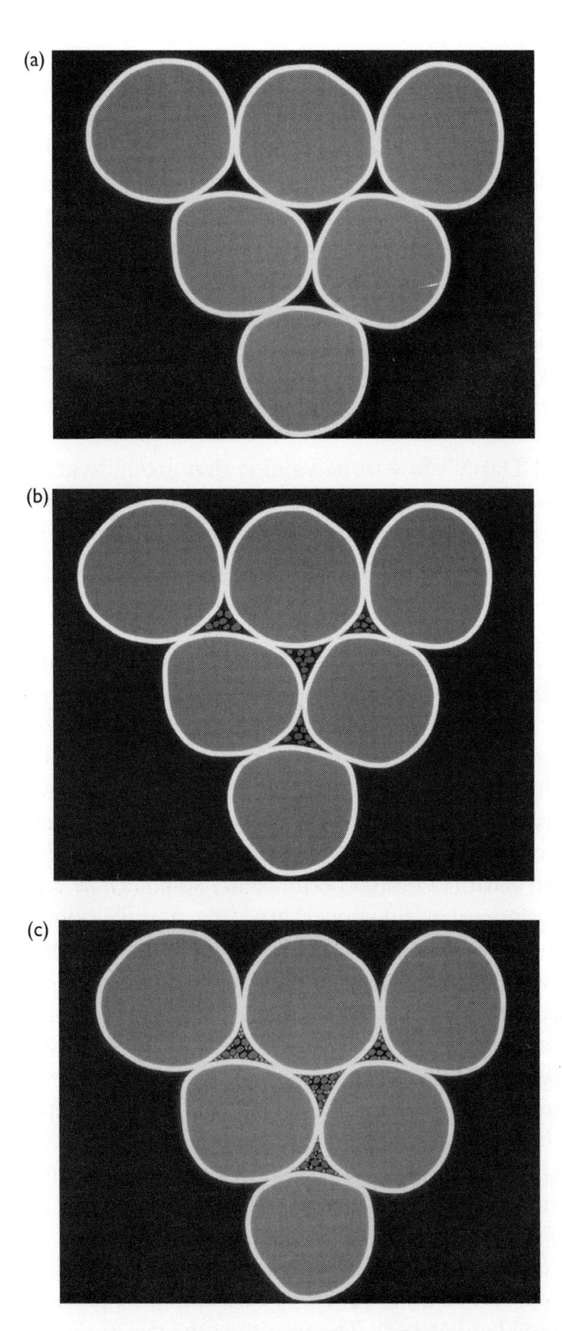

Figure 3.4 Soil structure and permeability. (a) High permeability, (b) medium permeability, (c) low permeability.

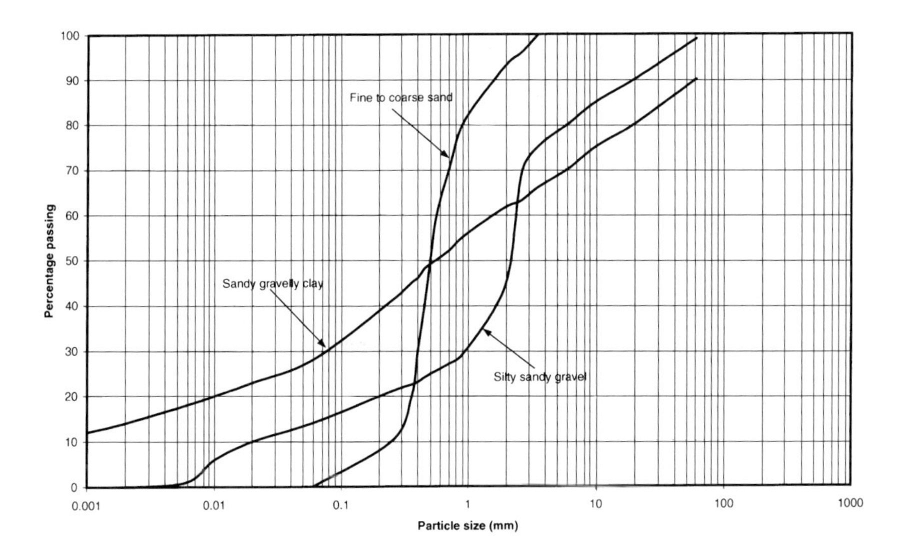

Figure 3.5 Particle size distribution.

sizes. Since there is a fairly intuitive link between particle size and permeability, over the last 100 years or so many researchers and practitioners have developed empirical correlations between permeability and certain particle size characteristics. Some of these methods are still in use today, and are described in Section 6.6, along with the drawbacks and limitations of such correlations.

As discussed earlier, the permeability is also dependent on the properties of the permeating fluid. Groundwater lowering is concerned exclusively with the flow of water, so this issue is not a major concern. It is worth noting, however, that the viscosity of water varies with temperature (and varies by a factor of two for temperatures between 20°C and 60°C), so in theory permeability will change with temperature. In practice, groundwater temperatures in temperate climates vary little and errors in permeability due to temperature effects tend to be small in comparison with other uncertainties.

3.2.2 *Typical values of permeability*

In practice, meaningful values of permeability are more difficult to visualize than in Darcy's experiment. Darcy's concept of permeability described above assumes that the soil permeability is homogenous (i.e. it is the same everywhere) and isotropic (i.e. it has the same properties in all directions).

Even a basic study of geology will show that many soils and rocks exhibit properties far removed from these assumptions. Many strata are more or less heterogeneous in character, so that even in an apparently consistent stratum the permeability may vary greatly between one part and another. This may result from inhomogeneities such as fissures, erosion features or sand and clay lenses. Also, the nature of deposition of soils can introduce anisotropy, where the soil permeability is not the same in all directions. Soils laid down in water may have a layered or laminated structure or 'fabric' – as a result the permeability in a horizontal direction is usually significantly greater than that in the vertical direction. The influence of soil fabric on permeability is discussed in detail by Rowe (1972).

Despite these complications, any rational attempt at groundwater lowering will require permeability to be investigated and assessed. Some of the design methods discussed in Chapter 7 assume, for simplicity, isotropic and homogeneous permeability conditions – on a theoretical basis this is clearly unrealistic. Yet these methods are established and field-proven methods and, if applied appropriately, can give useful results. This highlights the principle that in ground engineering it is often necessary to tolerate some simplification to make a problem more amenable to analysis. The key is to apply critical judgement to the parameters used in analysis, and not to blindly accept the results of the analysis until they are corroborated or validated by other data or experience.

As will be described in the section on site investigation (Chapter 6), tests to estimate realistic values of permeability can be problematic. As an engineering parameter permeability is unusual because of the tremendous range of possible values for natural soils and rocks; there can be a factor of perhaps 10^{10} between the most permeable gravels and almost impermeable intact clays. It is not unknown for the results of permeability tests to be in error by a factor of 10, 100, 1,000 or even more. This often stems purely from the limitations of the test and interpretation methods themselves, and can occur even if the tests are carried out in an exemplary fashion by experienced personnel.

Table 3.1 provides some general guidance on the permeability of typical soil types. This table is intended to be used for comparison purposes with permeability test results, to look for inconsistencies between soil descriptions (from borehole logs) and reported test results. Nevertheless, this table should be used with caution, especially in mixed soil types where the soil fabric (which is sometimes not adequately indicated in the soil description) may play a dominant role.

The permeability values used throughout this book will be reported in m/s (metres per second), which is the convention in engineering and soil mechanics. This is in contrast to hydrogeological references, which generally report permeability (or hydraulic conductivity) in m/d (metres per day). Conversion factors between various units are given at the end of this book.

Table 3.1 Typical values of soil permeability

Soil type	*Typical classification of permeability*	*Permeability (m/s)*
Clean gravels	High	$>1 \times 10^{-3}$
Clean sand and sand/gravel mixtures	High to moderate	1×10^{-3} to 5×10^{-4}
Fine and medium sands	Moderate to low	5×10^{-4} to 1×10^{-4}
Silty sands	Low	1×10^{-4} to 1×10^{-6}
Sandy silts, very silty fine sands and laminated or mixed strata of silt/sand/clay	Low to very low	1×10^{-5} to 1×10^{-8}
Fissured or laminated clays	Very low	1×10^{-7} to 1×10^{-9}
Intact clays	Practically impermeable	$<1 \times 10^{-9}$

3.3 Aquifers, aquitards and aquicludes

'Aquifer' is a useful term that appears whenever groundwater is discussed. As used by hydrogeologists an aquifer might be defined as 'A stratum of soil or rock which can yield groundwater in economic or productive quantities'. Almost all wells used for water supply purposes are drilled into, and pump from, aquifers. Examples of aquifers in the United Kingdom include the Chalk or Sherwood Sandstone. By this definition strata which yield water, at flow rates too small to be used for supply are not aquifers, and might be considered 'non-aquifers'. Examples of non-aquifers might include alluvial silts, glacial lake deposits or fissured mudstones.

From a groundwater lowering point of view, the hydrogeologists' definition of aquifer in terms of 'productive quantities' is unhelpful. The groundwater in many strata that yield just a little water (and so are non-aquifers) can cause severe problems for excavation stability (see Chapter 4). For the purposes of this book the definitions of CIRIA Report C515 (Preene *et al.* 2000) will be adopted, where the reference to productive quantities is omitted. This definition is given below, together with those for aquiclude and aquitard. The relationship between these strata types is discussed in the following sections.

Aquifer. Soil or rock forming a stratum, group of strata or part or stratum that is water-bearing (i.e. saturated and permeable).

Aquiclude. Soil or rock forming a stratum, group of strata or part or stratum of very low permeability, which acts as a barrier to groundwater flow.

Aquitard. Soil or rock forming a stratum, group of strata or part or stratum of intermediate to low permeability, which yields only very small groundwater flows.

3.3.1 Unconfined aquifers

An unconfined aquifer is probably the simplest for most people to visualize and understand. The soil or rock of the aquifer contains voids, pores or fissures. These voids are saturated (i.e. full of groundwater) up to a certain level known colloquially as the 'water table' (Fig. 3.6(a)), which is open to the atmosphere. An observation well (see Section 6.5) installed into the saturated part of the aquifer will show a water level equivalent to the water table. The analogy that can easily be drawn is that of digging a hole in the sand on a beach. Water does not enter the hole until the water table is reached, at which point water will enter the hole and stay at that level unless water was pumped out.

The water table can be defined as level in the aquifer at which the pore water pressure is zero (i.e. equal to atmospheric). Below the water table the soil voids are at positive pore water pressures and are saturated. Above the water table the pressure in the voids will be negative (i.e. less than atmospheric), and, depending on the nature of the soil or rock, may be unsaturated and contain both water and air. In most unconfined aquifers, the great majority of groundwater flow occurs below the water table. When the water table is lowered by pumping the saturated aquifer thickness will reduce.

If groundwater is pumped (or abstracted) from an unconfined aquifer, it is intuitively apparent that the water table will be lowered locally around the well and a 'drawdown curve' will be created (Fig. 3.6(b)). In simple terms the drawdown curve is the new, curved shape of the water table. Pore water will drain out of the soil above the new lowered water table, and will be replaced by air – this soil will become unsaturated. The amount of water contained in a soil will depend on the porosity of the soil, but it is important to note that not all of the water in an unconfined aquifer will drain out when the water table is lowered. Some water will be retained in the smaller soil pores by capillary forces. The proportion of water which can drain from an unconfined aquifer is described by the specific yield S_y, which is generally less than the soil porosity.

The zone above the water table (sometimes known as the vadose zone) is worthy of further consideration. Generally, the soil in this zone above standing water table will be moist, due to the presence of capillary water in the interconnected pores between soil particles. This capillary water differs from groundwater below the water table in that it is hardly noticeable in a borehole or excavation since it is at below atmospheric pressure and cannot flow into the excavation. It is held in place by capillary tension, at negative pore water pressures (Fig. 3.7).

The soil is saturated with capillary water for a height above the water table (this is the capillary saturated zone). The height of the capillary saturated zone above the water table is dependent on the effective size of the pores – the smaller the pores the greater the height. At the upper surface of the capillary

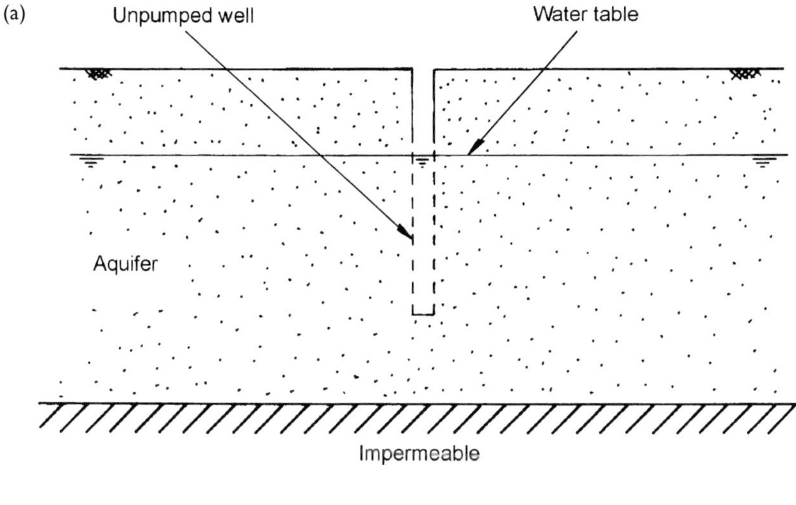

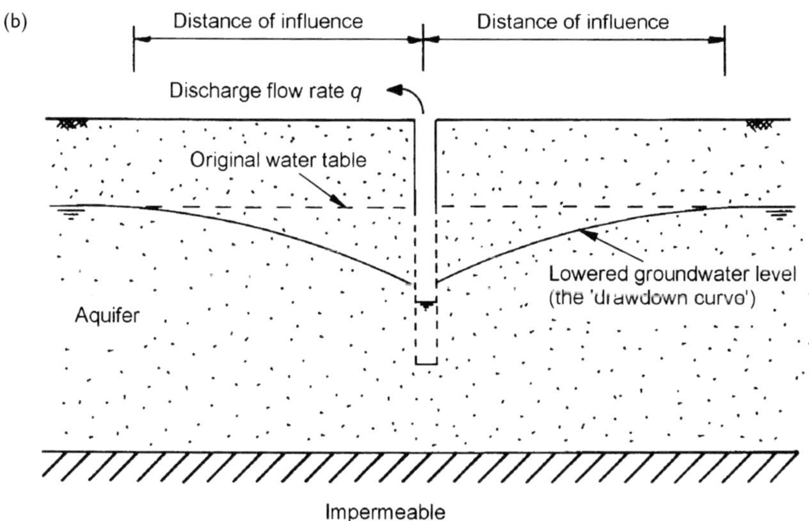

Figure 3.6 Unconfined aquifer. (a) Unpumped conditions, (b) during pumping.

saturated zone the surface tension forces between the water molecules and the soil particles are only just sufficient to prevent air being drawn into the soil pores. Above the top of the capillary saturated zone air can enter the soil, which becomes unsaturated. Typical heights of the capillary zone might be a few centimetres for coarse sand, or several metres for finer-grained soils (such as silts and clays) where the effective pore size is smaller.

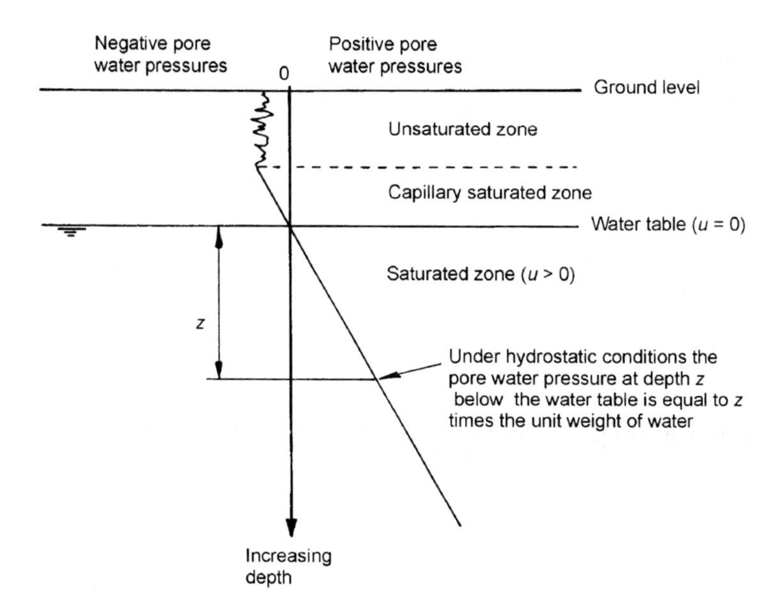

Figure 3.7 Groundwater conditions above the water table.

3.3.2 *Confined aquifers*

The distinction between unconfined and confined aquifers is important because they behave in quite different ways when pumped. In contrast to an unconfined aquifer, where the top of the aquifer is open to the atmosphere and an unsaturated zone may exist above the water table, a confined aquifer is overlain by a very low permeability layer known as an 'aquiclude' which forms a confining bed. A confined aquifer is saturated throughout, because the water pressure everywhere in the aquifer is above atmospheric. An observation well drilled into the aquifer would initially be dry when drilled through the confining bed. When the borehole penetrates the aquifer water will enter the borehole and rise to a level above the top of the aquifer. Because the pore water pressures are everywhere above atmospheric a confined aquifer does not have a water table. Instead its pressure distribution is described in terms of the 'piezometric level', which represents the height to which water levels will rise in observation wells installed into the aquifer (Fig. 3.8(a)).

If a confined aquifer is pumped, the piezometric level will be lowered to form a drawdown curve, which represents the new, lower, pressure distribution in response to pumping (Fig. 3.8(b)). Provided the piezometric level is not drawndown below the top of the aquifer (i.e. the base of the confining bed) the aquifer will remain saturated. Water will not drain out of the soil pores to be replaced by air in the manner of an unconfined aquifer. Instead

(a)

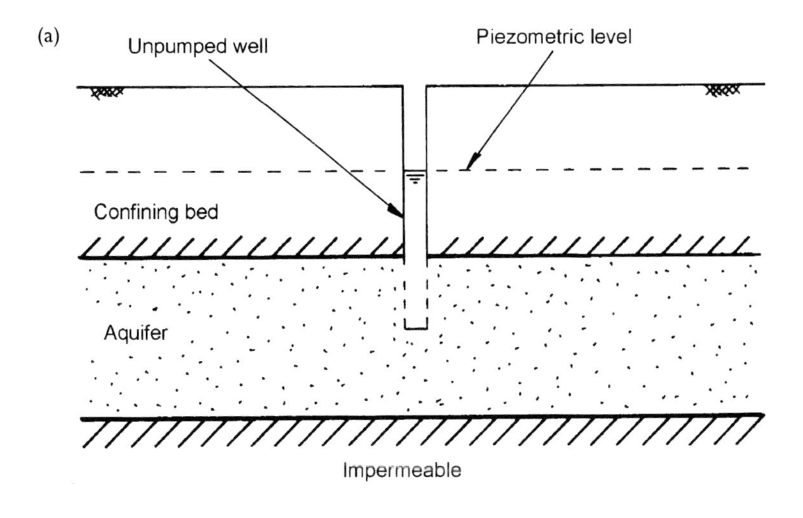

(b)

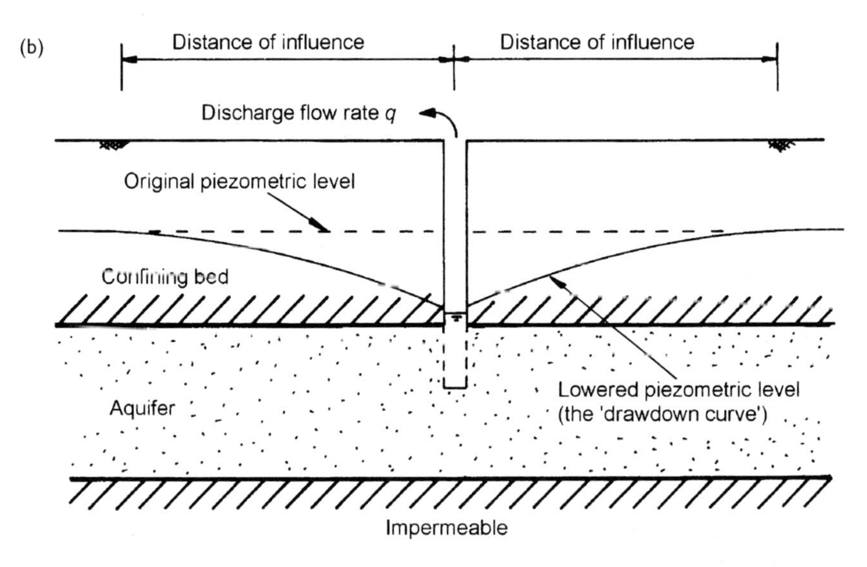

Figure 3.8 Confined aquifer. (a) Unpumped conditions, (b) during pumping.

a confined aquifer yields water by compression of the aquifer structure (reducing pore space) and expansion of the pore water in response to the pressure reduction. The proportion of water that can be released from a confined aquifer is described by the storage coefficient S.

If the water pressure in the confined aquifer is sufficiently high, a well drilled through the confining bed into the aquifer will be able to overflow naturally at ground level and yield water without pumping (Fig. 3.9). This is

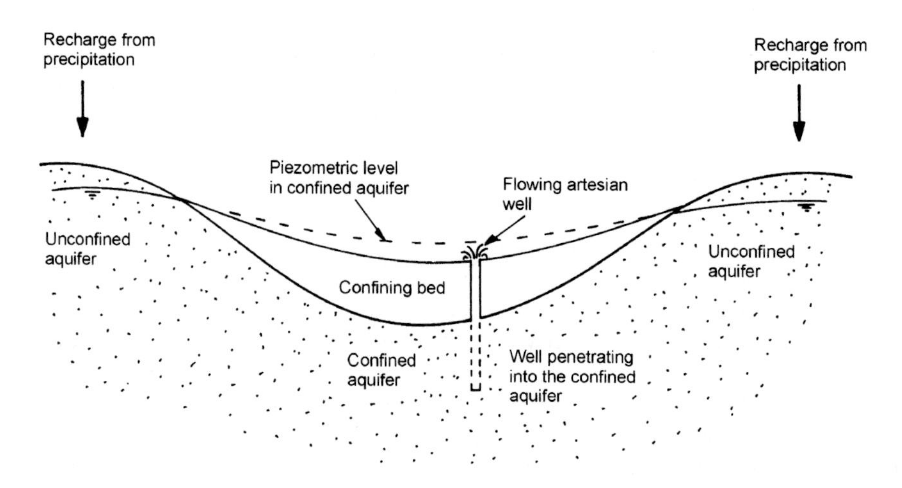

Figure 3.9 Flowing artesian conditions.

known as the flowing artesian condition – artesian is named after the Artois region of France where such conditions were first recorded. Flowing artesian conditions are possible from wells drilled in low-lying areas into a confined aquifer that is recharged from surrounding high ground. Flowing artesian aquifers are a special case of confined aquifers – confined aquifers that do not exhibit flowing conditions are sometimes known as artesian aquifers or occasionally (and incorrectly) as sub-artesian aquifers.

3.3.3 Aquicludes

An aquiclude is a very low permeability layer that will effectively act as a barrier to groundwater flow, for example, as a confining bed above an aquifer. It need not be completely 'impermeable' but should be of sufficiently low permeability that during the life of the pumping system only negligible amounts of groundwater will flow through it. The most common forms of aquiclude are layers of relatively unfissured clay or rock with permeabilities of 10^{-9} m/s or less.

In addition to having a low permeability, a stratum should meet two other criteria before it can be considered to act as an aquiclude.

i It must be continuous across the area affected by pumping; otherwise water may be able to bypass the aquiclude.
ii It must be of significant thickness. A thin layer of extremely low permeability material may be less effective as an aquiclude compared to a much thicker layer of greater, but still low, permeability. The thicker a layer of clay or rock is, the more likely it may act as an aquiclude.

3.3.4 Aquitards and leaky aquifers

An aquitard is a stratum of intermediate permeability between an aquifer and an aquiclude. In other words, it is of such low permeability that it is unlikely that anyone would consider installing a well to yield water, but the permeability is not so low that it can be considered effectively impermeable. Soil types that may form aquitards include silts, laminated clays/sands/silts, and certain clays and rocks that, while being relatively impermeable in themselves, contain a more permeable fabric of fissures or laminations.

Aquitards are of interest to hydrogeologists because they form part of 'leaky aquifer' systems. Such a system (also known as a semi-confined aquifer) consists of a confined aquifer where the confining layer is not an aquiclude, but is an aquitard (Fig. 3.10). When the aquifer is pumped water will flow vertically downward from the aquitard and 'leak' into the aquifer, ultimately contributing to the discharge flow rate from the well. It is apparent that the term 'leaky aquifer' is a misnomer since it is the aquitard that is actually doing the leaking.

Aquitards are relevant to the dewatering practitioner for the following reasons:

(a) If aquitards leak into underlying pumped aquifers the effective stress (see Section 4.2) will increase, leading to consolidation settlements (see Section 13.1). Analysis of the behaviour of any aquitards present is important when assessing the risk of damaging settlements.
(b) While aquitards may not yield enough water to form a supply, construction excavations into aquitards are likely to encounter small but problematic seepages and instability problems. Many applications of

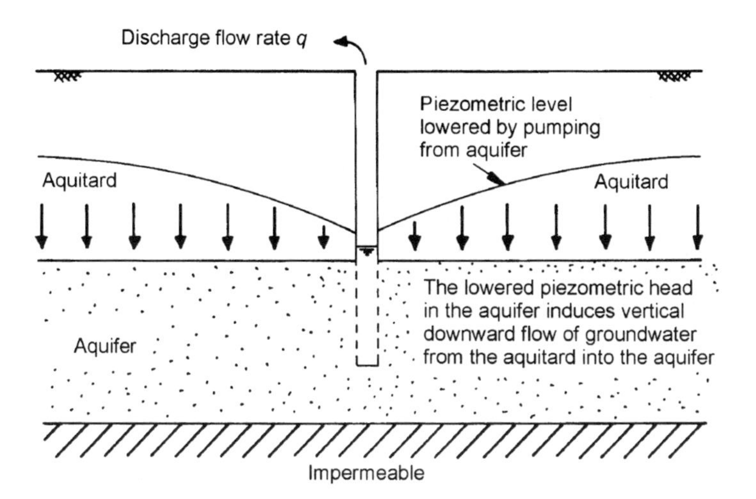

Figure 3.10 Leaky aquifer system.

pore water pressure control systems using some form of vacuum wells (see Section 5.4) are carried out in soils that would be classified as aquitards.

3.3.5 Aquifer parameters

The concept of permeability k has been introduced earlier (see Section 3.2), and is an important parameter used to describe aquifer properties. The thickness D of an aquifer is also important, since thicker aquifers of a given permeability will yield more water than thinner ones. These two terms can be combined into the hydrogeological term 'transmissivity' T:

$$T = kD \tag{3.3}$$

For SI units k and D will be in m/s and metres respectively, so T will have units of m^2/s. Results of pumping tests (see Section 6.6) are sometimes reported in terms of transmissivity; equation (3.3) allows these to be converted to permeability. In unconfined aquifers where the aquifer thickness reduces as a result of pumping transmissivity will reduce in a similar manner.

The amount of water released from an aquifer as a result of pumping is described by the storage coefficient. This is defined as the volume of water released from storage, per unit area of aquifer, per unit reduction in head; it is a dimensionless ratio. Because of the different way that water is yielded by confined and unconfined aquifers, storage coefficient is dealt with differently for each.

For an unconfined aquifer the storage coefficient is termed the specific yield S_y. This indicates how much water will drain out of the soil, to be replaced by air, under the action of gravity. Coarse-grained aquifers, such as sands and gravels, yield water easily from their pores when the water table is lowered. Finer-grained soils such as silty sands have smaller pores where capillary forces may retain much of the pore water even when the water table is lowered; S_y may be much lower than for gravels. Typical values of S_y are given in Table 3.2. Surface tension forces may also mean that the water may not drain out of the pores instantaneously when the water table is lowered; it may drain out slowly with time. This phenomenon is known as 'delayed yield' and can affect drawdown responses in unconfined aquifers.

In a confined aquifer, since the aquifer remains saturated, there is no specific yield and the storage coefficient S is used to describe the aquifer behaviour. Since water is only released by compression of the aquifer and expansion of the pore water typical values of S will be small, perhaps of the order of 0.0005–0.001. More compressible confined aquifers will yield more water under a given drawdown, and tend to have a greater storage coefficient, than stiffer aquifers. If the piezometric level in the aquifer is lowered sufficiently that it falls below the top of the aquifer, unconfined conditions will develop and a value of S_y will apply to the unconfined part of the aquifer.

Table 3.2 Typical values of specific yield

Aquifer	*Specific yield*
Gravel	0.15–0.30
Sand and gravel	0.15–0.25
Sand	0.10–0.30
Chalk	0.01–0.04
Sandstone	0.05–0.15
Limestone	0.005–0.05

Note: Based on data from Driscoll (1986) and Oakes (1986).

Most of the design methods presented later in this book (see Chapter 7) are based on simple steady-state methods commonly used for temporary works construction applications. The storage coefficient does not appear in those calculations because by the time steady-state has occurred all the water will have been released from storage. For relatively small construction excavations this is a reasonable assumption, because steady-state is generally achieved within a few days or weeks, so the volumes from storage release are only a concern during the initial drawdown period. Storage volumes may be more of a concern for large quarrying or opencast mining projects when steady-state may take a much longer time to develop.

3.4 Flow to wells

Our interest in groundwater is more than purely academic; we need to be able to control groundwater conditions to allow construction to proceed. The principal approach is to install a series of wells and to pump or abstract water from these wells – this is the essence of groundwater lowering or dewatering. Hence flow toward wells must be understood.

Some definitions will be useful. For the purposes of this book a 'well' is any drilled or jetted device that is designed and constructed to allow water to be pumped or abstracted. These definitions are, again, slightly different from those used by some hydrogeologists. In hydrogeology, a 'well' is often taken to mean a large diameter (greater than, say, 1.8 m) well or shaft, such as may be dug by hand in developing countries. A smaller diameter well, constructed by a drilling rig, is often termed a 'borehole', or in developing countries a 'tube well'.

A well will have a well screen (a perforated or slotted section that allows water to enter from certain strata) and a well casing (which, conversely, prevents water entering from certain strata), see Fig. 3.11.

In dewatering terminology wellpoints (Chapter 9), deep wells (Chapter 10) and ejectors (Section 11.1) are all types of wells, categorized by their

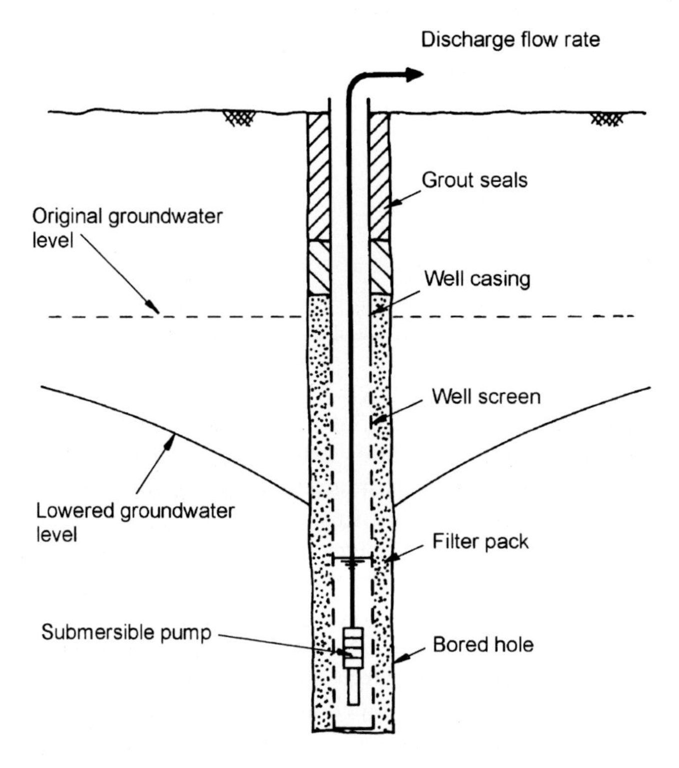

Figure 3.11 Principal components of a groundwater lowering well.

method of pumping. A 'sump' (Chapter 8) is not considered a well, since it is effectively a more or less crude pit to allow the collection of water.

3.4.1 Radial flow to wells

One of the defining features of flow through an aquifer toward a well is that flow converges radially as it passes through an ever smaller cross-sectional area of aquifer, resulting in a corresponding increase in flow velocity as the well is approached (Fig. 7.3). A drawdown curve is generated around the well as a result of pumping; in three dimensions around the well the drawdown curve describes a conic shape known as the 'cone of depression'. The limit of the cone of depression defines the 'zone of influence of a well'.

The effect of converging flow is compounded in unconfined aquifers because the drawdown at the well effectively reduces the aquifer thickness at the well (Fig. 3.6), further reducing the area available for groundwater flow. This is one of the reasons why flow rate equations for unconfined aquifers are more complex than for confined aquifers.

Convergence of flow, and the associated high groundwater velocities, leads to the phenomenon of 'well loss' where the water level inside a pumped well may be significantly lower than in the aquifer immediately outside. In some circumstances large well losses can be a sign of poorly designed or constructed wells.

3.4.2 Zone of influence

The zone of influence is a theoretical concept used to visualize how a well is affecting the surrounding aquifer. Imagine a well penetrating an aquifer that has an initial water table or piezometric level at the same elevation everywhere (Fig. 3.12). When water is first pumped from the well the water level in the well will be lowered, and flow will occur from the aquifer into the well. This water will be released from storage in the aquifer around the well. As time passes the cone of depression will expand away from the well, releasing further water from storage. The zone of influence will continue to increase with time, but at a diminishing rate until either an aquifer boundary is reached or the infiltration into the aquifer within the zone of influence is sufficient to supply the yield from the well. An idealized zone of influence is perfectly circular. The size of the zone is described by the 'distance of influence', which is the distance from the centre of the well to the outside edge of the zone where drawdown occurs.

Even if no source of recharge is encountered by the expanding zone of influence, after some time of pumping it will be expanding so slowly that it is effectively constant and is said to have reached a quasi steady-state. This is the approach assumed in several of the design methods described in Chapter 7. This simplifies all the sources of water (whether from storage or external recharge) as being equivalent to a single circular source of water at distance R_0 from the well. R_0 is the distance of influence for radial flow, sometimes known as the 'radius of influence'. The distance of influence for plane flow (e.g. for flow to line of wellpoints) is given the symbol L_0.

The distance of influence is an important parameter when estimating the discharge flow rate from a well or dewatering system. All other things being equal, a small distance of influence will predict a higher flow rate than a large distance of influence. It is vital that realistic values of distance of influence are used in calculations – poor selection of this parameter is one of the prime causes of errors in flow rate calculations. Important points to consider are:

i The steady-state distance of influence used in many calculations is a theoretical concept only. The true distance of influence is zero when pumping begins and increases with time.
ii The distance of influence will generally be greater in high permeability soils than in low permeability soils.

iii The distance of influence will generally be greater for larger well drawdowns than for small drawdowns.

It may be possible to determine the true distance of influence from appropriately analysed well pumping tests (see Section 6.6). There are however a

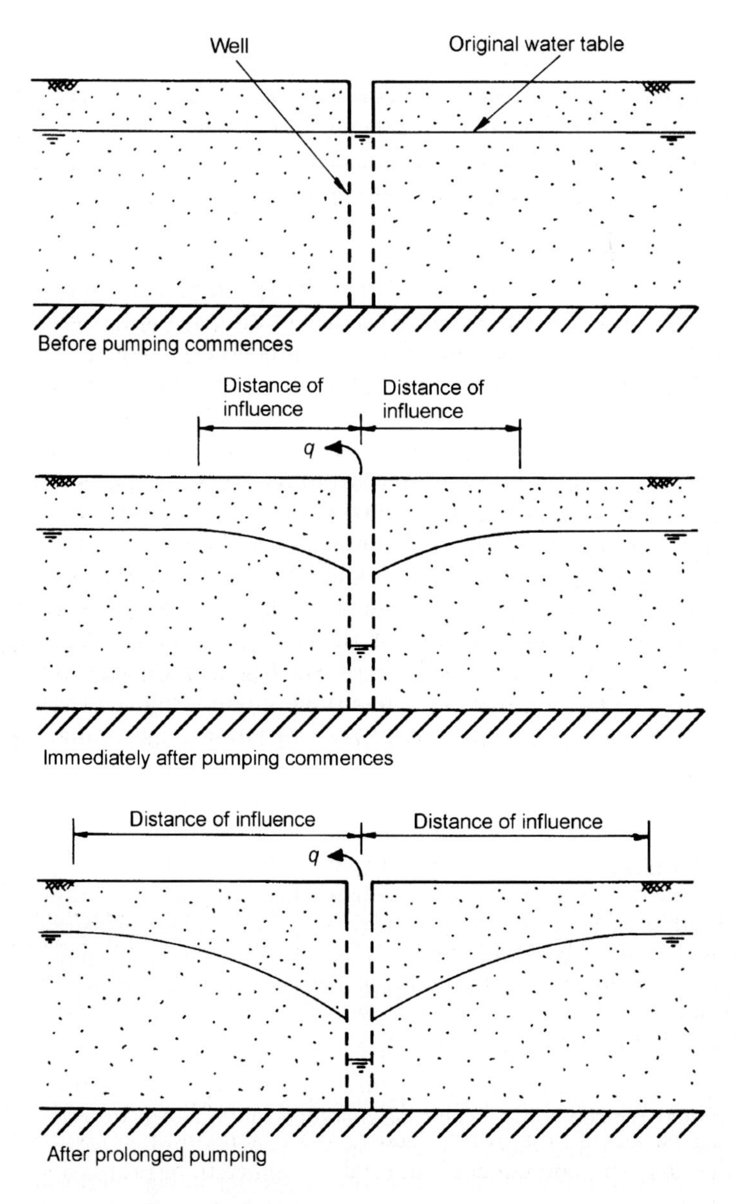

Figure 3.12 Zone of influence.

number of equations available to estimate R_0 and L_0. Two of the most commonly used are given below.

The empirical formula developed by Weber, but commonly known as Sichardt's formula, allows R_0 to be estimated from the drawdown s and aquifer permeability k:

$$R_0 = Cs\sqrt{k} \tag{3.4}$$

where C is an empirical factor. For radial flow to a well, if s is in metres and k is in m/s, R_0 can be obtained in metres using a C value of 3,000. The Sichardt estimate of R_0 is a quasi steady-state value and does not take into account the time period of pumping. The time dependent development of R_0 is described by the formula of Cooper and Jacob (1946):

$$R_0 = \sqrt{\frac{2.25kDt}{S}} \tag{3.5}$$

where D is the aquifer thickness, k and S are the aquifer permeability and storage coefficient respectively, and t is the time since pumping began.

When using distance of influence in design calculations, it is important to be vigilant for unrealistic values of R_0 and L_0, especially very small or very large values. In the authors' experience values of less than around 30 m or more than 5,000 m are rare and should be viewed with caution. It may be appropriate to carry out sensitivity analyses using a range of distance of influence values to see the effect on calculated flow rates.

Another problem can occur if the system is designed using the long-term R_0 or L_0. This will predict a much lower flow rate than may be generated during the initial period of pumping when R_0 is small as the cone of depression is expanding. It may be appropriate to design for a smaller, short-term R_0, or to design using the long-term R_0, but provide spare pumping capacity, over and above that predicted, to deal with the higher flows during the initial drawdown period.

3.4.3 Well losses

In general, the water level inside the screen or casing of a pumped well will be lower than in the aquifer immediately outside the well. This difference in level is known as the 'well loss' and results from the energy lost as the flow of groundwater converges toward the well. The drawdown observed in a pumped well has two components (Fig. 3.13).

i The drawdown in the aquifer. This is the drawdown resulting from laminar (smooth) flow in the aquifer, and is sometimes known as the aquifer loss. This component is normally assumed to be proportional to flow rate q.

ii The well loss. This is the drawdown resulting from resistance to turbulent flow in the aquifer immediately outside the well, and through the well screen and filter pack. This component is normally assumed to be proportional to q^n, and so increases rapidly as the flow velocity increases.

Well loss means that the water levels observed in a pumped well are unlikely to be representative of the aquifer in the vicinity of the well. Drawdowns estimated from pumped wells will tend to over-estimate drawdowns. It is preferable to have dedicated monitoring or observation wells, or at least to observe drawdown in some of the wells that are not being pumped.

Jacob (1946) suggested that in many cases $n = 2$. This is now widely accepted and allows the drawdown s_w in a well to be expressed as

$$s_w = Bq + Cq^2 \tag{3.6}$$

where B and C are calibration coefficients (which can be determined from the results of step drawdown tests, Clark 1977). The Bq term is the aquifer drawdown, and Cq^2 is the well loss.

If a well experiences high well loss its yield will be limited. At high flow rates the drawdown inside the well will be large, but drawdown in the aquifer (which is the aim of any groundwater lowering system) may be much smaller. In poorly performing wells it is not unusual for the drawdown outside a well to be less than half that inside the well. Well losses can

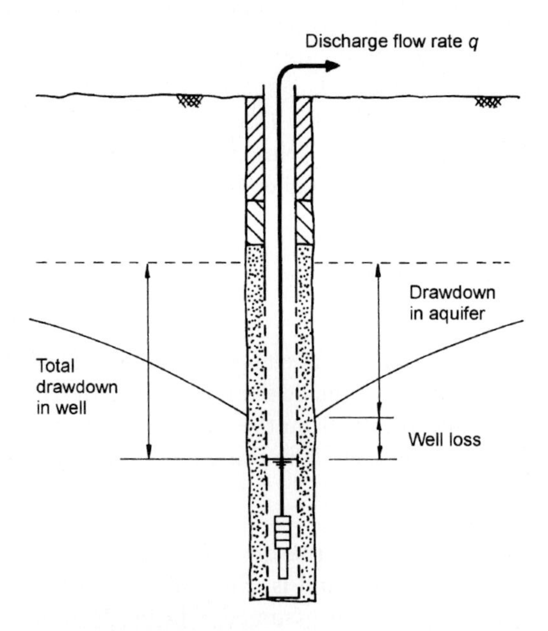

Figure 3.13 Well losses.

be minimized by designing the well with sufficient 'wetted screen length' and with a low screen entrance velocity (see Section 10.2). Ensuring that the well is adequately developed (see Section 10.6) can help reduce well losses.

3.4.4 *Effect of diameter of well*

Common sense suggests that, other things being equal, a larger diameter well will have the ability to yield more water than a smaller diameter well. This is because the well will have more contact area with the aquifer, and also because the flowing water has to converge less to reach a larger well.

However, the relationship between yield and diameter is not straightforward. It is worth stating that doubling the diameter of the well will not double the potential yield from the well, and may only increase the yield by a small proportion. Table 3.3, based on work by Ineson (1959) shows the estimated relation between yield and bored diameter (the drilled diameter through the aquifer, not the diameter of the well screen). This indicates, for example, that in an homogeneous aquifer increasing the well diameter by 50 per cent from approximately 200 mm to approximately 300 mm will increase the yield by only 11 per cent. The increase in yield is more marked in fissured aquifers, because a larger diameter well has an increased chance of intercepting water-bearing fissures. The true relationship may be more complex than this, especially for wells with high losses. In those cases the increase in diameter, which will reduce groundwater flow velocity, may help reduce well losses and attainable well yield may increase rather more than shown in Table 3.3.

In practice, the economics of drilling and lining wells means that temporary works groundwater lowering wells are rarely constructed at diameters of greater than 300–450 mm. Well diameters are often chosen primarily to ensure that the proposed size of electro-submersible pump can be installed in the well (see Section 10.2).

3.4.5 *Equivalent wells and lines of wells*

The foregoing discussions have concentrated on a single pumped well in isolation. In reality, most groundwater lowering is carried out using several

Table 3.3 Effect of well diameter on yield

Bored diameter of well through aquifer	*203 mm*	*305 mm*	*406 mm*	*457 mm*	*610 mm*
Homogeneous aquifer (intergranular flow)	1.00	1.11	1.21	1.23	1.32
Fissured aquifer	1.00	1.29	1.52	1.61	1.84

Note: Based on data from Ineson (1959).

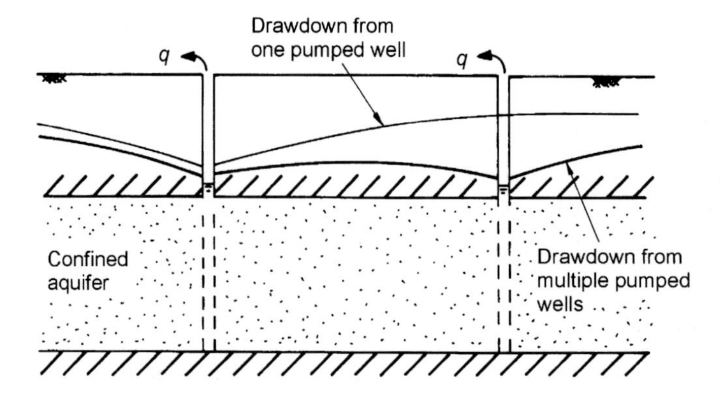

Figure 3.14 Superposition of drawdown.

closely spaced pumped wells (be they wellpoints, deep wells or ejectors) *acting in concert*. With this approach the cone of depression of each well overlaps (or interferes), creating additional drawdown. This effect is known as the superposition of drawdown and can create groundwater lowering over wide areas, which is ideal to allow excavations to be made below original groundwater levels (Fig. 3.14).

Wells are normally installed either as rings around an excavation or as lines alongside one or more sides of the dig. It is probably theoretically possible to consider the influence of each well individually and determine the complex interaction between the wells, but this would be tedious, especially when large numbers of wells are involved. A more practical, and empirically very effective, approach is to consider a dewatering system of many wells to act on a gross scale as a large equivalent well or slot.

Circular or rectangular rings of wells can be thought of as large equivalent wells (see Fig. 7.4) to which flow is, on a gross scale, radial. Similarly, lines of closely spaced wells (such as wellpoints alongside a trench) can be thought of as equivalent slots (Fig. 7.5) to which flow is predominantly plane to the sides. Once this conceptual leap is made, the well and slot formula given in Chapter 7 can be used to model the overall behaviour of dewatering systems, without having to consider each well individually. Formulae for estimating the size of an equivalent well from the dimensions of a dewatering system are given in Section 7.3.

3.5 Aquifers and geological structure

The aquifer types described in the foregoing sections are a theoretical ideal. At many sites more than one aquifer may be present (perhaps separated by aquicludes or aquitards), or the aquifers may be of finite extent and be influenced by their boundaries. The particular construction problems resulting from the presence of aquifers at a site will be strongly influenced by the soil stratification and geological structure.

This section will illustrate the importance of an appreciation of geological structure to the execution of groundwater lowering works. Two case histories will be presented: the London basin shows how large-scale geological structure can allow multiple aquifer systems to exist; and a problem of base heave in a small trench excavation illustrates the importance of smaller-scale geological details.

3.5.1 Multiple aquifers beneath London

The city of London is founded on river gravels and alluvial deposits associated with the River Thames, which are underlain by the very low permeability London Clay. These gravels form a shallow (generally less than 10 m thick) aquifer. Construction of utility pipelines, basements and other shallow structures often requires groundwater lowering to be employed; wellpointing and deep wells have proved to be effective expedients in these conditions. However, the geology beneath London allows another, deeper, aquifer to exist below the city, largely isolated from the shallow aquifer.

Beneath the London Clay lie a series of sands and clays comprising the Lambeth Group stratum (formerly known as the Woolwich and Reading Beds) and the Thanet Sand stratum. These are underlain by the Chalk, a fissured white or grey limestone, which rests on the very low permeability Gault Clay. The overall geological structure is a syncline forming what is often called the 'London Basin'. The Chalk, Thanet Sand and parts of the Lambeth Group together form an aquifer. The upper 60–100 m of the Chalk are probably the dominant part of the aquifer, where significant fissure networks readily yield water to wells. The sands are of moderate permeability and generally do not yield as much water as the Chalk. The overlying London Clay acts as an aquiclude or confining bed, effectively separating the deep aquifer from the shallow gravel aquifer.

The Chalk has a wide exposure on the North Downs to the South of London and on the Chilterns to the North and occurs as a continuous layer beneath the Thames Valley (Fig. 3.15). Rain falling on central London may ultimately reach the gravel aquifer, but the London Clay prevents it from percolating down to the Chalk. The Chalk obtains its recharge from rain falling on the North Downs and the Chilterns many miles from the city. Ultimately this water forms part of the reservoir of water in the chalk aquifer. If the recharge exceeds the discharge from the aquifer (either from wells, or natural discharge to springs and the River Thames) the water pressure in the aquifer will rise slowly. If discharges exceed recharge the water pressure will fall.

Before London developed as a city, the natural rates of recharge and discharge meant that the deep aquifer had sufficient water pressure for it to act

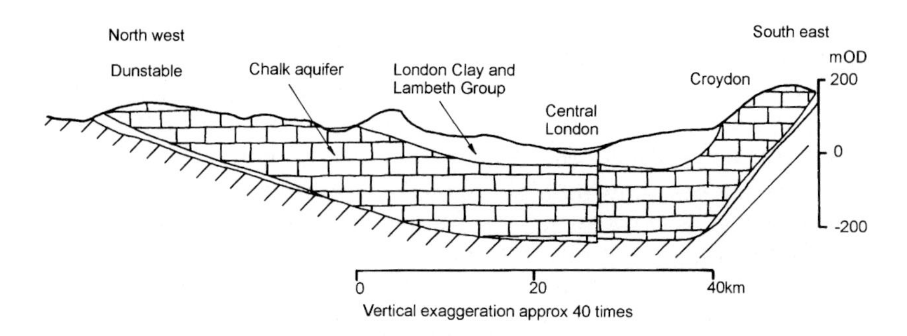

Figure 3.15 Chalk aquifer beneath London (after Sumbler 1996). The chalk aquifer extends beneath the London basin and receives recharge from the unconfined areas to the north and south. The London Clay deposits act as a confining layer beneath central London. Prior to the twentieth century flowing artesian conditions existing in many parts of the city.

as a confined aquifer (see Section 3.3). In the lower lying areas of the city there was originally sufficient pressure in the aquifer to allow a well drilled through the London Clay into the Chalk to overflow naturally as a flowing artesian well. In fact, in central London there are still a few public houses called the *Artesian Well*, indicating that in earlier days the locals were probably supplied with water from a flowing well.

This availability of groundwater led to a large number of wells being drilled into the deep aquifer (where the water quality was more 'wholesome' than in the gravel aquifer). Rates of groundwater pumping increased during the eighteen, nineteenth and early part of the twentieth centuries. This resulted in a significant decline in the piezometric level of the deep aquifer. Artesian wells ceased to flow, pumps had to be installed to allow water to continue to be obtained and over the years the pumps had to be installed lower and lower to avoid running dry. By the 1960s the water level in wells in some areas of London was 90 m below the ground surface – a huge drop relative to the original artesian conditions. In some locations the water pressure was reduced below the base of the London Clay, so the formerly confined aquifer became unconfined. The deeper water levels increased pumping costs and made well supplies less cost-effective compared with mains water. This, together with a general re-location of large water-using industries away from central London, has resulted in a significant reduction in groundwater abstraction. As a result the piezometric level in the aquifer has recovered since then (at more than 1 m/year at some locations in the 1980s). By the 1990s the piezometric level in many areas was within 55 m of ground level.

This continuing rise of water pressures is a major concern because much of the deep infrastructure beneath London (deep basements, railway and utility tunnels) was built during the first half of the twentieth century when water pressures were at an all-time low. Several studies (see e.g. Simpson *et al.* 1989) have addressed the risk of flooding or overstressing of existing deep structures if water levels continue to rise. The management of water levels beneath London (and, indeed, beneath other major cities around the world) is an important challenge to be faced by groundwater specialists during the first half of the twenty-first century.

In a construction context, an appreciation of the aquifer system is vital to ensure deep structures are provided with suitable temporary works dewatering. Fig. 3.16 shows a typical arrangement as might be used for a deep

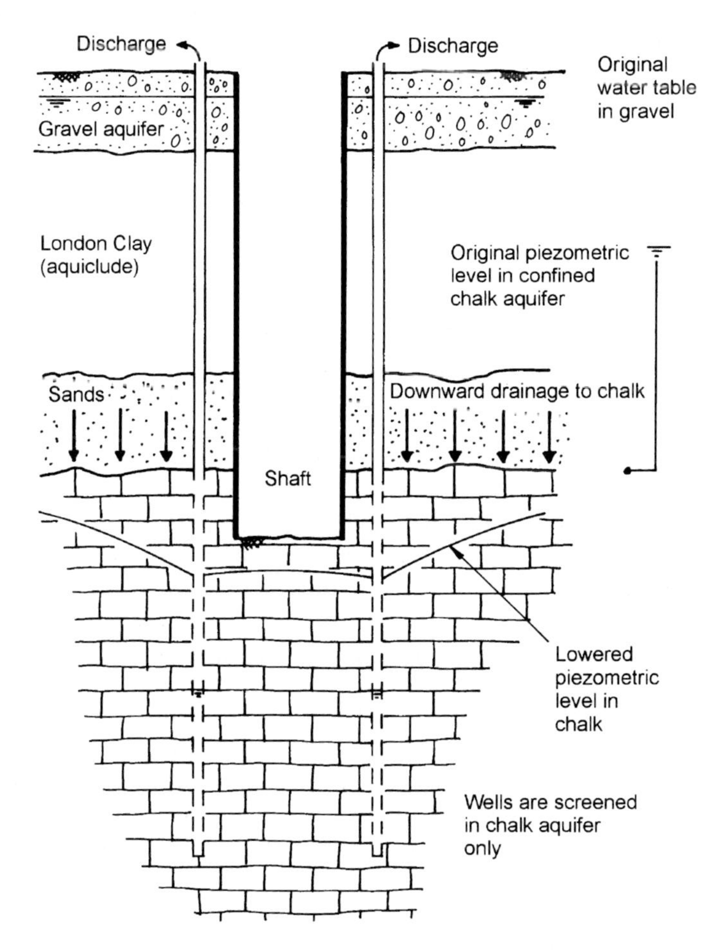

Figure 3.16 Groundwater control in multiple aquifers.

shaft structure in central London. Important points to note are:

i The structure penetrates two aquifers, separated by an aquiclude. Groundwater will need to be dealt with separately in each aquifer.
ii In London it is common to deal with the upper aquifer by constructing a cut-off wall, penetrating to the London Clay, to exclude the shallow groundwater. This is possible because the London Clay is at relatively shallow depth. If clay was present only at greater depth any cut-off would need to be deeper and it may be more economic to dewater the upper aquifer.
iii Wells are used to pump from the deep aquifer to lower the piezometric level to a suitable distance below the excavation. Because the wells must be relatively deep (perhaps up to 100 m), and therefore costly, it is important to design the wells and pumps to have the maximum yield possible, so that the number of wells can be minimized.
iv Although the lower aquifer consists of both the Chalk and the various sand layers between the top of the Chalk and the base of the London Clay, wells are often designed to be screened in the Chalk only, and are sealed from the sands using casing. This is because it can be difficult to construct effective well filters in the sands (which are fine-grained and variable), yet wells screened in the Chalk are simpler to construct and can be very efficient, especially if developed by acidization (see Section 10.6). This is the approach successfully adopted for some structures on the London Underground Jubilee Line Extension project described by Linney and Withers (1998). Excavations in the Thanet Sand were dewatered without pumping directly from the sand, but by pumping purely from the underlying Chalk. This might seem a rather contradictory approach, but is an example of the 'underdrainage' method. This is a way of using the geological structure to advantage by pumping from a more permeable layer beneath the layer that needs to be dewatered; the upper poorly draining layer will drain down into the more permeable layer (see Section 7.3).

3.5.2 Water pressures trapped beneath a trench excavation

Dr W. H. Ward (1957) reported some construction difficulties encountered by a contractor excavating a pipeline trench near Southampton. The trench excavation was made through an unconfined aquifer of sandy gravel overlying the clays of the Bracklesham Beds. The contractor dealt with the water in the sandy gravel by using steel sheet-piling to form a cut-off on either side of the trench and exclude the groundwater. The clay in the base of the excavation was not yielding water, so dewatering measures were not adopted.

The trench was approximately 6.1 m deep, to allow placement of a 760 mm diameter pipe which was laid on a 150 mm thick concrete slab in the base of the excavation. The construction difficulties encountered consisted of uplift of the bottom of the trench (called 'base heave', see Section 4.3), often occurring overnight whilst the trench was open. At one location the trench formation rose by almost 150 mm before the concrete slab was cast, and a further 50 mm after casting.

When Dr Ward and his colleagues at the Building Research Station were consulted, they suggested that the problem might be due to a high groundwater pressure in a water-bearing stratum below the base of the trench. This was proved to be the case when a small borehole was drilled in the base of the trench. This borehole overflowed into the trench with the flowing water bringing fine sand with it. The water pressure in the borehole was later determined to be at least 1.3 m above trench formation level, but it is likely that the original piezometric level was even higher, since the flowing discharge from the borehole may have reduced pressures somewhat. Once the problem had been identified, the contractor was able to complete the works satisfactorily by installing a system of gravel-filled relief wells (see Section 11.3) in the base of the trench to bleed off the excess groundwater pressures (Fig. 3.17).

This case history illustrates the importance of identifying the small-scale geological structure around an excavation. It appears that in this case the

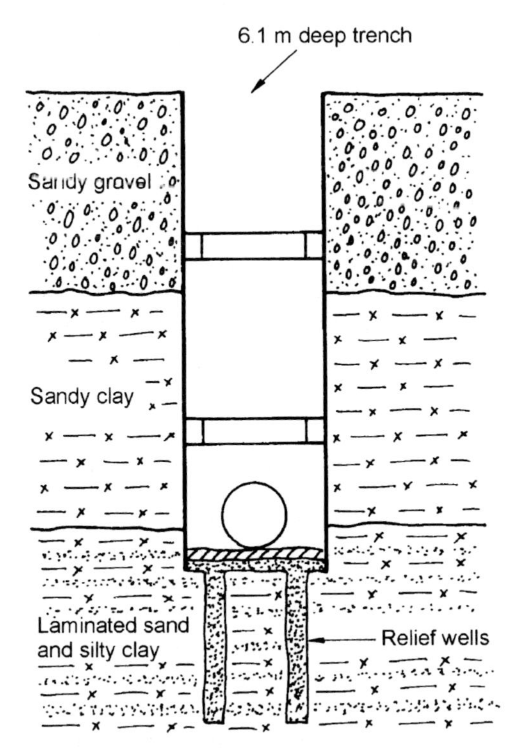

Figure 3.17 Use of simple relief wells to maintain base stability (after Ward 1957).

trench was excavated through the upper aquifer (the sandy gravel), which was dealt with using a sheet-pile cut-off, and the base of the trench was dug into low permeability clay, which is effectively an aquiclude. Problems occurred because there was a separate confined aquifer beneath the aquiclude, which contained sufficient water pressure to lift the trench formation. Once identified and understood, the problem was solved easily using relief wells. However, because the problem was not identified in the site investigation before work started, time and money was wasted in changing the temporary works, as well as making good the damaged pipelines. A classic failing of site investigations for groundwater lowering projects is that the boreholes are not taken deep enough to identify any confined aquifers which may exist beneath the proposed excavations. Guidelines on suitable depths for boreholes are given in Section 6.4.

3.6 Aquifer boundaries

Civil and geotechnical engineers can readily appreciate the importance of permeability in the design of groundwater lowering systems – permeability is a traditional numerical engineering value. However, engineers are sometimes less proficient in recognizing the less quantifiable effects that aquifer boundary conditions have on groundwater flow to excavations. The following sections will outline some important aquifer boundary conditions.

3.6.1 Interaction between aquifers and surface water

It is obvious from the hydrological cycle (see Section 3.1) that groundwater is inextricably linked with surface waters such as rivers, streams and lakes. The significance of the link between groundwater and surface water at a given site depends on the geological and hydrological setting.

Where surface water flows across or sits on top of an aquifer, water will flow from one to the other – the direction of flow will depend on the relative hydraulic head. The magnitude of the flow will be controlled by Darcy's law, and will be affected by the permeability and thickness of any bed sediment, and the head difference between the aquifer and the surface water. Bodies of surface water that are quiet or slow flowing may have low permeability bed sediments, which may dramatically reduce flow between surface and groundwater. Similarly, if the surface water is sitting on an aquitard or aquiclude it may be effectively isolated from groundwater in underlying aquifers.

Where watercourses (such as rivers and streams) are connected to aquifers they commonly receive water from the aquifer (the water entering the river from groundwater is termed 'baseflow'). Such a river is said to be a 'gaining' river (Fig. 3.18(a)). This is perhaps contrary to many people's expectations, that rivers should feed groundwater, rather than vice versa. 'Losing' rivers do sometimes exist, especially in unconfined aquifers with

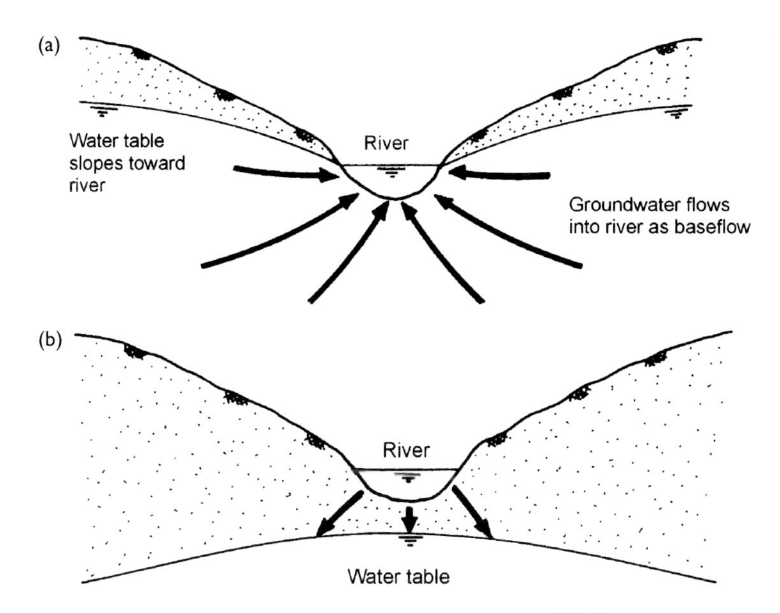

Figure 3.18 Interaction between rivers and aquifers. (a) Cross-section through a gaining river, (b) cross-section through a losing river.

deep water tables (Fig. 3.18(b)). If groundwater lowering is carried out near a 'gaining' river the hydraulic gradients may be reversed, changing a normally 'gaining' river to a 'losing' one. Even if the hydraulic gradients do not reverse, groundwater lowering may reduce baseflow to a gaining river. If pumping continues for an extended period this may reduce river flow, and perhaps result in environmental concerns.

Lakes often have low permeability silt beds, reducing the link with groundwater, although wave action near the shore may remove sediment, allowing increased flow into or out of the groundwater body. Man-made lagoons or dock structures may have silt beds (especially if they are relatively old) or may have linings or walls of some sort – however, just because linings exist, it does not mean that they do not leak! Analysis of a pumping test (Section 6.6) can be an effective way of determining whether groundwater lowering will be significantly influenced by any local bodies of surface water.

3.6.2 *Interaction between aquifers*

In the same way that there can be flow between groundwater and surface water, groundwater can flow between aquifers if hydraulic head differences exist between them.

Many aquifer systems exist under 'hydrostatic conditions' – this is the case when the hydraulic head is constant with depth, so that observation wells installed at different depths show the same level (Fig. 3.19(a)). In this case there would be no flow of groundwater between shallow and deep aquifers separated by an aquitard. But sometimes non-hydrostatic conditions exist. In Fig. 3.19(b), the hydraulic head in the deep aquifer is greater than the shallow aquifer, so water will flow upwards from the deep to the shallow aquifer. The flow may be very slow if the aquitard is of low permeability; if the stratum between the aquifers is an aquiclude, the flow will be so small that it is often ignored in the analysis of short-term groundwater lowering installations.

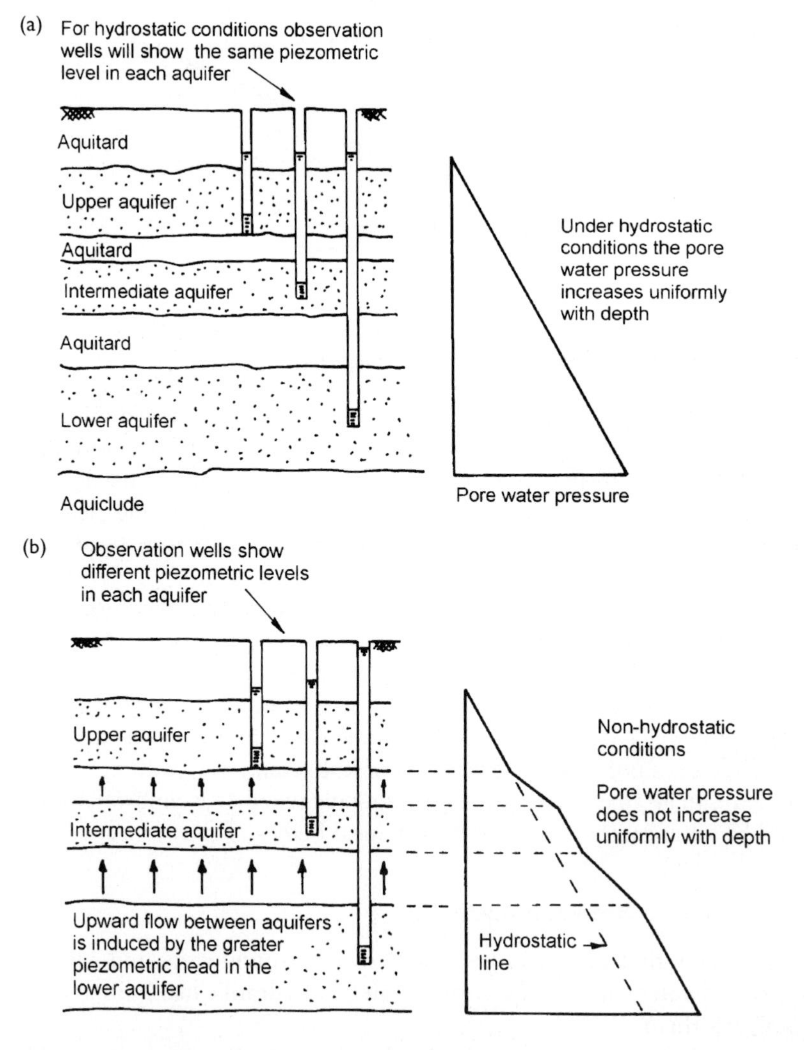

Figure 3.19 Interaction between aquifers. (a) hydrostatic conditions, (b) flow between aquifers.

Inter-aquifer flow may be a long-term phenomenon, perhaps sustained by different recharge sources for each aquifer. On the other hand, it may be a short-term condition resulting from temporary groundwater lowering operations disturbing the groundwater regime (this artificially induced flow between aquifers is one possible side effect of dewatering, see Section 13.3).

3.6.3 Recharge boundaries

Zones or features where water can flow into an aquifer are termed 'recharge boundaries', some commonly occurring examples of which are shown in Fig. 3.20. If recharge boundaries exist within the distance of influence they can have a significant effect on the behaviour of dewatering schemes. They may cause the cone of depression to become asymmetric (since the extent of the cone will be curtailed where it meets the recharge source). The flow rate that must be pumped by a dewatering system will

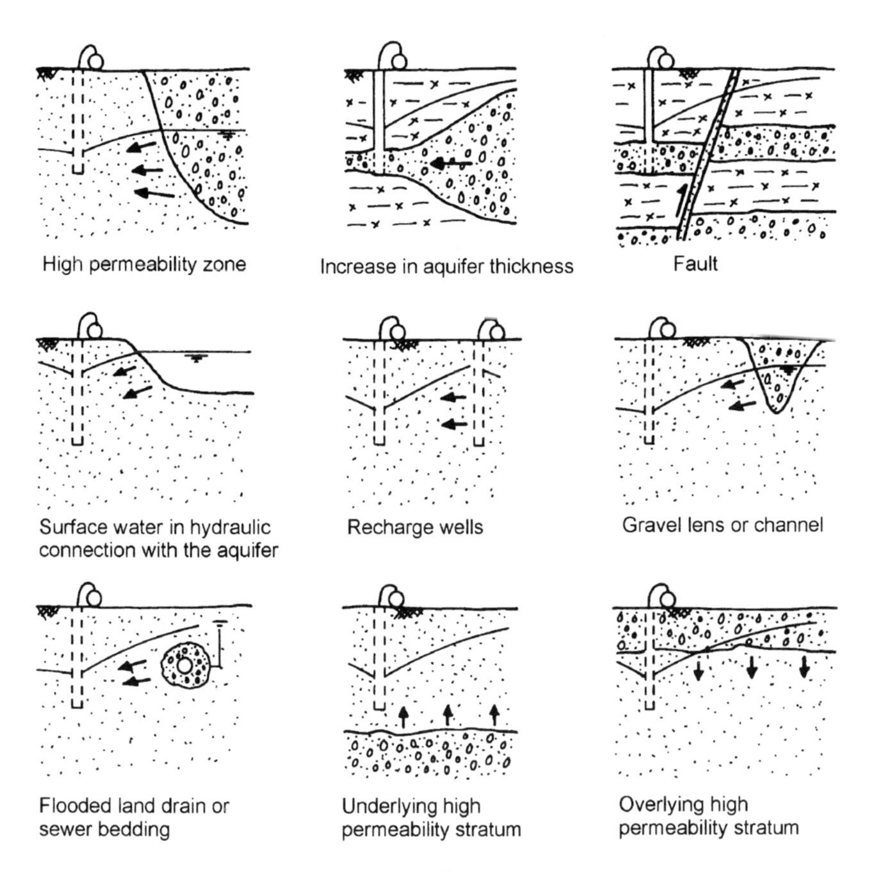

Figure 3.20 Potential aquifer recharge boundaries.

often be increased by the presence of a recharge boundary. It is essential that any potential recharge boundaries be considered during the investigation and design of dewatering works.

3.6.4 Barrier boundaries

Real aquifers are rarely of infinite extent and may be bounded by features which form barriers to groundwater flow. Fig. 3.21 shows some commonly occurring barrier boundaries. The presence of barrier boundaries will tend to reduce the pumped flow rate necessary to achieve the required drawdown.

3.6.5 Discharge boundaries

Water can sometimes be discharged naturally from aquifers. Water will flow from an unconfined aquifer if the water table intersects the ground surface. Diffuse discharges are called seepages or, if the flow is very localized (perhaps at a fault or fissure), the discharge is termed a spring. Flowing artesian aquifers can also discharge if faults or fissures allow water a path to the surface. Water flowing between aquifers may also constitute a discharge boundary condition.

Man's influence, in the form of pumping from wells (either for supply or for groundwater lowering for construction, quarrying, mining, etc.) can also create discharge boundaries.

Fig. 3.22 shows some examples of discharge boundaries.

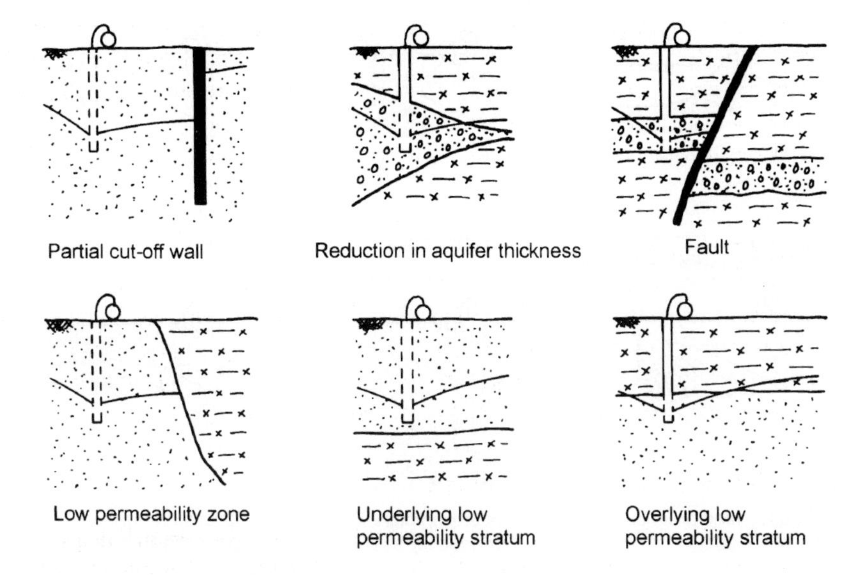

Figure 3.21 Potential aquifer barrier boundaries.

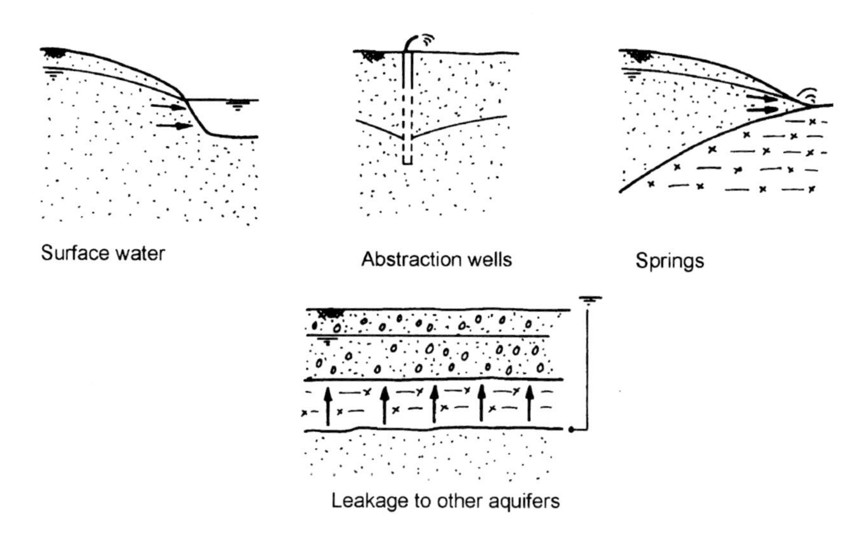

Figure 3.22 Potential aquifer discharge boundaries.

3.7 Using geological structure to advantage

Previous sections have described how the boundaries and structure of aquifers and aquitards can complicate the flow of groundwater. The presence of recharge boundaries or multiple layered aquifers will have a significant effect on the groundwater lowering requirements for an excavation. It is essential that, as the first step in any design process, a 'conceptual model' of groundwater conditions is developed. The background to conceptual models is described in Section 7.2.

Although geological structures and boundaries can make dewatering more difficult, conversely a designer can try to use these features to advantage. For example, if the conceptual model highlights the presence of a more permeable layer within the aquifer sequence by far the most effective approach is to abstract water directly from the permeable layer. This will maximize well yields and will induce the adjacent, less permeable, layers to drain to the permeable layer. Where the permeable layer is at depth, this approach of pumping preferentially from the most permeable layer is known as underdrainage (see Section 7.3 and Fig. 7.8).

3.8 Groundwater chemistry

The study of the chemical composition of groundwater is a key area of hydrogeology, helping to ensure that water abstracted for potable use is safe to drink or fit for its intended use. The water abstracted for groundwater lowering purposes is rarely put to any use, and, subject to the necessary permissions (see Chapter 15), is typically discharged to waste. Accordingly, water chemistry has often been thought of as an irrelevance to the dewatering practitioner.

However, a basic knowledge of groundwater chemistry can be useful for the following reasons:

i Discharge consents and permissions. In many locations permission to dispose of the discharge water to a sewer or surface watercourse will only be granted if it can be demonstrated that the water is of adequate quality.
ii Corrosion and encrustation. Water chemistry will influence well clogging, encrustation, corrosion and biofouling (see Section 14.8) that may affect systems in long-term operation. Clogging and encrustation can be a particular problem for artificial recharge systems (see Section 13.6).
iii The effect of groundwater lowering on water quality. Abstracting water from an aquifer may affect existing groundwater quality. Monitoring may be necessary to determine if these effects are occurring, and to control any mitigation measures (see Section 13.3).

If a project is likely to require detailed study of water chemistry, specialist advice should be obtained. This section is intended only to highlight some of the important practical issues of water chemistry relevant to groundwater lowering. Readers interested in the field are commended to Lloyd and Heathcote (1985) as an introductory text.

3.8.1 Chemical composition of groundwater

Water is a powerful solvent, and a wide range of substances will dissolve in it to some degree. Almost all water in the hydrological cycle will contain some dissolved minerals or other substances. Even rainwater contains dissolved carbon dioxide and some sodium chloride lifted into the atmosphere from the oceans.

Substances dissolved in water exist as electrically charged atoms or molecules known as ions. Positively charged ions are known as anions and negatively charged ions are known as cations. For example, when sodium chloride (common salt, NaCl, which is highly soluble) dissolves in water it exists as sodium anions (Na^{+}) and chloride cations (Cl^{-}).

There are certain substances which commonly exist at substantial concentrations (several milligrams per litre, mg/l) in natural groundwater. These are known as the major anions and cations (Table 3.4), which are routinely tested for, in groundwater analyses. A variety of trace metals are also present (called trace because they are generally present in much smaller concentrations, perhaps only a fraction of a mg/l). For groundwater lowering purposes the most important trace metals are iron and manganese because they influence the severity of encrustation and biofouling (see Section 14.8).

Table 3.4 Major anions and cations in groundwater

Major anions	*Major cations*
Sodium (Na^+)	Bicarbonate (HCO_3^-)
Potassium (K^+)	Carbonate (CO_3^-)
Calcium (Ca^{2+})	Sulphate (SO_4^{2-})
Magnesium (Mg^{2+})	Chloride (Cl^-)
	Nitrate (NO_3^-)
	Nitrite (NO_2^-)
	Ammonia (NH_3)
	Phosphate (PO_4^{2-})

In addition to dissolved natural minerals and compounds, if groundwater contamination has occurred other, potentially harmful, substances may be present. Many of the most problematic substances are said to be 'organic compounds'; this does not mean that they result from natural growth, but means that they are based around molecules formed from carbon atoms, the building blocks of organic life. Some of the organic compounds (which include pesticides) are toxic even at extremely low levels (below one microgram per litre, μg/l). At these low concentrations detection of these compounds requires the highest quality of sampling and testing; specialist advice is essential. Further details on contamination problems can be found in Fetter (1993).

It is apparent that almost no groundwater is 'pure', if by pure we mean containing nothing but H_2O! Nevertheless, water that contains relatively little dissolved material is said to be 'fresh'. Most groundwater abstracted for drinking use is classified as fresh, and requires little treatment other than basic sterilization to kill any harmful bacteria present. Even water suitable for industrial use (such as steam raising in a boiler) tends to be relatively fresh, because the lower mineral content reduces the build up of scale deposits within the pipework. In many developed countries permissible limits are set for the chemical composition of water that can be used for human consumption; guidelines are also available for water to be used for cultivation of crops or livestock (see Lloyd and Heathcote 1985, Chapter 10; Brassington 1995, Chapter 6).

3.8.2 *Field monitoring of groundwater chemistry*

In principle, the most obvious way to investigate groundwater chemistry is to obtain a sample of groundwater, seal it into a bottle and send it to a laboratory for testing. It might be appropriate to test for major anions and cations, selected trace metals and any other substances of interest at the site in question.

For groundwater lowering systems, obtaining a sample is relatively straightforward. It may be possible to fill a sample bottle directly at the discharge tank. However, when taking groundwater samples from the discharge flow the following factors should be considered.

- i Try to minimize the exposure of the sample to the atmosphere. Try and obtain it directly from the discharge of the pump or dewatering system. Totally fill the bottle and try and avoid leaving any air inside when it is sealed. If the pump discharge is 'cascading' before the sampling point the water will become aerated and oxidation may occur. The discharge arrangements should be altered so as the sample can be obtained before aeration occurs.
- ii Samples may degrade between sampling and testing. The samples should be tested as soon as possible after they are taken and, ideally, should be refrigerated in the meantime. The bottles used for sampling should be clean with a good seal. However, the sample may degrade while in the bottle (e.g. by trace metals oxidizing and precipitating out of solution). Specialists may be able to advise on the addition of suitable preservatives to prevent this occurring. The choice of sample bottle (glass or plastic) should also be discussed with the laboratory since some test results can be influenced by the material of the sample bottle.
- iii Use an accredited, experienced laboratory.

Sometimes groundwater samples may be required from a site when there is no dewatering pumping taking place – perhaps for pre or post construction background monitoring. In that case a sampling pump will have to be used to obtain a sample from an observation well. The water in the well has been exposed to the atmosphere and is unlikely to represent the true aquifer water chemistry. Therefore, it is vital to fully 'purge' the well before taking a sample. Purging involves pumping the observation well at a steady rate until at least three 'well volumes' of water have been removed (a 'well volume' is the volume of water originally contained inside the well liner). Specialist sampling pumps should be used in preference to airlifting, since the latter method may aerate the sample, increasing the risk of oxidation of trace metals and other substances.

If samples are taken to an off-site laboratory, it may be several days before the results are ready. Even if the tests are rushed through as priority work, some of the actual procedures may take a week or more. Comprehensive testing of samples is not cheap, either; a reasonably complete suite of testing may cost several hundred pounds per sample at 2001 prices. A good way of reducing costs, and obtaining rapid results, is to take and test water samples on a periodic basis (perhaps monthly), but carry out daily (or, using a datalogger, continuously) monitoring of the 'wellhead chemistry'.

Wellhead chemistry is a hydrogeological term used to describe certain parameters, which are best measured as soon as the water is pumped from the well – that is at the wellhead. It is best to measure these parameters here as they are likely to change during sampling and storage, which makes laboratory determined values less representative. Typical wellhead chemistry parameters include:

(a) Specific conductivity, EC.
(b) Water temperature.
(c) pH.
(d) Redox potential, E_H.
(e) Dissolved oxygen, DO.

Perhaps the most commonly measured wellhead parameter for groundwater lowering systems is specific conductivity, EC. Specific conductivity is a measure of the ability of the water to conduct electricity and is a function of the concentration and charge of the dissolved ions; it is reported in units of microseimens per centimetre (μS/cm) corrected to a reference temperature of 25°C. EC is useful in that it can be related to the amount of total dissolved solids TDS of the water (Lloyd and Heathcote 1985):

$$TDS = k_e EC \tag{3.7}$$

where: TDS is the total dissolved solids in mg/l; EC is the specific conductivity in μS/cm at 25°C; and k_e is the calibration factor with values between 0.55 and 0.80 depending on the ionic composition of the water.

TDS and EC have been related to each other for various water classifications in Table 3.5. Fresh water will have a low TDS (most water supply boreholes for potable use produce water with a TDS of no more than a few 100 mg/l). The higher the TDS, the less fresh the water. The term 'saline' is used for convention's sake and does not necessarily imply a high TDS is the result of saline intrusion (see Section 13.3). A high TDS may be an indicator of highly mineralized waters that have been resident in the ground for very long periods, slowly leaching minerals from the soils and rocks.

Table 3.5 Classification of groundwater based on total dissolved solids (TDS)

Classification of groundwater	*Total dissolved solids, TDS (mg/l)*	*Specific conductivity, EC (μS/cm)*
Fresh	< 1,000	< 1,300–1,700[a]
Brackish	1,000–10,000	1,300–1,700 to 13,000–17,000[a]
Saline	10,000–100,000	13,000–17,000 to 130,000–170,000[a]

Note: [a] Approximate correlation; precise value depends on ionic composition of water.

Daily monitoring of *EC* using a conductivity probe has proved useful on sites where there was concern that water of poorer quality (e.g. from saline intrusion or from deeper parts of the aquifer) might be drawn towards the pumping wells. Readings of conductivity were taken every day, and reviewed for any sudden or gradual changes in *EC*, which would have indicated that poor quality water was reaching the wells. In effect, the *EC* monitoring was used as an early warning or trigger to determine when more detailed water testing was required.

References

Blyth, F. G. H. and De Freitas, M. H. (1984). *A Geology for Engineers*, 7th edition. Edward Arnold, London.

Brassington, R. (1995). *Finding Water: A Guide to the Construction and Maintenance of Private Water Supplies*. Wiley, Chichester.

Clark, L. (1977). The analysis and planning of step drawdown tests. *Quarterly Journal of Engineering Geology*, **10**, 125–143.

Cooper, H. H. and Jacob, C. E. (1946). A generalised graphical method for evaluating formation constants and summarising well field history. *Transactions of the American Geophysical Union*, **27**, 526–534.

Darcy, H. (1856). *Les Fontaines Publiques de le Ville de Dijon*. Dalmont, Paris.

Driscoll, F. G. (1986). *Groundwater and Wells*. Johnson Division. Saint Paul, Minnesota.

Fetter, C. W. (1993). *Contaminant Hydrogeology*. Macmillan, New York.

Fetter, C. W. (1994). *Applied Hydrogeology*, 3rd edition. MacMillan, New York.

Head, K. H. (1982). *Manual of Soil Laboratory Testing*. Pentech Press, London (3 volumes).

Ineson, J. (1959). The relation between the yield of a discharging well at equilibrium and its diameter, with particular reference to a chalk well. *Proceedings of the Institution of Civil Engineers*, **13**, 299–316.

Jacob, C. E. (1946). Drawdown test to determine effective radius of artesian well. *Proceedings of the American Society of Civil Engineers*, **72**, 629–46.

Linney, L. F. and Withers, A. D. (1998). Dewatering the Thanet beds in SE London: three case histories. *Quarterly Journal of Engineering Geology*, **31**, 115–122.

Lloyd, J. W. and Heathcote, J. A. (1985). *Natural Inorganic Hydrochemistry in Relation to Groundwater: An Introduction*. Clarendon Press, Oxford.

Oakes, D. B. (1986). Theory of groundwater flow. *Groundwater, Occurrence, Development and Protection* (Brandon, T. W., ed.). Institution of Water Engineers and Scientists, Water Practice Manual No. 5, London, 109–134.

Peck, R. B. (1969). Advantages and limitations of the observational method in applied soil mechanics. *Géotechnique*, **19**(2), 171–187.

Powrie, W. (1997). *Soil Mechanics: Concepts and Applications*. Spon, London.

Preene, M. Roberts, T. O. L., Powrie, W. and Dyer, M. R. (2000). *Groundwater Control – Design and Practice*. Construction Industry Research and Information Association, CIRIA Report C515, London.

Price, M. (1996). *Introducing Groundwater*, 2nd edition. Chapman & Hall, London.

Rowe, P. W. (1972). The relevance of soil fabric to site investigation practice. *Géotechnique*, **22**(2), 195–300.

Simpson, B., Blower, T., Craig, R. N. and Wilkinson, W. B. (1989). *The Engineering Implications of Rising Groundwater in the Deep Aquifer Below London*. Construction Industry Research and Information Association, CIRIA Special Publication 69, London.

Sumbler, M. G. (1996). *British Regional Geology: London and the Thames Valley*, 4th edition. HMSO, London.

Todd, D. K. (1980). *Groundwater Hydrology*, 2nd edition. Wiley, New York.

Ward, W. H. (1957). The use of simple relief wells in reducing water pressure beneath a trench excavation. *Géotechnique*, 7(3), 134–139.

Chapter 4

Groundwater effects on the stability of excavations

4.0 Introduction

To avoid troublesome conditions during excavation and construction, measures must be taken to control groundwater flows and pore water pressures in water-bearing soils. Surface water run-off must also be effectively managed. An understanding of how an excavation may be affected will assist in assessing which groundwater control measures are necessary to ensure stability.

This chapter discusses the various circumstances by which inadequately controlled groundwater can allow unstable ground conditions to develop. The interaction between pore water pressures, effective stress and stability is introduced. Various large-scale and small-scale stability problems are presented and suitable approaches to prevent or control such occurrences are described. The particular groundwater problems associated with the presence of surface water and with excavations in rock are also discussed. The various methods of groundwater control available to stabilize excavations will be outlined in Chapter 5.

4.1 Groundwater control – the objectives

The fundamental requirement of groundwater control measures is that they should ensure stable and workable conditions throughout, so that excavation and construction can take place economically and under safe conditions at all times.

Groundwater may be controlled by: exclusion methods (see Section 5.3); one or more types of pumping systems (see Section 5.4) commonly termed 'dewatering'; or a combination of pumping plus exclusion techniques. This book is primarily concerned with groundwater control by pumping or dewatering.

A correctly designed, installed and operated dewatering system ensures that construction work can be executed safely and economically by:

i Local lowering of the water table or groundwater level and interception of any seepages due to perched water tables which might otherwise emerge on the exposed slopes or at the base of the excavation(s).

ii Improving the stability of the excavation slopes and preventing material being removed from them by erosion due to seepage. Effective control of groundwater may allow slopes to be steepened and the area of excavation reduced.
iii Preventing heave, blow-outs or quick conditions from developing in the floor of the excavation with associated detrimental effects on the bearing capacity at formation level.
iv Draining the soil to improve excavation, haulage, trafficking and other characteristics of the soils involved.
v Reducing the lateral loads on temporary support systems such as sheet-piling.

The practical upshot of objective (i) is that, by lowering the groundwater level and controlling any local seepages the excavation will not be flooded by groundwater, and might be said to be 'dewatered'. This is perhaps the most obvious aim of groundwater control – prevention of flooding – but objectives (ii)–(v) can be equally important in improving the stability of excavations. This is discussed further in the following sections.

The reader may note that the word 'dry' does not appear anywhere in the objectives of groundwater control outlined above. The authors are not in favour of any specifications or contracts which state that the aim of dewatering is to provide 'dry' conditions. After all, a heavy rain shower may cause 'wet' conditions, irrespective of how well groundwater is controlled! A much more useful target is for the dewatering measures to provide 'workable conditions'.

4.2 Groundwater, effective stress and instability

The concept of effective stress is fundamental to understanding the interrelation between groundwater and soil strength and stability. The principle of effective stress was proposed by Karl Terzaghi in 1920s, and is described in detail in soil mechanics texts such as Powrie (1997). As was described in Section 3.2 soil has a skeletal structure of solid material with an interconnecting system of pores. The pores may be saturated (wholly filled with water) – in confined aquifers or below the water table in unconfined aquifers – or unsaturated (filled with a mixture of water and air) – above the water table.

As a saturated soil mass is loaded, the total stress is carried by both the soil skeleton and the pore water. The pore water pressure acts with equal intensity in all directions. The stress carried by the soil skeleton alone (by inter-particle friction) is thus the difference between the total applied stress σ and the water pressure set up in the pores u. This is termed the effective stress σ' and is expressed by Terzaghi's equation

$$\sigma' = \sigma - u \quad (4.1)$$

Because water has no significant shear strength, the soil skeleton may deform while the pore water is displaced (a process known as pore water pressure dissipation). This action continues until the resistance of the soil structure is in equilibrium with the external forces. The rate of dissipation of the pore water will be dependent on the permeability of the soil mass and the physical drainage conditions.

The shear strength τ of soil is primarily from interparticle friction, and so is dependent on the effective stress. The shear strength at failure τ_f can be expressed by the Mohr–Coulomb failure criterion:

$$\tau_f = \sigma' \tan \phi' \quad (4.2)$$

where ϕ' is the angle of shearing resistance of the soil.

These two simple equations show that reducing pore water pressures (as a result of drawing down the groundwater level) will increase the effective stress within the soil in the area affected. This produces a corresponding increase in the shear strength of the soil, which will improve the stability of the soils around and beneath an excavation.

The increase in effective stresses that results from groundwater lowering is also an important factor in the potential for ground settlements around dewatering systems. This is discussed in Section 13.1.

4.3 Large-scale instability caused by groundwater

Inadequately controlled groundwater can cause large-scale stability problems in a variety of ways, the mechanisms likely to be prevalent at a given site largely being controlled by ground conditions. These large-scale mechanisms can be explained in terms of effective stress and pore water pressures.

It is worth considering two classic cases of ground conditions. First, an excavation with battered side slopes made into a sandy soil forming an unconfined aquifer. Second, an excavation made through a low permeability stratum overlying a confined aquifer. The mechanisms of potential instability will be quite different in each case.

4.3.1 Unconfined aquifers

Consider a hole with sloping sides dug in a bed of silty fine sand below the standing groundwater level. If the inflow water is pumped from a sump within the excavation, the sides will slump in when a depth of about 0.5–1.0 m below original standing water level is reached. As digging proceeds the situation will get progressively worse, and the edges of the excavation will recede. The bottom will soon fill with a sand slurry in an almost liquid condition which will be constantly renewed by material slumping from the side slopes. The collapse of the side slopes results from the presence of positive pore water and seepage pressures which are developed in the ground by the flow of water to the pumping sump of the open excavation (Figure 4.1).

The mechanisms causing this unstable condition can be explained in terms of effective stress.

Above the water table, assuming that the soil is dry (i.e. has zero pore water pressure) then it can stand in stable slopes of up to ϕ' to the horizontal. In reality the pore water pressure above the water table may not be zero; in fine-grained soils such as silty sands negative pore water pressures may exist temporarily due to capillary effects. As a result, stable slopes even steeper than ϕ' may temporarily be possible. Figure 9.12 shows a near vertical excavation face in a silty fine sand where groundwater has been lowered by wellpoints. However, in time the negative pore water pressure will decay as the capillary water exposed on and near the face dries out. After some finite period of time (which may be no more than a few hours) the over-steepened slope will eventually crumble to that of the long-term stable slope angle.

Where seepage emerges from the slope there will be positive pore water pressures. Positive pore water pressures will reduce effective stress, the shear strength of the soil will reduce in turn and the soil will not stand at slopes as steep as in dry soils. This is the mechanism leading to the 'slumping' effect seen when excavating below the groundwater level.

Considering Figure 4.1 in detail, flow line 'A' represents the flow line formed by the water table or phreatic surface. Seepage into the excavation will cause a slight lowering of the water table so flow line 'A' curves downward and emerges almost parallel to the surface of the excavation. Immediately

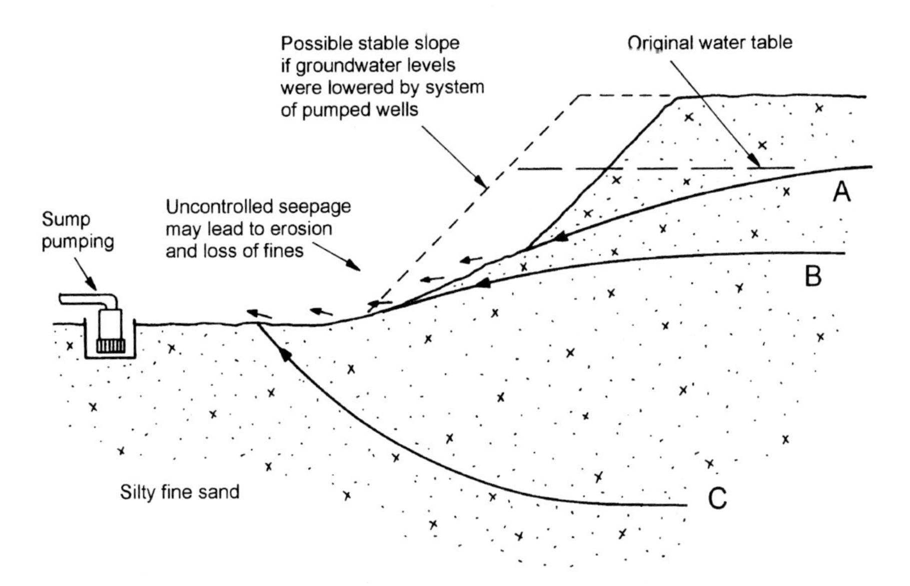

Figure 4.1 Instability due to seepage into an excavation in an unconfined aquifer.

below its point of emergence the positive pore water pressures generated by the seepage mean that the soil can no longer support a slope of ϕ'. Below the emergence of the seepage line the soil slope will slump to form shallower angles. At the point of emergence of the almost horizontal flow line 'B' the sand will stand at $\frac{1}{2}\phi'$ or less. Where there is upward seepage into the excavation base (flow line 'C') the effective stress may approach zero and the soil cannot sustain any slope at all. In these circumstances the soil may 'boil' or 'fluidize' and lose its ability to support anything placed on it – this is the so-called 'quicksand' case.

'Running sand' is another term used to describe conditions when a granular soil becomes so weak that it cannot support any slope or cut face, and becomes an almost liquid slurry. The term is often used as if it were a property of the sand itself. Actually, it is the flow of groundwater through the soil, and the resultant low values of effective stress which cause this condition. Effective groundwater lowering can change 'running sand' into a stable and workable material.

In addition to the loss of strength, seepage of groundwater through slopes may cause erosion and undermining of the excavation slopes. This is especially a problem in fine-grained sandy soils, and is discussed in Section 4.4.

Any solution to stability problems in unconfined aquifers will need to stabilize both the sides and base of the excavation. The most commonly used expedient is to install a system of dewatering wells (see Section 5.4) around the excavation to lower the groundwater level to below the base of the excavation, and to ensure that seepage does not emerge from the side slopes. A typical target is to lower the groundwater level a short distance (say 0.5 m) below the deepest excavation formation level.

If the excavation is only penetrating a short distance (say less than 1.5 m) below the original groundwater level, a sump pumping system (Chapter 8) used in combination with slope drainage (see Section 4.4) might also be effective. Extreme care must be taken when using sump pumping in this way, to avoid destabilizing seepages into the excavation; the risk of this increases for excavations further below original groundwater level.

If the sides of the excavation are supported by physical cut-off walls (see Section 5.3) then it is possible to sump pump from within the excavation, without the risk of instability of side slopes. The risk of fluidization of the base due to upward seepage remains (Figure 4.2). Fluidization will theoretically occur when the upward hydraulic gradient exceeds a critical value i_{crit}

$$i_{crit} = \frac{(\gamma_s - \gamma_w)}{\gamma_w} \tag{4.3}$$

where γ_s is the unit weight of soil, and γ_w is the unit weight of water. In general the density of soil is about twice the density of water (peat being a notable exception), so generally $i_{crit} \approx 1$. This is the hydraulic gradient at which fluidization will occur, so it is important that this value is not

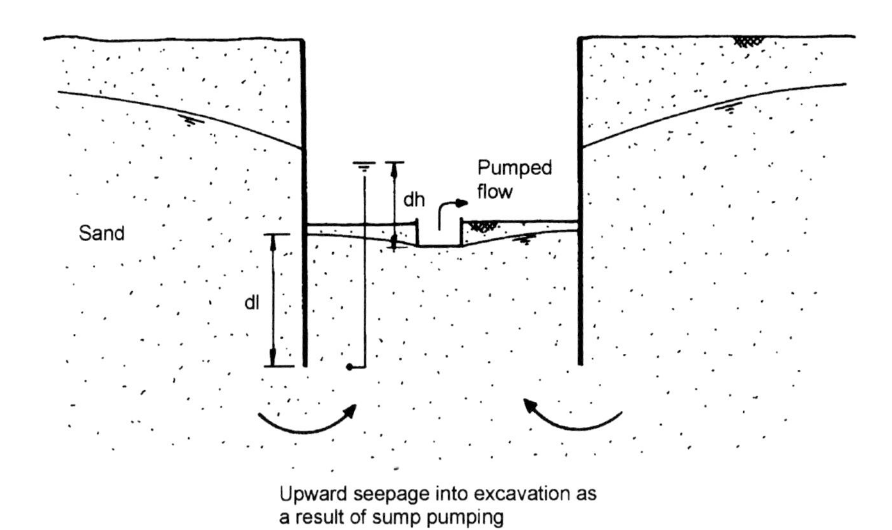

Figure 4.2 Fluidization of excavation base due to upward seepage in unconfined aquifer. Upward hydraulic gradient below excavation floor = dh/dl.

approached. In design it is normal to limit the predicted hydraulic gradients to be less than i_{crit}/F, where F is a factor of safety of 1.2 or greater.

The hydraulic gradient for upward seepage is defined as the head difference dh divided by the flow path length dl (see Figure 4.2). The hydraulic gradient can be controlled in two ways.

(a) By increasing dl. This can be achieved by ensuring the cut-off walls penetrate a sufficient depth below formation level. This is an important part of the design of cut-off walls in unconfined aquifers (see Williams and Waite 1995).

(b) By reducing dh. This requires reducing the groundwater head difference between the inside and the outside of the excavation enclosed within the cut-off walls. The most obvious way of doing this is to use a system of dewatering wells outside the excavation to lower the external groundwater level and so reduce the head difference to an acceptable level, while still sump pumping from within the excavation. This approach might be taken a step further and the external wells be used to lower the groundwater level to below formation level, and avoid the need for sump pumping altogether. An alternative approach to reducing dh is to keep the excavation partly or fully topped up with water and carry out excavation underwater (e.g. using grabs) and then place a tremie concrete plug in the base. This method is sometimes used for the construction of shafts and cofferdams.

4.3.2 *Confined aquifers*

The classic confined aquifer case is shown in Figure 4.3. It shows an excavation through a clay stratum, with the sides of the excavation supported by walls of some sort. The formation level of the excavation is in a very low permeability clay stratum that forms a confining layer above a permeable confined aquifer. The piezometric level in the aquifer is considerably above the base of the clay layer.

When assessing the stability of this case, the critical horizon is the interface between the confined aquifer and the underside of the overlying confining bed. Consider the two separate sets of pressures acting at this level. The excavation will be stable and safe provided the downward pressure from the weight of the residual 'plug' of unexcavated clay is sufficiently greater than the upward pressure of pore water confined in the aquifer. This is assuming that the confining stratum is consistently of very low permeability, and is competent and unpunctured (e.g. it is not penetrated by any poorly sealed investigation boreholes, which could form water pathways).

If excavation within the cofferdam is sunk further into the clay without any reduction of pore water pressures in the sand aquifer, there will come a time when the upward water pressure in the aquifer will exceed the weight of the clay plug. There will then be an upward movement or 'heave' of the excavation formation. If the clay plug heaves sufficiently the clay may rupture, allowing an uprush of water and sand (sometimes known as a 'blow') from

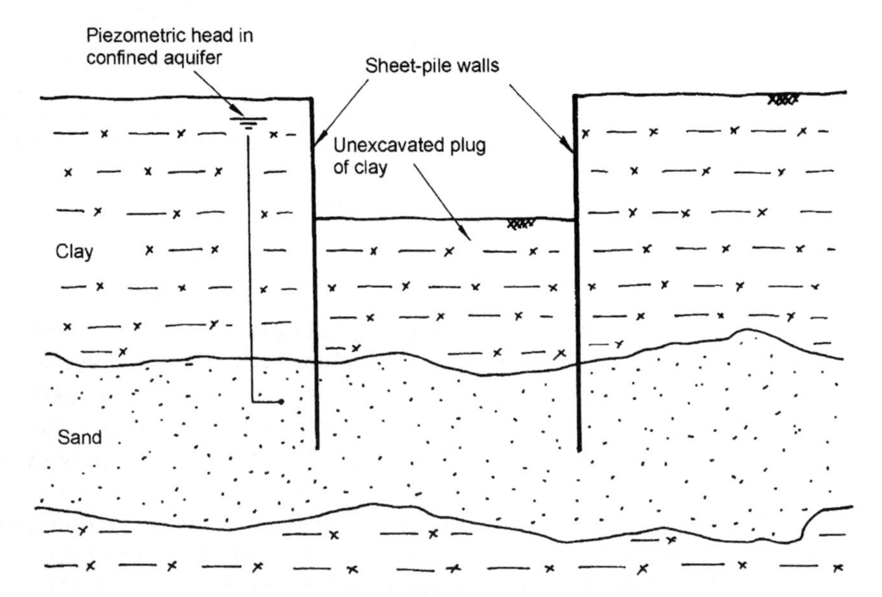

Figure 4.3 Excavation in very low permeability soil overlying a confined aquifer – stable condition. For a relatively shallow excavation the weight of the unexcavated plug of clay is greater than the upward water pressure from the confined aquifer.

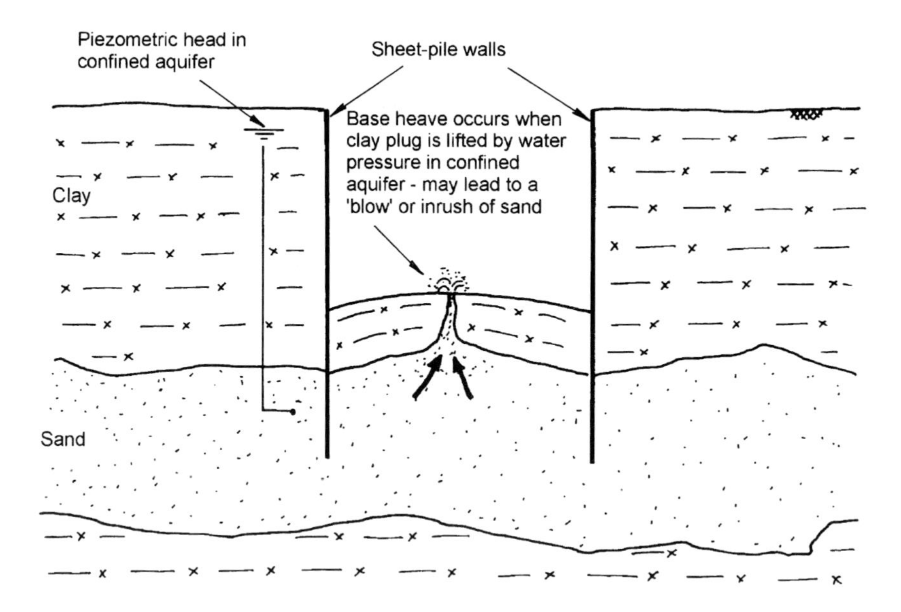

Figure 4.4 Excavation in very low permeability soil overlying a confined aquifer – unstable condition. For a deeper excavation the upward water pressure from the confined aquifer exceeds the weight of the clay plug, leading to base heave and, ultimately a 'blow'.

the aquifer which may even lead to the collapse of the excavation (see Figure 4.4).

The 'heave' situation can be avoided either by:

(a) Adequately reducing the pore water pressure in the confined aquifer by pumping from suitable dewatering wells inside the cofferdam. Alternatively, the wells may be sited outside the cofferdam but will be less hydraulically efficient. The wells should lower the aquifer water pressure so that the downward forces exceed the upward pressures by a suitable factor of safety.
(b) Increasing the depth of the physical cut-off wall sufficient to penetrate below the base of the confined aquifer. Provided the cut-off walls are water tight this will prevent further recharge, thus leaving only the water pressure contained in the aquifer within the cut-off walls to be dealt with. This can be effected by installing relief wells (see Section 11.3) prior to excavation. The economics of this alternative will depend primarily on the depth needed to secure a seal and the effectiveness of that seal.
(c) Increasing the downward pressure on the base of the excavation by keeping it partly topped up with water during the deeper stages of work. Excavation is made underwater, and a tremie concrete plug used to seal the base on completion.

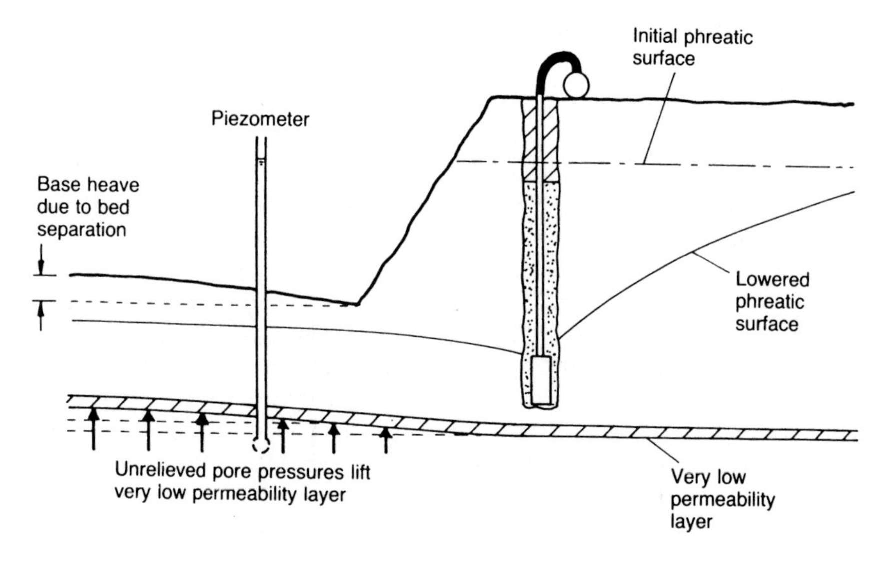

Figure 4.5 Pore water pressures not controlled beneath thin very low permeability layer – unstable condition (from Preene and Powrie 1994, with permission).

The above description describes the principal features of the 'base heave' failure in confined aquifers. For relatively narrow excavations the shear resistance between the clay plug and surrounding ground may be significant and should be considered when estimating factors of safety. Hartwell and Nisbet (1987) discuss the problem further.

Base heave can also sometimes occur in nominally unconfined aquifers that contain discrete (but perhaps very thin) very low permeability clay layers. Figure 4.5 shows a case where dewatering wells are used to lower groundwater levels around an excavation, but do not penetrate the clay layer below the excavation. The dewatering system has lowered water pressures above the clay layer, but the original high pressures remain beneath. These high pressures can cause base heave. This mechanism is sometimes known as 'bed separation', because as the clay layer moves upward a reservoir of water will develop beneath. This form of base heave can be avoided by using deeper dewatering wells to control pore water pressures at depth.

4.4 Localized groundwater problems

In addition to groundwater effects that can cause large-scale failure of an excavation, groundwater can also cause more localized problems in excavations. The effects of localized problems can vary greatly in scope, from minor inconveniences caused by a perched water table to severe problems resulting from the collapse of side slopes due to spring sapping – localized groundwater problems should not be ignored. A number of possible mechanisms are described in the following sections.

4.4.1 Drainage of slopes and formation

The slopes of open cut excavations should be designed so that instability due to seepage and associated continuous removal of fines does not occur. Where the slope is too steep or the hydraulic head too large, seepage can daylight on the slope and cause slope failure (see Figure 4.6). The term daylight is another expression for local trouble spots – but they may not be only local!

Often the solution is to have a flatter slope, sometimes together with a toe drain backfilled with filter media (see Figure 4.7) to prevent emerging water continuously removing fines and so causing damage leading to instability. It is advantageous to return the filter material back up the slope a short distance and so prevent seepage in that area from transporting fines. Periodic maintenance may be needed to ensure toe drains remain efficient.

An alternative solution is to adopt a slightly steeper slope angle with a properly backfilled drainage trench at the toe of the slope, in conjunction

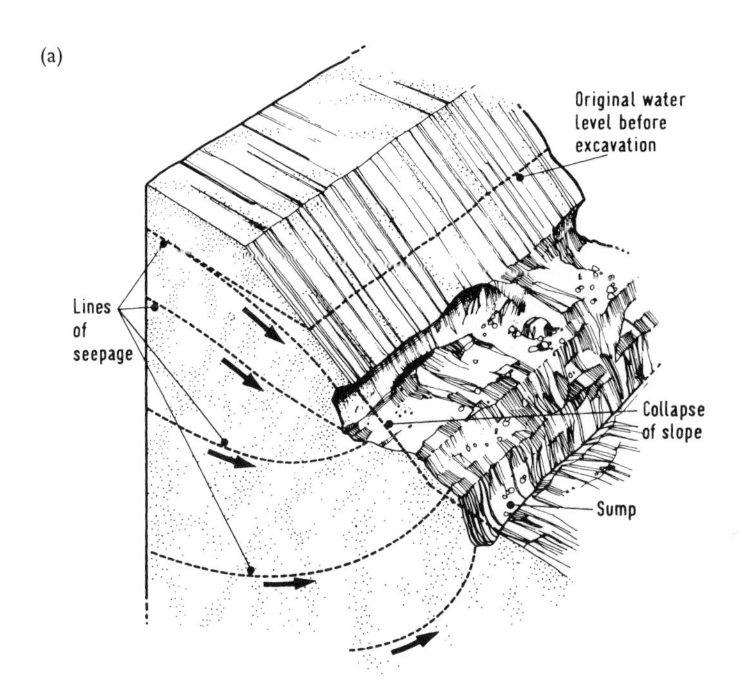

Figure 4.6 Effect of seepage on excavation. (a) Collapse of excavation caused by seepage from steep slopes (from Somerville, 1986: reproduced by kind permission of CIRIA), (b) unstable excavation slope resulting from seepage. Seepage has eaten back into the slope, and an outwash fan of saturated washed out material has formed at the base of the slope.

(b)

Figure 4.6 Continued.

with sand-bagging of the toe of the slope or weighting with graded filter material. Alternatively, the fill can be weighted with loose-laid timbers and sacking, with straw or hay padded behind them. Modern practice also allows the use of one of the various proprietary geotextile fabrics. Sand-bagging or a geotextile membrane may be needed where local perched water tables are revealed. The materials to be used will depend on local availability in the country concerned.

The gradients within the excavated floor areas should be formed so that the run-off water is directed to the collector drains and the risk of ponding is assiduously avoided. Run-off from slopes should be collected by the surface water drainage system, suitably sited at each berm and/or toe of slope, and thence be directed to pumping sumps for ultimate disposal off site.

Sumps are usually sited at the corners of excavations, below the general excavation level, and made big enough to hold sufficient water for pumping

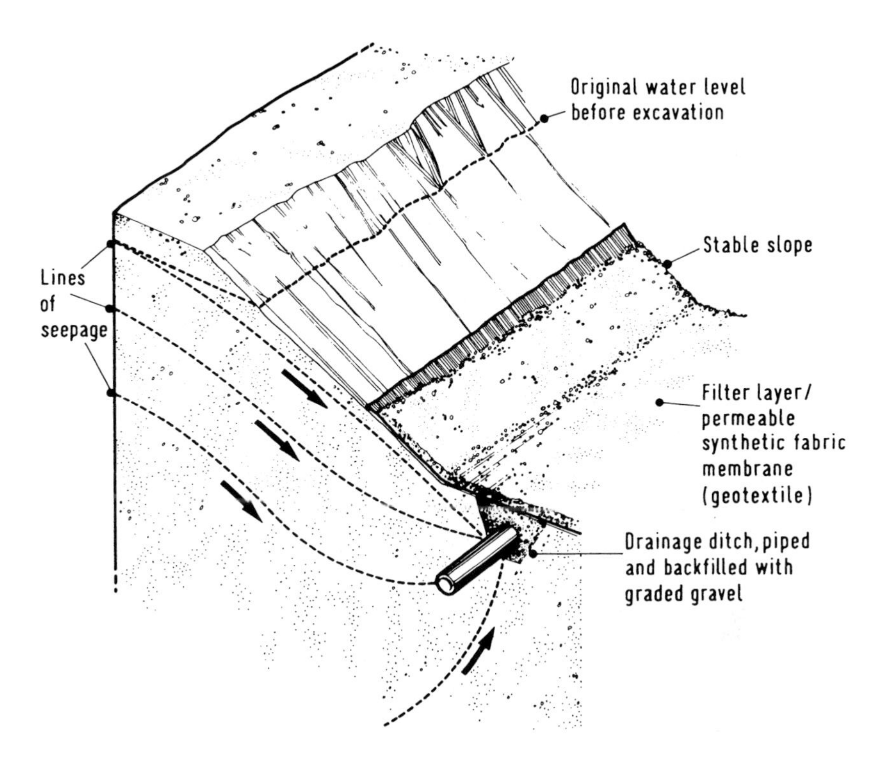

Figure 4.7 Stable excavated slopes resulting from flat gradient and provision of drainage trench (from Somerville 1986: reproduced by kind permission of CIRIA).

and keep the excavation floor relatively dry. A pump is provided for each sump and connected to a discharge pipe.

The drains leading to the sumps should be so arranged as to allow drainage of the whole excavation and have sufficient fall to prevent silting up. Also, the drains should be cleaned out from time to time. Ditches should be sufficiently wide to allow a water velocity low enough to prevent erosion. This may be achieved by constructing check weirs at intervals along the ditch. Additionally, the ditches can be improved by laying rough blinding or paving material or laying porous open-jointed pipes or other agricultural land drainage piping surrounded by filter material.

Observance of good practice applicable to slope grades and filter design criteria will help achieve stable slopes but it is essential to be on the look out for local trouble spots, generally due to variations in soil conditions.

4.4.2 Perched water table

Often within permeable granular strata random lenses of significantly lower permeability occur, locally inhibiting downward drainage of water.

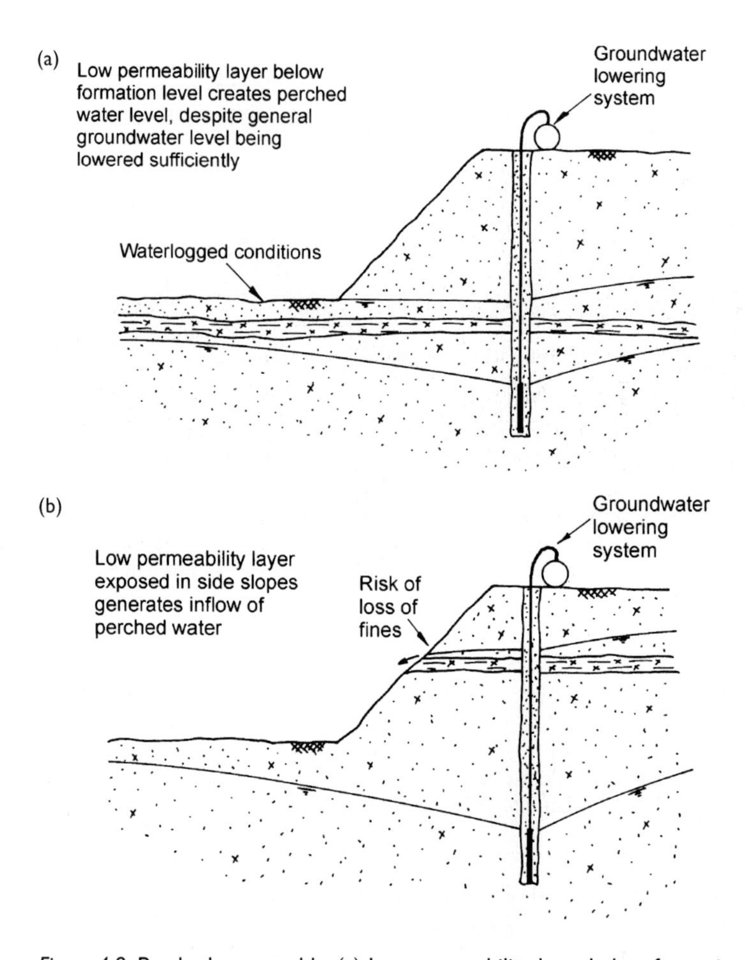

Figure 4.8 Perched water table. (a) Low permeability layer below formation level, (b) low permeability layer above formation level and exposed in side slope.

A 'perched water table' is likely to result from this phenomenon (Figure 4.8). This means a residual body of water remains just above the lens of low permeability soil even though the groundwater level in the permeable strata beneath has been drawn down lower. This troublesome condition can be dealt with effectively by forming vertical drainage holes through the low permeability lens. This can sometimes be done by the simple expedient of digging through the layer with an excavator bucket. Alternatively, relatively small diameter sand drains may be appropriate, perhaps jetted in using wellpoint equipment (see Section 9.7).

If the seepage from perched water tables continues once the excavation has passed through the perched water table this is known as residual seepage or

Figure 4.9 Perched water table (courtesy of Soil Mechanics). The groundwater level monitored in observation wells was several metres below the standing water visible in the photograph. When the clay layers below this level were punctured by excavating, the perched water drained away.

'overbleed'. If the rates of seepage are low, the water will merely constitute a nuisance, rather than lead to a serious risk of flooding. Nevertheless, even at small rates of seepage, if fines are being transported by the water this should be counteracted immediately. Possible measures include the placement of sand bags, a granular drainage blanket or geotextile mesh to act as filters.

An extensive perched water table in an unconfined aquifer of gravelly sand is shown in Figure 4.9. Readings in several observation wells indicated that the pumping from a system of deep wells had drawn the groundwater level down to the target level, which was several metres below the perched water level visible in the figure. The contractor was advised to continue excavation by dragline at one location. Within a few hours of continuous digging the water started to swirl and there was a loud gurgling sound as the perched water drained away.

4.4.3 High permeability zones

Sometimes a lens of significantly more permeable water-bearing soil, not previously detected, may be revealed during excavation. This will act as a source of copious recharge requiring additional pumping capacity to be installed to abstract groundwater from the lens. Such permeable gravel lenses are not uncommon in fluvio-glacial or alluvial soils which were laid down in water. The permeable zones may result from paths of former streams or flow channels within the geological deposits. Like modern day watercourses these gravel lenses may follow an irregular, difficult to predict path. Long, thin gravel lenses are known as 'shoestrings'; these may easily pass undetected during site investigation.

The same recharge effect as from a natural gravel lens can result from man-made features such as the permeable gravel bedding or sub-drains associated with sewers. There have been cases where dewatering adjacent to existing sewers has proved difficult, until the flow path formed by the permeable material around the sewer was blocked, or the flow intercepted directly by sump pumping from the gravel bedding.

Figure 4.10 shows a case described by Preene and Powrie (1994) where an ejector system for a small shaft excavation lowered the general groundwater level but failed to prevent persistent troublesome seepage on one side of the shaft. Additional ejector wells installed around the shaft (and logged as site investigation boreholes) revealed the presence of a shoestring of permeable gravel present within the main stratum of fluvio-glacial sand.

Additional wells were installed into the shoestring to intercept the water before it reached the excavation, the seepage was eliminated and the shaft

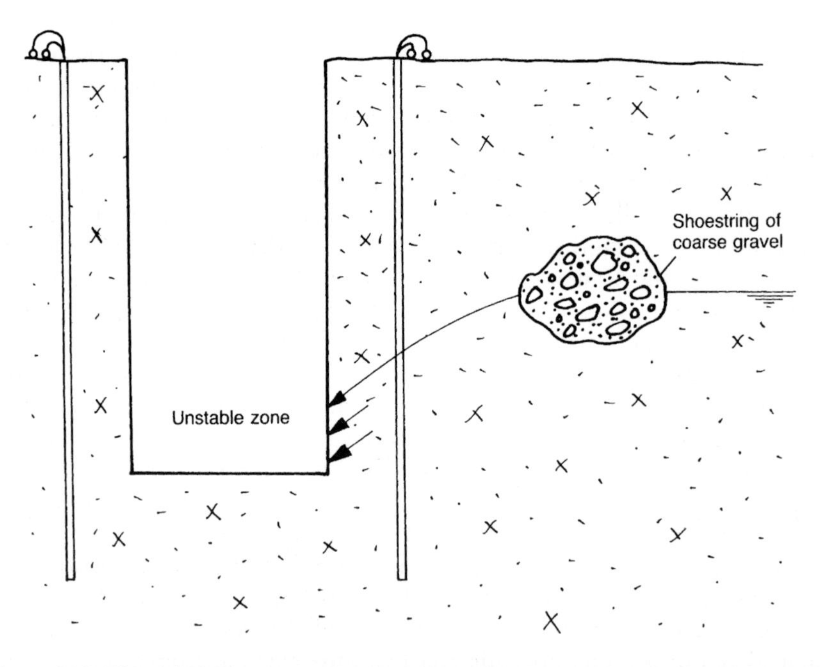

Figure 4.10 Effect of close source of recharge on effectiveness of groundwater lowering for shaft excavation (from Preene and Powrie 1994, with permission). A permeable gravel shoestring acts as a close source of recharge and concentrates seepage on one side of the shaft, leading to local instability.

successfully completed. This sounds very straightforward, but this was far from the true story. Because the location of the shoestring was not known precisely, the dewatering contractor had to keep drilling wells until a few of them directly intercepted the permeable zone. Up until that time, even when many additional wells had been installed and pumped, the seepage persisted. The total number of wells increased from five to twenty-two! In reality, only the handful pumping directly from the shoestring were truly effective.

The principle to be remembered is that whenever a local source of more copious recharge is encountered, additional wells or sumps should ideally draw groundwater direct from the recharge source (i.e. the permeable zone). It is preferable to intercept the flow upstream of the exposed excavation but sometimes this may not be practicable. In such circumstances the additional wells or sumps must be installed as close as possible to the area of water emergence into the work.

4.4.4 Spring sapping and internal erosion

This mechanism, which can occur in sands and silty fine sands, is not uncommon, but is not well understood. The sides of an excavation in silty fine sand may show gully structures from which sand slurry can be seen to flow into the bottom of the excavation. This is due to internal erosion. The water flowing along the upper surface of the lowered phreatic line carries sand slurry with it and thus tunnels backwards beneath the overlying sand, which has some strength owing to capillary tension. Before this tunnelling effect has gone far, the top collapses and the collapsed material is also carried away. This tunnelling phenomenon has often been observed in fine sands and especially in silty fine sands. It tends to occur immediately above a thin and less permeable layer of finer material which has caused a perched water table, though this condition may only be temporary. This phenomenon is quite well known in geology under the name of spring sapping or seepage erosion.

In the 1960s, an interesting experiment into this phenomenon was carried out at Imperial College, London. A slope of fine sand was formed in a glass-sided tank and a small source of water was allowed to flow in through the base of the tank. The base of the tank was at a flat angle. A very small flow of water was introduced beneath the sand and erosion rapidly started at the point of exit. One would intuitively expect a cone of eroded material to build up, and for the process of erosion to come to an end, but in fact the flow of sand was continuous – as it was in a case described by Ward (1948). It seems that this flow can only happen if excess pore pressures exist in eroded material, but their cause is still unexplained and it is hard to see why they are not quickly dissipated.

The site studied by Ward (1948) was near Newhaven in Sussex, at a point where Woolwich and Reading Beds (now known more correctly as the Lambeth Group) overlie the Thanet Sand and Chalk. A thin bed of fine sand in the Woolwich and Reading Beds carried a small flow of groundwater towards the sea, and although the hydraulic gradient was only about one-in-ten, internal erosion on a large scale had continued for centuries, and debris flows remained active and carried the sand and the collapsed mass of the overlying clay down to the beach some 55 m below. This state of affairs was eventually controlled by a simple system of filters and drains.

The curious thing about this process is that it takes place when the flow of water is very small, and the gradient very low, yet because it is continuous, in time very large quantities of material are removed.

Figure 4.11 shows a close up view of a small example of spring sapping. A thin lens of predominantly shells was present within an extensive stratum of fine sand. The levels recorded in nearby observation wells indicated that the water had been drawn down to some 4 m below this lens but because the permeability of the lens of shelly material was significantly greater there was still water flowing, continuously transporting fines and so causing local spring sapping. This was evidenced by the two cavities in the centre of the photograph and the outwash fans of almost liquid sand in the foreground.

Variable soil conditions may be the cause of local trouble spots being revealed as excavation proceeds. Drift deposits are usually heterogeneous

Figure 4.11 Spring sapping. Local spring sapping has caused the two cavities in the centre of the photograph with outwash fans of almost liquid sand in the foreground.

and so the possibility of exposing local trouble spots in the sides of excavations is ever present. Often these trouble spots are due to a layer or lens of slightly lower permeability material within stratum being dewatered. Due to the difference in permeability the pore pressures in the layer or lens are higher than in the surrounding stratum and despite the general effectiveness of the installed groundwater lowering system, there is some residual flow from the lens, which causes transportation of fines. If this condition is allowed to persist it will result in instability due to seepage erosion.

4.5 Excavations in rock

The preceding sections have described the effect of groundwater on excavations on uncemented soil or 'drift' deposits. Groundwater lowering is most commonly employed in these material types, but is also occasionally used for excavations in 'rock'. For the purposes of groundwater control the defining features of rock is that the rock mass is relatively strong and often of low permeability. As a result, groundwater tends not to flow through the whole rock mass, but tends to be concentrated locally along fissures, fractures, joints, solution fissures and other discontinuities.

In general, groundwater-induced instability is much less of a problem for excavations in rocks compared with excavations in soils. An excavation in rock below the water table will encounter localized inflows where joints or fissures are intercepted. Provided the rock around the fissures is competent, these inflows may not cause stability problems. In such cases the aims to prevent the inflows from affecting construction operations, and sump pumping is often used to manage water within the excavations.

There are a number of potential problems which may result from groundwater inflows into rock excavations.

- i The concentration of flow where water flows from a fissure into the excavation may create high flow velocities, and an associated risk of internal erosion. There have been cases in relatively soft mudstones and sandstones where initial inflows were modest, but increased significantly with time as the fissures were enlarged by erosion. One solution would be to install a system of external dewatering wells to reduce the groundwater head driving flow into the excavation. This will reduce the potential for erosion.
- ii Weathered beds or zones may exist within otherwise competent rock. These zones may behave like drift deposits and may slump inward as groundwater flows through them. These problems may be avoided by the use of external dewatering wells or, if very localized, by the use of drainage measures similar to those described in Section 4.4.
- iii Groundwater will emerge into the excavation at locations dictated by the rock structure, bedding, fissuring, etc. These locations will often be

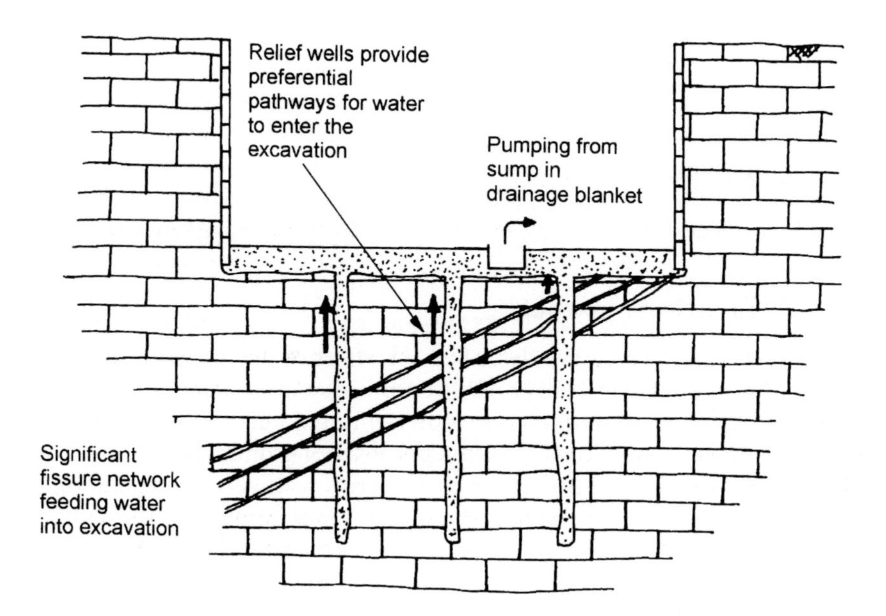

Figure 4.12 Relief wells used in rock excavations to intercept flow in discrete fissures.

inconvenient for construction operations – trying to place high quality structural concrete on top of a gushing fissure is unlikely to be efficient! This problem can be dealt with by constructing trench drains (see Section 5.2) to carry the water away from the working area to a sump. Alternatively, an array of relief wells (see Section 11.3) could be drilled through the base of the excavation to try and intercept fissures and to provide a preferential pathway to allow groundwater to enter the excavation at a more convenient location (Figure 4.12).

iv If the rock possesses bedding that is roughly horizontal, pore water pressures may remain trapped beneath lower permeability layers leading to the 'base heave' type problem described in Section 4.3. This sometimes occurs where the rock consists of alternating layers of mudstone (relatively impermeable) and sandstone (relatively permeable); such conditions exist in the Coal Measures rock in northern England. The installation of relief wells within the excavation will prevent this by allowing the water trapped in the more permeable zones to flow freely into the excavation.

v In very large rock excavations there may be a risk of 'block failure' where part of the rock mass slides or moves into the excavation, often moving along joints or fissures that slope into the excavation. The pore water pressures acting in the rock joints has a critical effect on the stability of such rock masses. Dewatering wells have occasionally been used to control pore water pressures as part of stabilization measures. This is a very specialized case, and it is essential that rock mechanics specialists are involved if this approach is being contemplated.

4.6 Surface water problems

This book is primarily concerned with the control of groundwater but, since the aim is to provide workable conditions for construction, the importance of surface water control must not be forgotten. Surface water may arise in many ways: as direct precipitation into the excavation; as precipitation run-off from surrounding areas; as waste water from construction operations such as concreting or even from washing down of plant and equipment. Since the excavation is obviously going to form a low point in the site, surface water, if given free rein, is likely to collect in the bottom of the excavation.

Whatever the source, if surface water is allowed to pond in the excavation it will impede efficient excavation and construction. Figure 4.13 shows an example of an excavation before and after surface water was suitably dealt with. Neglecting the management of surface water will undoubtedly cost the project time and money. Methods for control of surface water are described in Sections 5.1 and 8.2.

4.7 Effect of climate and weather

Depending upon climatic conditions it may be prudent to make major or minor contingency plans. For instance, in some parts of the world it is known that monsoons, typhoons and hurricanes will occur and the consequential high volume surface run-off may be very significant. Hence, planning for such areas would dictate that appropriate measures to deal with sudden influxes of surface water, prevent slope erosion and deal with short-term rises in groundwater level must be incorporated into the excavation plan at design stage. Particularly when working in shallow unconfined aquifers, intense rainfall events can have a major effect on lowered water levels. Figure 4.14 shows groundwater level and rainfall data taken during the rainy season at a site in South-East Asia. A wellpoint system was used to lower water levels in a very silty sand and had achieved a drawdown of several metres. Whenever there was heavy rain the water levels rose rapidly and only fell back to their original levels slowly under the influence of pumping.

Exposed excavated slopes may be protected from monsoon type erosion by laying an appropriate geotextile membrane. A judgement should to be made at design stage whether to cater for the once in ten-year occurrence, or the fifty-year occurrence or more onerous conditions.

(a)

(b)

Figure 4.13 Control of surface water in excavations. (a) Without adequate surface water control, (b) with effective surface water control.

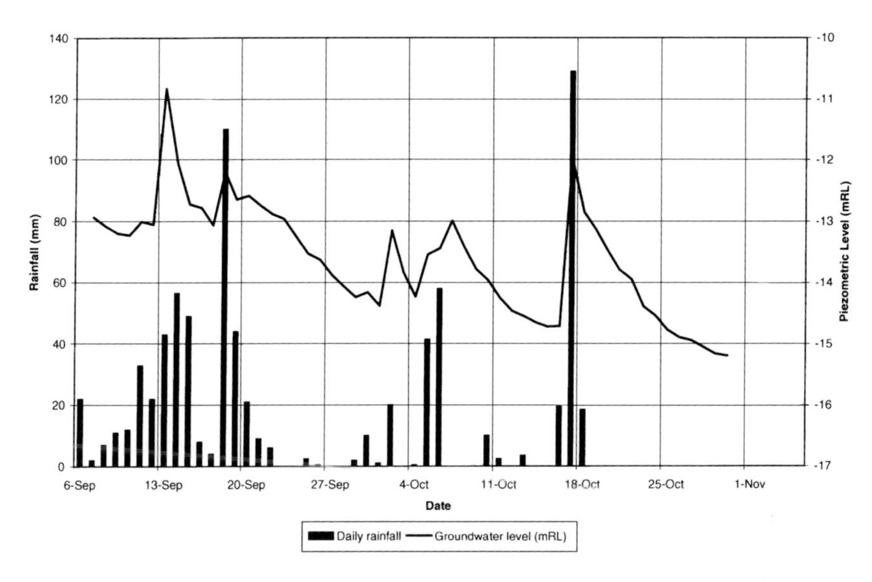

Figure 4.14 Effect of extreme rainfall events on groundwater levels. Data are from monitoring of a wellpoint system operating on a site in South-East Asia during the rainy season.

References

Hartwell, D. J. and Nisbet, R. M. (1987). Groundwater problems associated with the construction of large pumping stations. *Groundwater Effects in Geotechnical Engineering* (Hanrahan, E. T., Orr, T. L. L. and Widdis, T. F., eds). Balkema, Rotterdam, pp 691–694.

Powrie, W. (1997). *Soil Mechanics: Concepts and Applications*. Spon, London.

Preene, M. and Powrie, W. (1994). Construction dewatering in low permeability soils: some problems and solutions. *Proceedings of the Institution of Civil Engineers, Geotechnical Engineering*, **107**, 17–26.

Somerville, S. H. (1986). *Control of Groundwater for Temporary Works*. Construction Industry Research and Information Association, CIRIA Report 113, London.

Ward, W. H. (1948). A coastal landslip. *Proceedings of the 2nd International Conference on Soil Mechanics and Foundation Engineering*, Rotterdam, **2**, 33–38.

Williams, B. P. and Waite, D. (1993). *The Design and Construction of Sheet-Piled Cofferdams*. Construction Industry Research and Information Association, CIRIA Special Publication 95, London.

Chapter 5

Methods for control of surface water and groundwater

5.0 Introduction

As has been discussed in Chapter 4, the presence of groundwater can lead to troublesome conditions when construction operations are to take place below the original groundwater level. Surface water run-off must also be effectively controlled to prevent interference with excavation and construction works. This chapter briefly outlines some available methods for control of surface water and groundwater.

Techniques for the control of groundwater can be divided into two principal types:

(a) Those that exclude water from the excavation (known as exclusion techniques).
(b) Those that deal with groundwater by pumping (known as dewatering techniques).

This chapter includes brief discussions and tabular summaries of the various available techniques of both groups, including their advantages and disadvantages. Apart from this summary, this book addresses only the dewatering techniques. The practicalities of the more commonly used pumping or dewatering methods are described in detail in subsequent chapters.

5.1 Control of surface water

In order to optimize efficiency (and profitability) it is essential that conditions within the area of excavation and construction be workable; mechanical plant should be able to operate efficiently without getting bogged down. To achieve this, in addition to lowering of groundwater levels, surface water must be controlled. Thus both surface water run-off and groundwater should be disposed of expeditiously.

On many construction sites the measures enacted to control surface water run-off are inadequate, and unnecessary waste of time and money results.

5.1.1 *Surface water run-off*

The basic rule of good practice is to collect and control the surface water run-off as soon as, or better still even before, it enters the area of work.

The drainage system should incorporate appropriate measures to collect run-off from land areas surrounding and adjacent to an excavation to prevent surface water run-off encroaching into the construction area. Adoption of this philosophy requires the installation of adequate interceptor drains sited uphill or upstream of the excavation at original ground level.

Rain water or discharges from other construction activities (such as concreting or washing down of plant) should be prevented from entering an excavation, particularly on a sloping site, by the simple expedient of digging collector ditches or drains on the high ground. The drain should lead the water away to discharge points (which could be pumped by sump pumps, see Chapter 8) lower down the slope. Collector drains should be lined with an impermeable membrane to avoid potential upstream recharge that might cause a rotational slip of the excavation slope. The open ditch method (see Fig. 5.1(a) should be used only where the presence of open ditches does not inhibit construction work. As an alternative, agricultural type drains can be used (see Fig. 5.1(b), but the surface must be regularly scarified to reduce the effect of clogging by suspended particles in the surface water. This cleaning dictum should be applied to both the open ditch and the agricultural type drains.

5.2 Methods of groundwater control

The techniques available for control of groundwater fall into two principal groups:

(a) Those that exclude water from the excavation (known as exclusion techniques).
(b) Those that deal with groundwater by pumping (known as dewatering techniques).

This book is primarily concerned with the second group, the dewatering methods. However, the essential features of groundwater control by exclusion are outlined in the following section.

5.3 Exclusion methods

The aim of groundwater control by exclusion is to prevent groundwater from entering the working area. The methods used can be grouped into three broad categories:

1 Methods where a very low permeability discrete wall or barrier is physically inserted or constructed in the ground (e.g. sheet-piling, diaphragm walls).

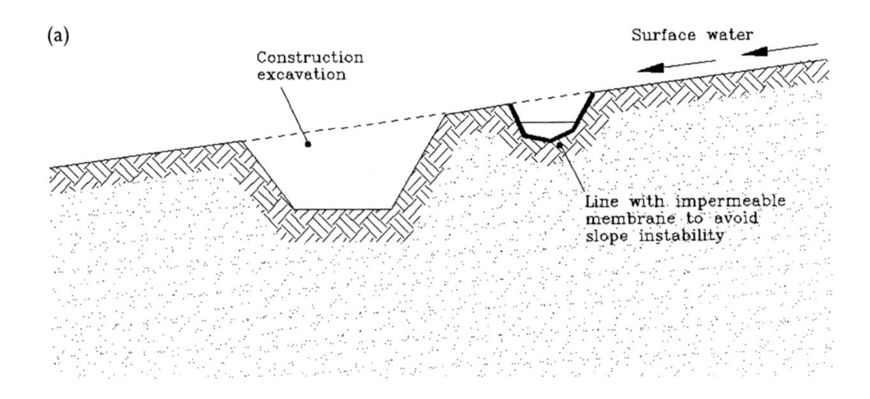

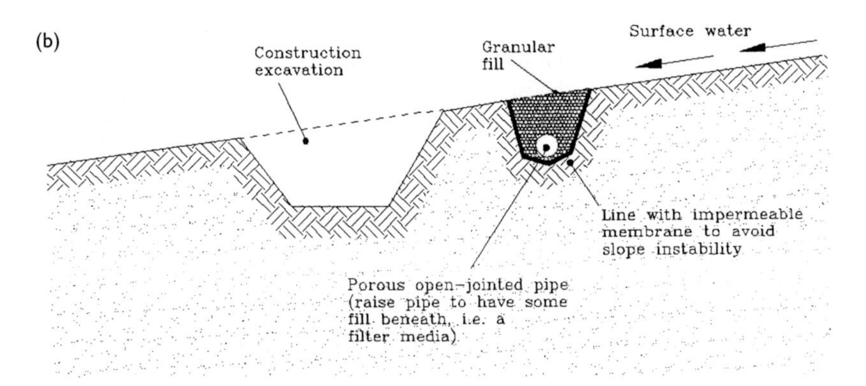

Figure 5.1 Drains for collection of surface water run-off (after Somerville 1986). (a) Open ditch, (b) agricultural drain.

2 Methods which reduce the permeability of the *in situ* ground (e.g. grouting methods, artificial ground freezing).
3 Methods which use a fluid pressure in confined chambers such as tunnels to counter balance groundwater pressures (e.g. compressed air, earth pressure balance tunnel boring machines).

Techniques used to exclude groundwater are listed in Table 5.1, which is based on information from Preene *et al.* (2000). Details of the various methods can be found in Bell and Mitchell (1986) or the further reading listed at the end of this chapter.

One of the most common applications of the exclusion method involves forming a notionally impermeable physical cut-off wall or barrier around the perimeter of the excavation to prevent groundwater from entering the working area. Typically, the cut-off is vertical and penetrates down to a

Table 5.1 Principal methods for groundwater control by exclusion

Method	*Typical applications*	*Notes*
Displacement barriers		
Steel sheet-piling	Open excavations in most soils, but obstructions such as boulders or timber baulks may impede installation	May be installed to form permanent cut-off, or used as temporary cut-off with piles removed at the end of construction. Rapid installation. Can support the sides of the excavation with suitable propping. Seal may not be perfect, especially if obstructions present. Vibration and noise of driving may be unacceptable on some sites, but 'silent' methods are available. Relatively cheap
Vibrated beam wall	Open excavations in silts and sands. Will not support the soil	Permanent. A vibrating H-pile is driven into the ground and then removed. As it is removed, grout is injected through nozzles at the toe of the pile to form a thin, low permeability membrane. Rapid installation. Relatively cheap, but costs increase greatly with depth
Excavated barriers		
Slurry trench cut-off wall using bentonite or native clay	Open excavations in silts, sands and gravels up to a permeability of about 5×10^{-3} m/s	Permanent. The slurry trench forms a low permeability curtain wall around the excavation. Quickly installed and relatively cheap, but cost increases rapidly with depth
Structural concrete diaphragm walls	Side walls of excavations and shafts in most soils and weak rocks, but presence of boulders may cause problems	Permanent. Support the sides of the excavation and often form the sidewalls of the finished construction. Can be keyed into rock. Minimum noise and vibration. High cost may make method uneconomical unless walls can be incorporated into permanent structure
Secant (interlocking) and contiguous bored piles	As diaphragm walls, but penetration through boulders may be costly and difficult	As diaphragm walls, but more likely to be economic for temporary works use. Sealing between contiguous piles can be difficult, and additional grouting or sealing of joints may be ncessary
Injection barriers		
Jet grouting	Open excavations in most soils and very weak rocks	Permanent. Typically forms a series of overlapping columns of

Table 5.1 Continued

Method	*Typical applications*	*Notes*
		soil/grout mixture. Inclined holes possible. Can be messy and create large volumes of slurry. Risk of ground heave if not carried out with care. Relatively expensive
Mix-in-place columns	Open excavations in most soils and very weak rocks	Permanent. Overlapping columns formed by *in-situ* mixing of soil and injected grout using auger-based equipment. Produces little spoil. Less flexible than jet grouting. Relatively expensive
Injection grouting using cementitious grouts	Tunnels and shafts in gravels and coarse sands, and fissured rocks	Permanent. The grout fills the pore spaces, preventing the flow of water through the soil. Equipment is simple and can be used in confined spaces. A comparatively thick zone needs to be treated to ensure a continuous barrier is formed. Multiple stages of treatment may be needed
Injection grouting using chemical and solution (acrylic) grouts	Tunnels and shafts in medium sands (chemical grouts), fine sands and silts (resin grouts)	As cementitious grouting, but materials(chemicals and resin) can be expensive. Silty soils are difficult and treatment may be incomplete, particularly if more permeable laminations or lenses are present
Other types		
Artificial ground freezing using brine or liquid nitrogen	Tunnels and shafts. May not work if groundwater flow velocities are excessive ($>$2 m/day for brine or $>$20 m/day for liquid nitrogen)	Temporary. A 'wall' of frozen ground (a freezewall) is formed, which can support the side of the excavation as well as excluding groundwater. Liquid nitrogen is expensive but quick; brine is cheaper but slower. Liquid nitrogen is to be preferred if groundwater velocities are relatively high. Plant costs are relatively high
Compressed air	Confined chambers such as tunnels, sealed shafts and caissons	Temporary. Increased air pressure (up to 3.5 bar) raises pore water pressure in the soil around the chamber, reducing the hydraulic gradient and limiting groundwater inflow. Potential health hazards to

Table 5.1 Continued

Method	*Typical applications*	*Notes*
		workers. Air losses may be significant in high permeability soils. High running and set-up costs
Earth pressure balance tunnel boring machine (TBM)	Tunnels in most soils and weak rocks	Temporary. The TBM excavates for the tunnel, and supports the soil and excludes groundwater by maintaining as balancing fluid pressure in the plenum chamber immediately behind the cutting head. The fluid is a mixture of soil cuttings, groundwater and conditioning agents (such as polymer or bentonite muds). TBMs need to be carefully selected to deal with given ground conditions; set-up and running costs may be high

very low permeability stratum that forms a basal seal for the excavation (Fig. 5.2(a)).

The costs and practicalities of constructing a physical cut-off wall are highly dependent on the depth and nature of any underlying permeable stratum. If a suitable very low permeability stratum does not exist, or is at great depth, then upward seepage may occur beneath the bottom of the cut-off wall, leading to a risk of base instability (see Section 4.3). In such circumstances dewatering

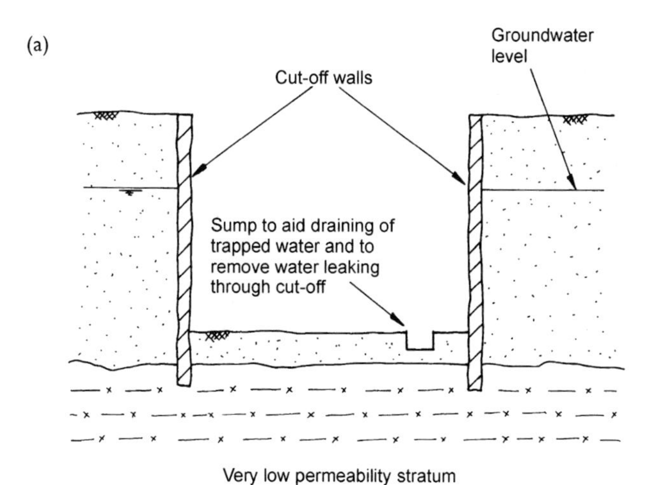

Figure 5.2 Groundwater control by exclusion using physical cut-offs. (a) Cut-off walls penetrate into very low permeability stratum, (b) cut-off walls used in combination with dewatering methods, (c) cut-off walls used with horizontal barrier to seal base.

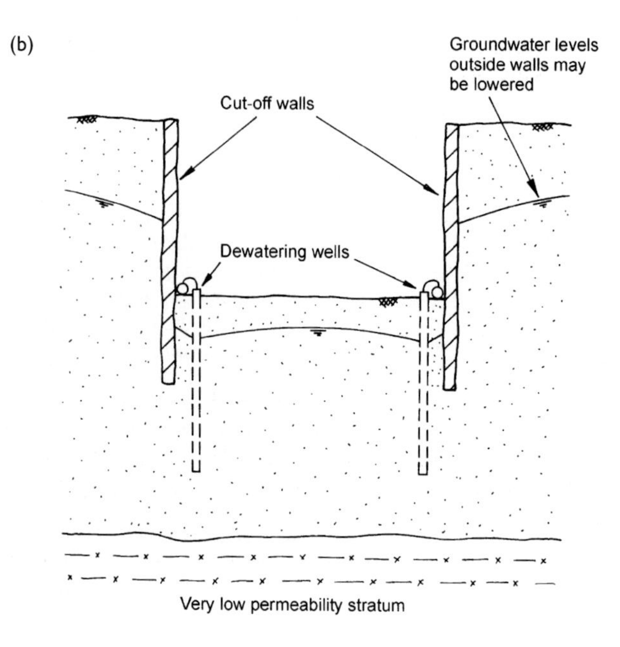

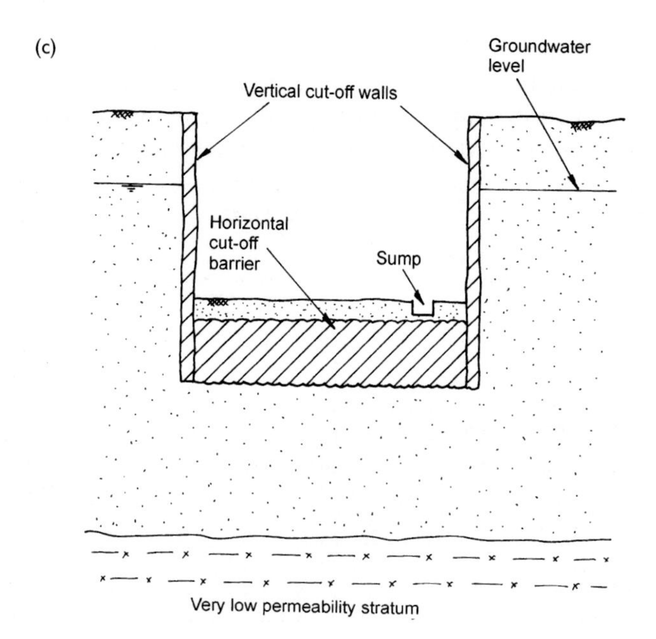

Figure 5.2 Continued.

methods may be used in combination with exclusion methods (Fig. 5.2(b)). Alternatively, it may be possible to form a horizontal barrier or 'floor' to the cut-off structure to prevent vertical seepage (Fig. 5.2(c)). The construction of horizontal barriers is relatively rare, but has been carried out using jet grouting, mix-in-place, grouting and artificial ground freezing techniques.

If a complete physical cut-off is achieved, some groundwater will be trapped inside the working area. This will need to be removed to allow work to proceed, either by sump pumping during excavation, or by pumping from wells or wellpoints prior to excavation.

One of the attractive characteristics of the exclusion technique is that it allows work to be carried out below groundwater level with minimal effects on groundwater levels outside the site. This means that any potential side effects of dewatering (see Chapter 13) are avoided. In particular, in urban areas exclusion methods are often used in preference to dewatering methods to reduce the risk of settlement damage resulting from lowering of groundwater levels. However, when considering using the exclusion technique to avoid groundwater lowering in areas outside the site, it is essential to remember that almost all walls will leak to some extent. Leakage may particularly occur through any joints (between columns, panels, piles, etc.) resulting from the method of formation.

Leakage of groundwater through cut-offs into the excavation or working area can cause a number of problems:

i During construction the seepages may interfere with site operations, necessitating the use of sump pump or surface water control methods to keep the working area free of water.
ii The leakage into the excavation may be significant enough to locally lower groundwater levels outside the site, creating the risk of settlement or other side effects.
iii If the cut-offs form part of a permanent structure (such as the walls of a deep basement) even very small seepages will be unsightly in the long term and may cause problems with any architectural finishes applied to the walls.

In many cases, the significant seepages that give rise to problems (i) and (ii) can be dealt with by grouting or other treatment. On the other hand, it can be very difficult to prevent or to seal the small seepages of (iii). David Greenwood (1994) has said:

> Water penetration is very difficult to oppose. It is comparatively easy to reduce torrents to trickles, but to eliminate trickles is difficult. If it is essential to have a completely dry or leakproof structure, costs rise steeply.

This is an important point to consider if cut-off walls are to be incorporated in the permanent works.

5.4 Dewatering methods

Dewatering methods control groundwater by pumping, affecting a local lowering of groundwater levels (Fig. 5.3). The aim of this approach is to lower groundwater levels to a short distance (say 0.5 m) below the deepest excavation formation level.

The two guiding principles for securing the stability of excavations and especially of slopes, by dewatering or groundwater lowering are:

i Do not hold back the groundwater. This may cause a build-up of pore water pressures that will eventually cause catastrophic movement of soil and groundwater.

ii Ensure that 'fines' are not continuously transported since this will result in erosion and consequent instability. Apply a suitable filter blanket to avoid any build-up of pore water pressures and prevent transportation of fines. As a general guide, if the permeability is less than about 1×10^{-7} m/s migration of fines ceases. (This is a helpful guide when water testing after grout treatment, say prior to shaft sinking or the like, to assess whether or not further grouting is desirable.)

These objectives can be achieved by lowering groundwater levels sufficiently to avoid groundwater seeping into the excavation. This is the basis of the so-called 'pre-drainage' methods such as wellpointing (Chapter 9), deep-wells (Chapter 10) and ejectors (Chapter 11) where groundwater levels are lowered *in advance* of excavation. Wells are typically installed outside the excavation (see Fig. 5.3) – this draws water away from the excavation, avoiding troublesome seepages and resulting instability.

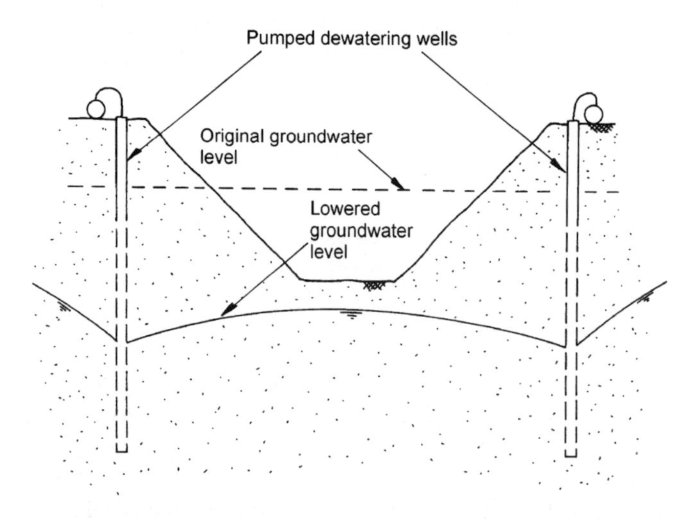

Figure 5.3 Groundwater control by pumping.

The alternative approach is to allow groundwater to enter the excavation and then deal with it by sump pumping (Chapter 8); this is less than ideal, since it draws water toward the excavation and may promote instability. Nevertheless, this can be acceptable, provided that fines are not removed from the slopes and base of the excavation. This philosophy requires pumping and disposal of the seepage water collected and the creation of a filter blanket and collector drainage system. The filter blanket may be a sand or gravel layer of adequate thickness (say 0.5 m) and appropriate grading, or a suitable geotextile membrane held in place by some weighting materials (say cobbles and boulders). Measures necessary to avoid the instability of excavations under these conditions are described in Section 4.3.

There are several techniques or methods available for controlling groundwater flow for a construction project. The selection of a technique or techniques appropriate to a particular project at a particular site or country will depend on many factors. However, the lithology and permeability(ies) of the soils will always be of paramount importance. Other factors to be considered are:

- Extent of the area of construction requiring dewatering.
- Depth of deepest formation level below existing ground level and the amount of lowering required.
- Proximity of existing structures, the nature of their foundations and the soil strata beneath them.

The various dewatering techniques are tabulated in Table 5.2, which is based on information from Preene *et al.* (2000).

There are projects where a single method is insufficient and a combination of methods is appropriate. Where the excavation is to penetrate a succession of soils of widely varying lithology this problem is more likely to arise.

It will be seen from study of the column of Table 5.2 headed 'Typical applications', that only a few methods are suitable for use in all types of soils. The ranges of soils that are suitable for treatment by the various dewatering methods are shown in Figure 5.4 in the traditional form of particle size distribution curves. These curves are taken from CIRIA Report 113 (Somerville 1986) and are based on the earlier work of Glossop and Skempton (1945) and others. Similarly the ranges of soils suitable for treatment by the various exclusion methods are shown in Figure 5.5. These are tentative economic and physical limits. The emphasis is on the word *tentative*.

An interesting and useful variation of Figure 5.4 was presented by Roberts and Preene (1994). Figure 5.6 is taken from CIRIA Report C515 (Preene *et al.* 2000) and is based on the Roberts and Preene (1994) original *Range of application of construction dewatering systems* paper but has been modified in the light of the joint experiences of the authors (Cashman 1994) and others. The practical achievements at the lower end of the permeability range are constrained by the physical limitations of applied vacuum. At the upper end

Table 5.2 Principal methods for groundwater control by pumping

Method	*Typical applications*	*Notes*
Drainage pipes or ditches (e.g. French drains)	Control of surface water run-off and shallow ground-water (including overbleed and perched water)	Simple methods of diverting or removing surface water from the working area. May obstruct construction traffic, and will not control groundwater at depth. Unlikely to be effective in reducing pore water pressures in fine-grained soil
Sump pumping	Shallow excavations in clean coarse soils, for control of groundwater and surface water	Cheap and simple. May not give sufficient drawdown to prevent seepage from emerging on the cut face of a slope, possible leading to loss of fines and instability. May generate silt or sediment laden discharge water, causing environmental problems
Wellpoints (including machine-laid horizontal wellpointing)	Generally shallow, open excavations in sandy gravels down to fine sands and possibly silty sands. Deeper excavations (requiring $>$5–6 m drawdown) will require multiple stages of wellpoints to be installed	Relatively cheap and flexible. Quick and easy to install in sands. Suitable for progressive trench excavations. Difficult to install in ground containing cobbles or boulders. Maximum drawdown is ~5–6 m for a single stage in sandy gravels and fine sands, but may only be ~4 m in silty sands. Horizontal wellpointing suitable for trench excavations outside urban areas, where very rapid installation is possible
Deep wells with electric submersible pumps	Deep excavations in sandy gravels to fine sands and water-bearing fissured rocks	No limit on drawdown in appropriate soil conditions. Installation costs of wells are significant, but fewer wells may be required compared with most other methods. Close control can be exercised over well screen and filter.
Deep wells with electric submersible pumps and vacuum	Deep excavations in silty fine sands, where drainage from the soil into the well may be slow	No limit on drawdown in appropriate soil conditions. More complex and expensive than ordinary deep wells because of the separate vacuum system. Number of wells may be dictated by the requirement to achieve an adequate drawdown between wells, rather than the flow rate, and an ejector system may be more economical
Shallow bored wells with suction pumps	Shallow excavations in sandy gravels to silty fine sands and water-bearing fissured rocks	Particularly suitable for coarse, high permeability materials where flow rates are likely to be high. Useful where correct filtering is important as closer control can be exercised over the well filter than with wellpoints. Drawdowns limited to ~4–7 m depending on soil conditions

Table 5.2 Continued

Method	*Typical applications*	*Notes*
Ejector system	Excavations in silty fine sands, silts or laminated or fissured clays in which pore water pressure control is required	In practice drawdowns generally limited to 20–50 m depending on equipment. Low energy efficiency, but this is not a problem if flow rates are low. In sealed wells a vacuum is applied to the soil, promoting drainage. Inclined holes possible
Passive relief wells and sand drains	Relief of pore water pressure in confined aquifers or sand lenses below the floor of the excavation to ensure basal stability	Cheap and simple. Create a vertical flowpath for water into the excavation; water must then be directed to a sump and pumped away
Collector wells	High permeability sands and gravels	Each collector well is expensive to install, but relatively few wells may produce large flow rates and be able to dewater large areas
Electro-osmosis	Very low permeability soils, e.g. clays, silts and some peats	Only generally used for pore water pressure control or ground improvement when considered as an alternative to artificial ground freezing. Installation and running costs are comparatively high

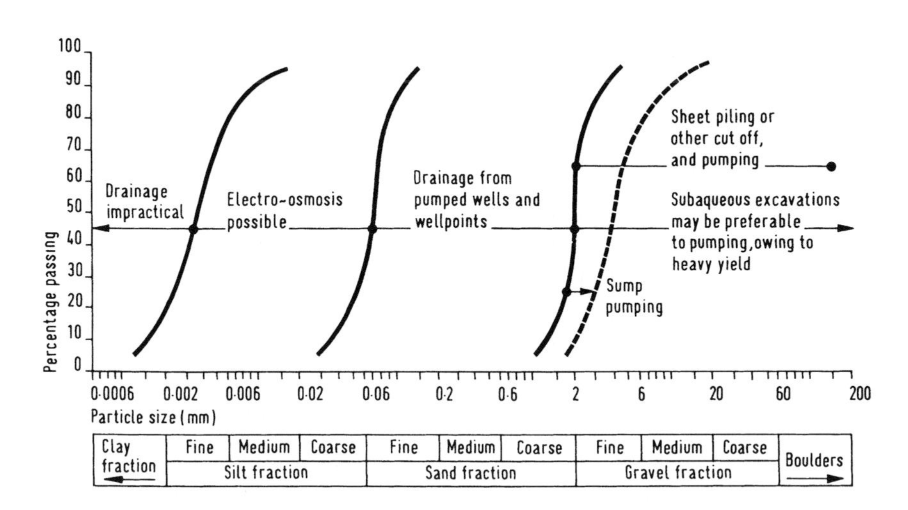

Figure 5.4 Tentative ranges for groundwater lowering methods (from Somerville 1986: reproduced by kind permission of CIRIA).

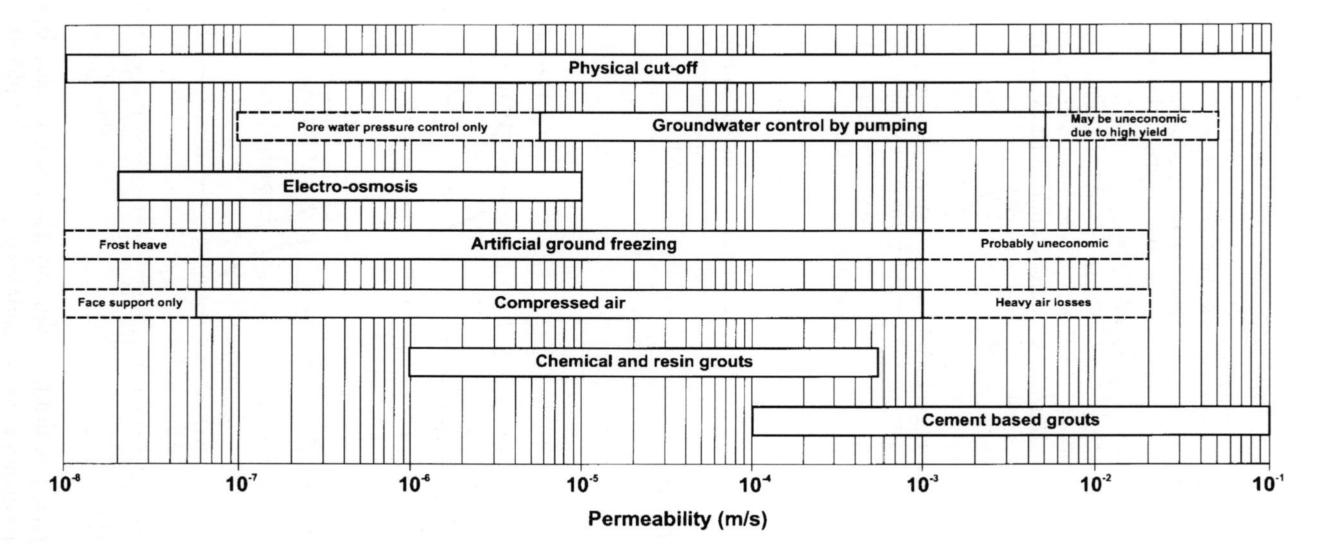

Figure 5.5 Tentative economic ranges for exclusion methods in soils (after Doran, Hartwell, Kofoed *et al.* 1995 and Preene *et al.* 2000).

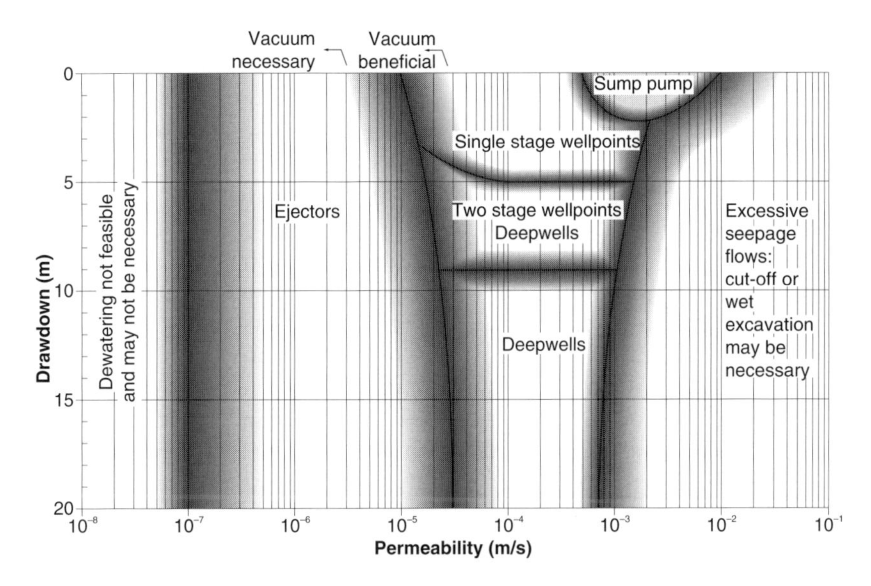

Figure 5.6 Range of application of pumped well groundwater control techniques – adapted from Roberts and Preene (1994), and modified after Cashman (1994) (from Preene *et al.* 2000: reproduced by kind permission of CIRIA).

of the range of permeability values, the cost of pumping the ensuing large volumes of water is the constraining factor. The greater the permeability, the greater the rate of pumping necessary to achieve the required lowering: the energy costs increase at an alarming rate.

The upper economic limit for the use of a deep well or wellpoint system is of the order of 5×10^{-3} m/s. The pumping costs of dealing with soils of greater permeability are generally uneconomic – except in the almost unheard of situation where fuel or electrical power are provided at no cost. In such high permeability soils exclusion methods may offer a more cost-effective expedient.

5.4.1 Pore water pressure control systems in fine-grained soils

When dewatering methods are applied to high and medium permeability soils, local lowering of groundwater levels occurs by gravity drainage in response to pumping. The period of pumping necessary to lower the water level is quite short as the pore water is rapidly replaced by air. In fine-grained soils of low and very low permeability, gravity drainage of the pore water is resisted by capillary tension. Such soils drain poorly and slowly by gravity drainage.

Because these soils do not drain easily, any excavation made below the groundwater level will encounter only minor seepages, and is unlikely to flood rapidly. Yet, even the small seepages encountered (perhaps less than 1 l/s even for a large excavation) can have a dramatic destabilizing effect. Side slopes may collapse or slump inwards and the base may become unstable or 'quick'. On site, people are often surprised that such small flow rates can be a problem. The theory of effective stress explains the mechanism of instability (see Chapter 4). The seepages imply the presence of high positive pore water pressures around and beneath the excavation. This implies low levels of effective stress, and hence low soil strength – instability is the natural result.

The solution to this problem is to abstract groundwater and so lower the pore water pressures around and beneath the excavation. This will maintain effective stresses at acceptable levels and prevent instability. The aim is not to totally drain the pore water from the soils – in any event this would be very difficult as capillary forces mean that fine-grained soils can remain saturated even at negative pore water pressures (see Fig. 3.7). Because the soil is not being literally 'dewatered', pumped well systems in fine-grained soils are more correctly referred to as pore water pressure control systems, rather than dewatering systems. The application of pumped well systems in low permeability soils is discussed further by Preene and Powrie (1994).

Glossop and Skempton (1945) and Terzaghi *et al.* (1996) indicated that gravity drainage will prevail in soils of permeability greater than about 5×10^{-5} m/s. Where wells are installed as part of pore water pressure control systems in soils of lesser permeability the well yields will be very low. This can make the continued operation of conventional wellpoint or deep well systems difficult, as the pumps are prone to overheating at low flow rates. However, if the top of the wells are sealed a partial vacuum can be applied to assist drainage. The increase in yield of a well due to the application of vacuum is likely to be of the order of 10 per cent, sometimes up to 15 per cent.

A sealed wellpoint system (see Section 9.5) operated by vacuum tank pump (see Section 12.1) may be effective in soils of permeability down to about 1×10^{-6} m/s. If a vacuum is applied to sealed deep wells (see Section 10.8) the lower limit of effectiveness of deep wells may be extended to about 1×10^{-5} m/s. Ejectors (see Section 11.1) installed in sealed wells will automatically generate a vacuum in the well when yields are low. Ejector systems can be effective in soils of permeability as low as 1×10^{-7} m/s.

Again, the permeability ranges quoted above are tentative. In fine-grained soils, structure and fabric has a great influence on the performance of vacuum well systems. If the soil structure consists of thin alternating layers or laminations of coarser and finer soils, the drainage will be more rapid. This is because the layers of coarse material will more rapidly drain the adjacent layers of finer-grained soil. There have also been cases where pore water pressure control systems have been effective in clays of very low permeability – the success of the method was attributed to the presence of a permeable fissure network in the clay.

5.4.2 *Some deep well and ejector projects deeper than 20 m*

The 20 m depth limit in Fig. 5.6 is not restrictive, but is merely a convenience for presentation of the diagram. However, there are insufficient reported case histories to have confidence in extending Fig. 5.6 data below 20 m depth but four known case histories are reported below to substantiate the view that, with care it is economically feasible to use wells and ejectors to depths of the order of 40 m or more.

At Dungeness A nuclear power station sited on the south coast of England, sixty deep wells were pumped for more than two years. The soil conditions at the site were:

(a) Gravel with cobbles, +5.5 mOD to −1.5 mOD, average permeability 3×10^{-3} m/s
(b) Original groundwater level from +2.4 mOD to 0.9 mOD
(c) Gravelly sand, −1.5 mOD to −9.1 mOD, average permeability 8×10^{-4} m/s
(d) Sand, −9.1 mOD to −33.5 mOD, permeability 1.5×10^{-4} m/s decreasing with depth to 5×10^{-5} m/s.

The construction excavations were encircled by a continuous girdle of interlocking sheet piles driven to −9.1 mOD level so as to exclude recharge from the overlying high permeability soils. The lowering for the excavations for the Turbine Hall and the Reactors was achieved by pumping twenty-nine wells 20 m deep. The lowering for the Cooling Water Pumphouse and the Syphon Recovery Chamber was achieved by pumping nineteen and twelve wells respectively, each 41 m deep.

Prior to the construction of the East Twin Dry dock in Northern Ireland, site investigation borings had revealed a confined aquifer of Triassic sandstone beneath alluvial and glacial deposits (both mainly clays). The depth to the upper surface of Triassic sandstone varied between 34 m below ground level and 21 m (which was 1 m below the formation level of the entrance structure). There was a high piezometric level confined in the sandstone – to about 3 m above original ground level. Twenty deep wells were installed to depths between 33.5 m and 43 m below ground level using reverse circulation rotary drilling methods. The wells were operated for over two years.

At the Mufulira mine No. 3 dump, in northern Zambia, about 200 twin pipe ejectors were installed to reduce the moisture content and thereby stabilize slimes lagoon deposits, which were of porridge like consistency and had broken through the roof of the main adit causing disastrous loss of life. The Ministry of Mines, Zambia required that the slimes deposits over the cave-in be stabilized before the mine could be permitted to reopen. After nine months of pumping of the ejector installation the phreatic surface was drawn down about 10 m. Eventually the phreatic surface was drawn down some 30 m and resulted in an acceptable increase in the shear strength of the slimes deposits over the cave-in. The pumping lift was in the range from 20 m to over 40 m.

At the Benutan dam site in Brunei single pipe ejectors were installed to stabilize alluvium comprising heterogeneous loose silty fine sands, very soft clays, thin layers of peat and buried timbers (see Cole *et al.* 1994). Brunei is in an earthquake zone of moderate seismic activity so it was judged that there was risk of liquefaction of these loose alluvial foundation soils beneath the proposed dam unless they were stabilized or replaced with other more stable materials. As at Mufulira, an ejector system was installed and operated to reduce the pore water pressures of the loose soft soils and thereby increase their stability. The depths to which the ejectors were installed ranged from about 10 m to 38.5 m. The ejector pumping was continuous on the deepest line for about two years. The dam height was 20 m above valley floor level.

5.5 Groundwater control for tunnels and shafts

There is some merit in considering groundwater control for tunnels and shafts separately from methods used for surface excavations (although cut-and-cover tunnels are effectively surface excavations and can be dewatered accordingly). Even though the great majority of tunnels are constructed below groundwater level, traditionally most tunnels through water-bearing ground are constructed without dewatering by pumping from wells.

Dewatering is sometimes used for shaft construction, although many shafts are constructed as flooded or 'wet' caissons. This method involves the shaft lining being constructed at ground level and sunk (by jacking or kentledge) into the ground, while material is excavated by grab from within the flooded caisson. This avoids the need to lower groundwater levels, but carries its own set of risks. It can be difficult to control excavation levels when grabbing through considerable depths of water, and problems have occurred when due to over-excavation, or when boulders are present.

Ground treatment is often used to exclude groundwater from tunnels. Methods used include jet or injection grouting and artificial ground freezing, carried out from within the tunnel or from the surface. Any residual seepage into the tunnel is dealt with by maintaining some sump pumping capacity at the tunnel face.

Because many tunnelling methods construct a relatively watertight lining immediately behind the working face, only a small area of tunnel face is exposed to water-bearing ground. This confined environment has allowed methods to be developed whereby groundwater is excluded by maintaining the tunnel face at a fluid pressure more or less equal to the groundwater pressure.

The traditional way of balancing groundwater pressure is by compressed air working (Fig. 5.7(a)). This method, in use since the nineteenth century,

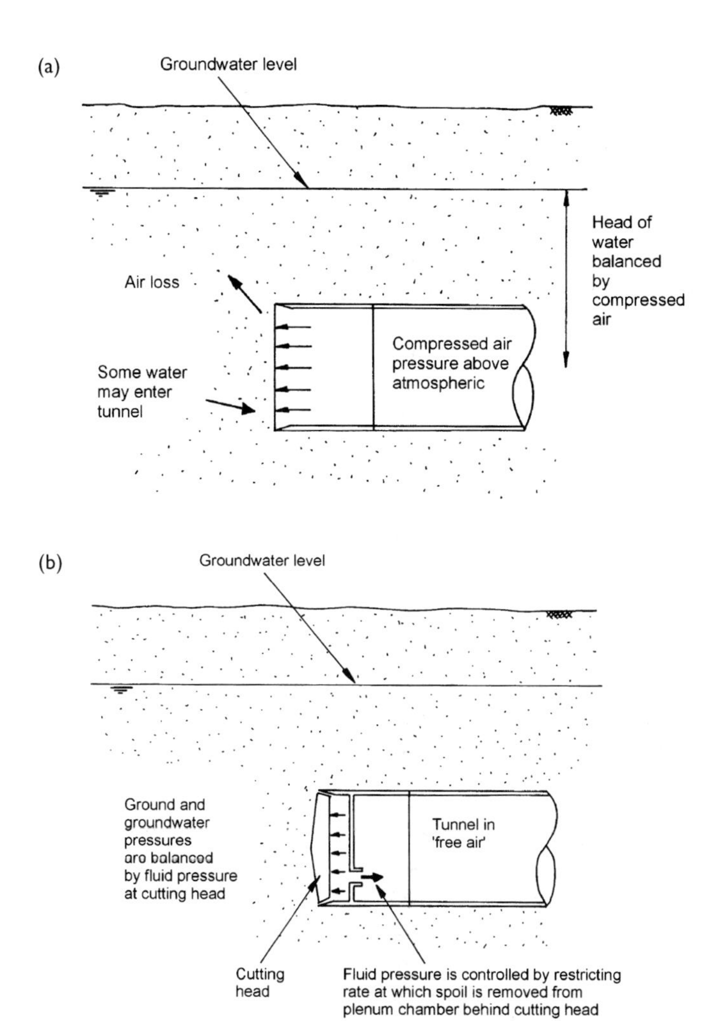

Figure 5.7 Pressure balancing techniques used to exclude groundwater from tunnels. (a) Compressed air, (b) earth pressure balance tunnel boring machine.

uses pressurized air in the tunnel face and working area (for some distance from the face) requiring the miners to work in air pressures above atmospheric, with concomitant health risks. Physiological effects on workers exposed to compressed air mean that the method can only be used up to 3.5 bar of pressure above atmospheric (equivalent to 35 m below water level), and only then with very careful medical controls on working arrangements. Most compressed air working is carried out at pressures of less than 0.75 bar above atmospheric (known as low pressure compressed air); costs rise considerably above 0.75 bar (high pressure compressed air) due to the additional medical constraints. Compressed air working can also be used for shaft construction, when an air deck and airlocks are used to seal the top of the shaft.

An alternative method of balancing ground and groundwater pressures at the tunnel face was developed in the late twentieth century – the earth pressure balance (EPB) tunnel boring machine (TBM). EPB machines are a form of full face TBM that cuts the soil from the tunnel face, and forms a fluid or 'earth paste' (consisting of soil cuttings, groundwater and conditioning agents such as polymer or bentonite muds) in a plenum chamber behind the face (Fig. 5.7(b)). By controlling the rate at which the fluid or paste is extruded from the plenum chamber as the TBM excavates and moves forward, the face can be supported by a balanced pressure. The TBM driver and miners work in a 'free air' (i.e. atmospheric) environment behind the plenum chamber, avoiding the health risks of compressed air working.

Cases do sometimes arise when groundwater control by pumping is used in tunnel construction. Hartwell (2001) describes several case histories. Applications of groundwater lowering for tunnelling include:

i For construction of shafts to launch or receive TBMs, and for construction of 'soft eyes' in shaft linings.
ii The entry or exit of TBMs into or out of shafts, portals, outfalls and other structures.
iii For construction of tunnel enlargements, step plate junctions, cross passages, adits or other connections where the tunnel lining has to be breached temporarily.
iv To lower groundwater levels to allow compressed air working at reduced pressures (ideally less than 1 bar), for example to allow access to the working head of a TBM for maintenance purposes. The dewatering wells should be located with care to avoid compressed air escaping through the ground to the wells.
v To reduce pore water pressures in fine-grained soils such as silts or very silty sands, to reduce the risk of the soils liquefying as a result of vibration from the machinery in the TBM.
vi To lower groundwater levels to below invert (to effectively 'dewater' the tunnel) to allow open face tunnelling methods to be used in otherwise unstable ground. This is sometimes necessary when a TBM designed for relatively stable soils present over most of the tunnel length has to traverse a short section of alluvial or glacial soils which may be present in a buried channel or other geological feature.
vii For recovery of damaged or inundated TBMs.
viii To control groundwater velocities to allow use of ground treatment methods (such as artificial ground freezing or grouting) in problematic conditions.

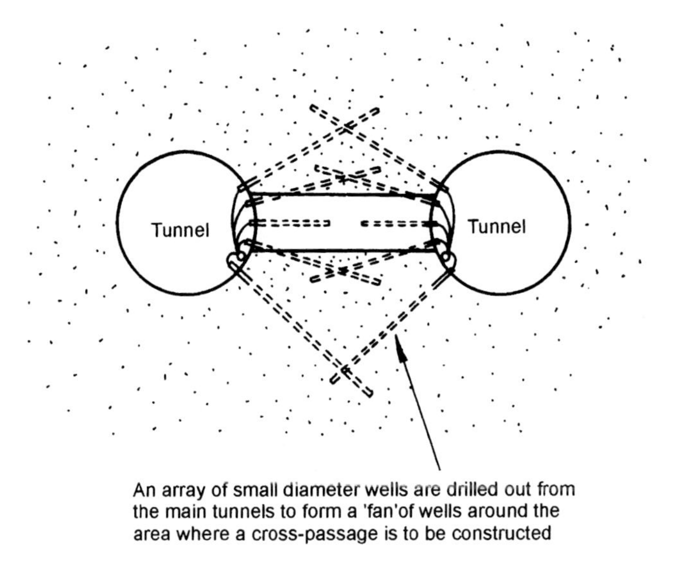

Figure 5.8 Typical pattern of dewatering wells drilled out from tunnel for localized depressurization to allow cross-passage construction (after Doran, Hartwell, Roberti *et al.* 1995).

Where dewatering is used for tunnel works, wells drilled from the surface can be used, provided that access is available above the tunnel, and that there are no intervening service pipes between ground level and the tunnel. If surface access is not available, it may be possible to drill small diameter wells radially out from the tunnel or shaft (or both). This technique was used in the 1990s on the London Underground Jubilee Line Extension and also on the Storebaelt railway tunnel in Scandinavia (Doran *et al.* 1995). An example of well arrangement is shown in Fig. 5.8.

Drilling out through the existing tunnel lining is a challenging task, and runs the risk of destabilizing or inundating the tunnel; it is essential that such works are meticulously planned and executed, and supervized by experienced personnel. Drilling must normally be carried out through stuffing boxes and blow-out preventers secured and sealed to the tunnel lining. Care must be taken to avoid loosing ground into the tunnel during drilling and subsequent pumping – this is a particular risk in fine-grained uniformly graded soils such as silts and sands. Experience suggests that in such soils if groundwater heads are in excess of around 10 m above tunnel invert, successful well installation will be very difficult.

References

Bell, F. G. and Mitchell, J. K. (1986). Control of groundwater by exclusion. *Groundwater in Engineering Geology* (Cripps, J. C., Bell, F. G. and Culshaw,

M. G., eds). Geological Society Engineering Geology Special Publication No. 3, London, pp 429–443.

Cashman, P. M. (1994). Discussion of Roberts and Preene (1994). *Groundwater Problems in Urban Areas* (Wilkinson, W. B., ed.). Thomas Telford, London, pp 446–450.

Cole, R. G., Carter, I. C. and Schofield, R. J. (1994). Staged construction at Benutan Dam assisted by vacuum eductor wells. *Proceedings of the 18th International Conference on Large Dams*, Durban, South Africa, 625–640.

Doran, S. R., Hartwell, D. J., Kofoed, N. and Warren, S. (1995). Storebælt Railway tunnel – Denmark: design of cross passage ground treatment. *Proceedings of the 11th European Conference on Soil Mechanics and Foundation Engineering*, Copenhagen, Denmark.

Doran, S. R., Hartwell, D. J., Roberti, P., Kofoed, N. and Warren, S. (1995). Storebælt Railway tunnel – Denmark: implementation of cross passage ground treatment. *Proceedings of the 11th European Conference on Soil Mechanics and Foundation Engineering*, Copenhagen, Denmark.

Glossop, R. and Skempton, A. W. (1945). Particle-size in silts and sands. *Journal of the Institution of Civil Engineers*, **25**, 81–105.

Greenwood, D. A. (1994). Engineering solutions to groundwater problems in urban areas. *Groundwater Problems in Urban Areas* (Wilkinson, W. B., ed.). Thomas Telford, London, pp 369–387.

Hartwell, D. J. (2001). Getting rid of the water. *Tunnels & Tunnelling International*, January, 40–42.

Preene, M. and Powrie, W. (1994). Construction dewatering in low permeability soils: some problems and solutions. *Proceedings of the Institution of Civil Engineers, Geotechnical Engineering*, **107**, 17–26.

Preene, M., Roberts, T. O. L., Powrie, W. and Dyer, M. R. (2000). *Groundwater Control – Design and Practice*. Construction Industry Research and Information Association, CIRIA Report C515, London.

Roberts, T. O. L., and Preene, M. (1994). Range of application of construction dewatering systems. *Groundwater Problems in Urban Areas* (Wilkinson, W. B., ed.). Thomas Telford, London, pp 415–423.

Somerville, S. H. (1986). *Control of Groundwater for Temporary Works*. Construction Industry Research and Information Association, CIRIA Report 113, London.

Terzaghi, K., Peck, R. B. and Mesri, G. (1996). *Soil Mechanics in Engineering Practice*, 3rd edition. Wiley, New York, p 305.

Further reading – exclusion methods

Steel sheet-piling

Williams, B. P. and Waite, D. (1993). *The Design and Construction of Sheet-Piled Cofferdams*. Construction Industry Research and Information Association, CIRIA Special Publication 95, London.

Vibrated beam walls

Privett, K. D., Matthews, S. C. and Hodges, R. A. (1996). *Barriers, Liners and Cover Systems for Containment and Control of Land Contamination*.

Construction Industry Research and Information Association, CIRIA Special Publication 124, London, pp 59–60.

Slurry trench walls

Jefferis, S. A. (1993). In-ground barriers. *Contaminated Land – Problems and Solutions* (Cairney, T., ed.), Blackie, London, pp 111–140.

Structural concrete diaphragm walls and secant pile walls

Puller, M. (1996). *Deep Excavations: A Practical Manual.* Thomas Telford, London, pp 97–116.

Jet grouting

Essler, R. D. (1995). Applications of jet grouting in civil engineering. *Engineering Geology of Construction* (Eddleston, M., Walthall, S., Cripps, J. C. and Culshaw, M. G., eds). Geological Society Engineering Geology Special Publication No. 10, London, pp 85–93.
Lunardi, P. (1997). Ground improvement by means of jet grouting. *Ground Improvement*, **1**, 65–85.

Mix-in-place columns

Blackwell, J. (1994). A case history of soil stabilisation using the mix-in-place technique for the construction of deep manhole shafts at Rochdale. *Grouting in the Ground* (Bell, A. L., ed.). Thomas Telford, London, pp 497–509.
Greenwood, D. A. (1989). Sub-structure techniques for excavation support. *Economic Construction Techniques*. Thomas Telford, London, pp 17–40.

Injection grouting

Bell, A. L., (ed.). (1994). *Grouting in the Ground*. Thomas Telford, London.
Little, A. L. (1975). Groundwater control by exclusion. *Methods of Treatment of Unstable Ground* (Bell, F. G., ed.). Butterworths, London, pp 37–68.

Artificial ground freezing

Harris, J. S. (1995). *Ground Freezing in Practice*. Thomas Telford, London.

Compressed air

Megaw, T. M. and Bartlett, J. V. (1981). *Tunnels: Planning, Design, Construction*, Volume 1. Ellis Horwood, Chicester, pp 125–156.

Chapter 6

Site investigation for groundwater lowering

6.0 Introduction

Site investigation was defined by Clayton *et al.* (1995) as:

> The process by which geological, geotechnical and other relevant information which might affect the construction or performance of a civil engineering or building project is acquired.

This chapter describes the information which needs to be gathered to allow the design of groundwater lowering systems. Good practice in the planning and execution of site investigations is outlined, together with brief details of methods used for boring, probing, testing, etc. The analysis of the data gathered during site investigation is described with the design methods in Chapter 7.

This chapter deals specifically with the site investigation requirements for dewatering projects. The successful design of such projects is highly dependent on obtaining realistic estimates of the permeability of the various soil and rock strata present beneath the site. Methods of permeability testing, their range of applicability and their relative merits and limitations are discussed in detail.

6.1 The purpose of site investigation

Site investigation is the essential starting point, without which the design of any construction or geotechnical process cannot progress. Groundwater lowering is no exception to this rule. To quote the fictional detective Sherlock Holmes:

> It is a capital mistake to theorize before one has data
> (Sir Arthur Conan Doyle, *A Scandal in Bohemia*)

The best site investigations are deliberately planned and executed processes of discovery, carefully matched to the characteristics of the site, and to the work to be carried out. Sadly, many investigations in the past have not met this challenging standard. The problems of poor or inadequate investigations were highlighted in two reports by the Institution of Civil Engineers: *Inadequate Site Investigation* (Institution of Civil Engineers 1991) and *Without Investigation Ground is a Hazard* (Site Investigation Steering Group 1993).

A particular problem for the designer of groundwater lowering works is that, in many cases, investigations are designed primarily to provide information for the design of the permanent works. The information needed for design of temporary works (including groundwater lowering) is often neglected. This problem can arise when the persons designing the site investigation do not have appropriate expertise or experience. Alternatively, poor communication may result in them not being informed of the likely need for dewatering, so they will not plan to gather the relevant information.

There is a wide and useful literature on site investigation, including Clayton *et al.* (1995), the *British Standard Code of Practice for Site Investigation* (BS5930: 1999) and the Institution of Civil Engineers' reports cited previously, and the reader should consult these for background on the subject. The remainder of this chapter will concentrate on the particular site investigation needs for projects where temporary works groundwater lowering schemes are to be employed.

6.2 Planning of site investigations

All site investigations need to be planned and designed in order that they provide the information needed by the various designers, estimators and construction managers. There is no such thing as a 'standard' site investigation, purely because 'standard' ground conditions have yet to be discovered, no matter how much we might wish for them!

According to the Institution of Civil Engineers (1991), the designers and planners of site investigations should attempt to answer the following questions:

(a) What is known about the site?
(b) What is not known about the site?
(c) What needs to be known?

On all but the smallest investigations, these questions cannot be answered by one person, and may require input from specialists in soil and rock mechanics, engineering geology, geophysics, archaeology and hydrogeology. For groundwater lowering projects, advice from a dewatering specialist can also aid the planning of investigations.

The planning, design, and ultimately procurement, of site investigations is highly specialized. It is essential that this work is guided by a suitably qualified and experienced person, who should be associated with the site investigation from conception to completion. Recommendations for the qualifications and experience required to act as geotechnical specialists and geotechnical advisors are given in Site Investigation Steering Group (1993).

Effective communication between all parties involved in the construction process is vital. This includes the site investigation as well, because without accurate and up-to-date information how can an investigation be designed to answer the questions listed earlier? The *Construction (Design and Management) Regulations 1994* (known as the CDM Regulations, see Chapter 15) formalize this ethos, and require clients and designers to work with all parties from an early stage so that safe methods of work can be planned and adopted. This means that clients and designers must provide the designer and manager of the site investigation with details of the proposed works (e.g. location, depth and size of excavation, support methods). Without these details it is difficult to plan an investigation on a rational basis.

6.3 Stages of site investigation

A site investigation includes all the activities required to gather the necessary data about the site and should consist of a number of stages, listed below:

i Desk study,
ii Site reconnaissance,
iii Ground investigation,
iv Reporting.

To many non-geotechnical specialists the ground investigation is perceived to be the 'essence' of a site investigation. In fact the ground investigation (which can involve trial pits, borings, *in situ* testing and laboratory testing) aims only to determine ground and groundwater conditions at the site. If other stages (especially the desk study) are neglected, inadequate investigations may result.

6.3.1 Desk study and site reconnaissance

The desk study and site reconnaissance (sometimes called a walk-over survey) are essential in any investigation – their importance cannot be overestimated. Unfortunately, they are sometimes overlooked or considered irrelevant.

The desk study is a review of all available information relevant to the proposed project including: geological and hydrogeological maps; aerial photographs (if available); records of construction on nearby sites or those where similar soil and groundwater conditions were encountered; records of nearby wells, boreholes and springs; and records of mining or previous use of the site. Sources of information for desk studies are given by Dumbleton and West (1976) and in Chapter 3 of Clayton *et al.* (1995).

The site reconnaissance is usually carried out after the desk study has been largely completed. It comprises a walk-over survey of the site to gather further information on the site surface conditions and access for the ground investigation, any exposed geological, groundwater or surface water features, and the nature of the areas surrounding the site.

Clayton *et al.* (1995) point out that both the desk study and site reconnaissance can provide large amounts of useful information at low cost – they are by far the most cost effective stages of site investigation. They are essential to allow efficient design of the ground investigation; failure to anticipate any predictable groundwater problems can result in poor or inadequate investigations, leading to potential problems during construction.

The desk study and site reconnaissance can also be used to gather information about the surroundings to the site, perhaps in areas where access for ground investigation cannot be obtained. For groundwater lowering projects this is particularly useful to help determine whether there will be any adverse side effects of dewatering (such as ground settlement or derogation of water supplies, see Chapter 13) outside the site. Brassington (1986) recommends that a survey of nearby groundwater supplies be carried out to aid the assessment of the risk of derogation to water users.

The problems which may result if the desk study and site reconnaissance are neglected can be illustrated by two case histories.

(a) On a project through variable glacial soils in northern England, a new sewer was laid in an open-cut trench to replace an existing sewer, constructed a few decades earlier. The new sewer was generally laid parallel to the old one, apart from one section where the old sewer took a circuitous 'dog-leg' route between manholes. The new sewer was to take the obvious straight-line route between the manholes. During construction of this section severe groundwater problems were encountered, including a flowing artesian aquifer (see Section 3.3) immediately beneath the base of the excavation. Work was further hampered when old abandoned sections of steel sheet-pile trench supports (probably dating from construction of the old sewer) were encountered. With hindsight, it is fairly obvious that the old sewer was originally intended to take the direct route between the manholes, but that groundwater problems forced them to abandon work and re-route the sewer. The new sewer was eventually completed, but a desk study of old construction records may have allowed different methods to be adopted from the start, thus avoiding cost and time delays.
(b) A small sewerage project involved a wellpoint system to allow construction of a manhole. The manhole itself was excavated in sand and

gravels, and was successfully dewatered. However, significant damage occurred to several neighbouring structures as a result of ground settlements. Subsequent investigations showed that the damaged buildings were founded on an extensive deposit of compressible peat. Although ground investigation at the manhole site did not encounter any peat, a desk study review of geological maps would have revealed the presence of peat beneath surrounding areas. This highlights the risk of damaging settlements, and would have allowed appropriate mitigation measures, to reduce the impact of dewatering on nearby properties (see Section 13.1), to be included in the project.

6.3.2 Ground investigation

There are numerous techniques available to physically investigate a site. It is rarely obvious at the start of an investigation which methods will be suitable to gather the information needed by the project designers. It is preferable, therefore, that all but the smallest ground investigations be carried out in phases. The first phase may consist of an initial pattern of boreholes and testing across the whole site. Preliminary results from the first phase will allow second and subsequent stages to be designed; such stages may include more closely spaced boreholes in areas where ground conditions are unclear, or perhaps specialist testing such as a pumping test.

Ground investigation usually involves the use of some combination of the following methods: boring, drilling, probing and trial pitting; *in situ* testing; geophysics; and laboratory testing. These ground investigation methods are briefly outlined below, mainly in relation to British practice as outlined in BS5930 (1999) and BS1377 (1990). Factors particularly relevant to investigations for dewatering projects are highlighted.

In Britain the most common form of boring is light cable percussion drilling, colloquially known as 'shell and auger' drilling (Fig. 6.1). Soil samples may be recovered from the borehole (for description of soil type), or certain types of *in situ* test may be performed within the borehole. A key point to note is that at various stages during drilling water may be added to the borehole by the driller, or may be removed by the action of the boring tools. This can lead to natural groundwater levels and inflows being masked during boring operations.

Rotary drilling (using a water or air-based flush medium with polymer or foam additives) is also widely used (Binns 1998), particularly in relatively intact rock strata, but also in uncemented drift deposits. Core samples can be recovered from the boreholes (for soil and rock description), and certain types of *in situ* tests carried out. Because the borehole is generally kept full of the flush medium, groundwater levels and inflows can be difficult to determine during drilling.

The older technique of wash boring is rarely used in Europe but is still employed in countries where labour is cheap. The basic rig is a winch tripod.

Figure 6.1 Light cable percussion boring rig (courtesy of WJ Groundwater Limited).

The associated equipment consists of an outer pipe with a chisel bit at the lower end and a swivel head at the upper end of the wash pipe, and incorporating a water pressure hose connection with a weight for driving the casing into the ground. A pump passes water down the wash pipe to slurify the soil at the bottom of the outer casing. The return washings are not regarded as reliable for identification of soil types – though recordings of wash-water colour changes should be noted. This method is suitable for use in sands and silts, but progress in clayey soils is likely to be slow. Groundwater levels and inflows are masked in a similar way to rotary drilling.

Hollow stem continuous flight augers are suitable for use in cohesive soils but are of limited use in water-bearing granular soils; indeed, in granular soils this technique is often unworkable. The drilling spoil brought to ground surface gives only an approximate indication of soil types and horizons. Drive-in samplers can be inserted through the hollow stem to obtain strata samples at convenient depth intervals.

As an alternative to boring or drilling, in recent years probing methods have been developed. A wide range of equipment exists and is used, but all have the objective of determining a profile of penetration resistance with depth. Most methods were developed as a low-cost and rapid alternative to drilling and boring. Two of the most commonly used methods are dynamic probing and static probing by the 'cone penetration test' (commonly known as the CPT).

Dynamic probing involves a percussive action to drive the probe into the ground, producing output in the form of blows per unit depth of penetration (see, e.g. Card and Roche 1989). Window sampling is a variant on the dynamic probing method that allows soil samples to be obtained via sampling tubes driven into the ground.

Static probing by CPT (also known as 'Dutch cone' testing after the country where the method was developed) is more sophisticated than dynamic probing. The cone is pushed continuously into the ground, using reaction from the test truck, producing an output of resistance against depth (see Meigh 1987; Lunne *et al.* 1997). Piezocone testing is a variant of the CPT method, where pore water pressures are measured in addition to resistance parameters; this can allow estimates of permeability to be obtained in low permeability soils.

Trial pitting is a simple and widely used method for investigation of shallow strata. A pit is dug, exposing the sub-soil for inspection and sampling. Groundwater inflows and seepages can normally be clearly identified. During excavation it may be possible to form an opinion as to appropriate methods of full-scale excavation, if relevant.

Trial pits are normally dug by mechanical excavator (Fig. 6.2). Small backhoe loaders can normally excavate to a depth of around 3 m, and larger excavators may be able to work to a maximum depth of around 5 m.

Figure 6.2 Trial pitting using a mechanical excavator.

Trial pitting is a potentially hazardous exercise. Trial pits of greater than 1.2 m depth should only be entered if adequately shored and supported. Even pits of less than 1.2 m depth may be unstable. Each pit should be assessed before entry, and if any doubt exists the pit should not be entered. Soils can be described from the surface, and samples taken from the spoil in the excavator bucket. In difficult locations or where ground disturbance must be kept to a minimum, it may be possible to excavate pits by hand excavation. However, this method is very slow, and such excavations must not be taken deeper than 1.2 m without employing timbering or some form of proprietary side support system. Safety in pits and trenches is discussed by Irvine and Smith (1992).

Further details on all these methods can be found in Clayton *et al.* (1995). Their relative merits are outlined in Table 6.1. The ground investigation stage

Table 6.1 Advantages and disadvantages of methods of boring, drilling, probing and trial pitting

Method	*Advantages*	*Disadvantages*
Drilling and boring	Suitable for a wide range of soils Allows soil and groundwater samples to be obtained Can allow *in situ* permeability tests to be carried out Allows observation wells (standpipes and standpipe piezometers) to be installed	Progress can be difficult if cobbles or boulders are present Some methods require specialist equipment
Probing	Provides information on soil profile Piezocone can provide information on soil permeability Can allow simple standpipe observation wells to be installed Some methods allow soil samples to be obtained	Does not provide soil or groundwater samples Some methods require specialist equipment Needs to be used in conjunction with boreholes to enable correlation of soil types Penetration depth is limited in stiff materials and coarse granular soils Does not allow installation of standpipe piezometers
Trial pitting	Suitable for a wide range of soils, including very coarse soils Allows stability and ease of excavation of soils to be directly observed Requires no specialist equipment Allows soil and groundwater samples to be obtained Can allow simple standpipe observation wells to be installed	Depth limited to around 5 m May be difficult to progress below groundwater level in unstable soils Does not allow installation of standpipe piezometers

also includes the installation and monitoring of groundwater observation wells; this is discussed further in Section 6.5.

Various methods of *in situ* testing can be carried out as part of boring, drilling, probing and trial pitting. The most relevant of these to groundwater lowering works are permeability tests; these are described in Section 6.6.

Geophysical methods are sometimes used in investigations for civil engineering works, but these methods are much more widely used in the oil, mineral extraction and water resource fields. In general, geophysical methods are used to provide information on changes in particular properties of strata beneath a site, and can be used to provide information between widely spaced boreholes. The use of boreholes in combination with geophysics is important, because the borehole data can be used to 'correlate' or 'calibrate' the geophysical results for the site in question. Geophysical methods used for civil engineering investigations are described in Clayton *et al.* (1995), Chapter 4 and McCann *et al.* (1997). Geophysical methods used in hydrogeological and water resource investigations are described by Barker (1986) and Beesley (1986). On a number of occasions when geophysics has been used in investigations the results have been perceived to be disappointing or inconclusive. This is probably a reflection on an inappropriate choice and specification of method, rather than a systematic drawback with the use of geophysics. To get the most out of geophysical surveying, it is essential that engineering geophysicists are involved at an early stage of planning and thereafter; otherwise the method will not achieve its potential.

Samples of soil or rock obtained from boreholes or trial pits may be tested in the laboratory. The purpose of testing can be as an aid to soil description and classification, or can be to determine soil properties for engineering design. Properties routinely tested for include strength, compressibility, permeability (see Section 6.6) and chemical characteristics. Samples of groundwater recovered during investigation may also be chemically tested in the laboratory.

6.3.3 Reporting

To be useful to the designers and managers of the project, the site investigation must be reported in an organized, concise and intelligible manner. Ideally, reporting should be carried out by geotechnical specialists who have been involved with the investigation since its inception. The minimum reporting requirement is for a 'factual report' which presents the data gathered during the desk study and site reconnaissance, and the borehole logs, trial pits logs, test results and groundwater monitoring data from the ground investigation. Such reports do not usually comment on the implications of the data gathered.

In many cases, particularly on large or complex projects, an 'interpretative report' is produced in addition to the factual report. Again, written by geotechnical specialists, this should review the ground and groundwater conditions at a site. It should include discussion of the effect of the anticipated conditions on the proposed design and construction methods. At the time the interpretative report is produced the project design may not be finalized, but the report should discuss the geotechnical aspects of the full range of design options current at that stage. If particular potential problems in design and construction are highlighted, one of the report's conclusions may be to recommend further or specialist investigations.

6.4 Determination of ground profile

Any investigation will need to identify the nature, depth, extent and orientation of the strata beneath the site. Collectively these parameters describe the 'ground profile', normally based on the results of trial pits, boreholes, geophysics and probing. Determination of the ground profile is an essential part of developing the groundwater conceptual model (see Section 7.2) needed to allow dewatering design.

It is essential that boreholes penetrate to adequate depth. The presence of confined aquifers or of localized zones of high permeability beneath excavations is a significant risk for many excavations. Boreholes in the area of the excavation should penetrate to a depth of 1.5–2 times the depth of the excavation. There have been several cases where excavations failed due to base heave (see Section 4.3) when investigation boreholes were not taken to adequate depths, but were terminated a few metres below proposed formation level. The failure was caused by confined aquifers below formation, undetected during investigation. Such problems are frustrating because if the boreholes had been deeper and had detected the aquifer, groundwater control (using relief wells) would have been simple and cost effective. As it was, major cost and time delays resulted. The only case when boreholes shallower than the recommendation can be tolerated is if the desk study clearly indicates that impermeable soils are present to considerable depth below formation level.

In addition to soil descriptions, groundwater level information and permeability test results should help identify which strata are water-bearing (and may act as aquifers) and which strata are of low permeability (and may act as aquitards and aquicludes). Compressibility test results may help identify any strata that may give rise to significant groundwater lowering induced settlements.

6.5 Determination of groundwater conditions

In many investigations the observations of groundwater conditions are totally inadequate, providing little concrete information on groundwater levels. Accurate knowledge of the likely range of groundwater levels and pore water pressures in the various strata is essential for the design of dewatering systems. This section will describe the types of groundwater observations that may be taken during investigation, and will discuss their various merits and limitations.

6.5.1 Water level observations in trial pits and borings

The easiest and most common form of groundwater level observations are those taken in trial pits and borings. Unfortunately, there are two important limitations to the accuracy of readings obtained in this way:

i The natural groundwater inflows and levels may be masked or hidden by the excavation or boring method, particularly if water is added or removed from the pit or borehole, or if drilling casing seals off inflows of water.

ii For a pit or borehole to show a representative groundwater level, sufficient water must flow into the pit or borehole to fill it up to the natural groundwater level. In soils of moderate or low permeability it can take a long time for the water level in the pit or borehole to come into equilibrium with the natural groundwater level. Most observations do not allow sufficient time for equilibrium, and so may report unrepresentative water levels.

Trial pits offer a simple way to observe shallow groundwater conditions. The size of the pit allows direct visual observation of inflows and seepages (it may be possible to categorize these as ‘slow’, ‘medium’ or ‘fast’ seepages on a subjective basis). The location of the seepages in relation to the soil fabric and layering can provide information on the relative permeability of various strata – this can be a useful way of identifying perched water tables. The disadvantage of trial pits is that, because of their relatively large volume, a significant volume of water has to enter the pit to fill it up to the natural groundwater level. This process of equalization may take several days in soils of moderate permeability. The pit could be left open and monitored daily but, for safety reasons, trial pits are rarely left open for long periods of time. In general, groundwater levels in trial pits may not be representative of natural groundwater levels.

In British site investigation practice, the most common form of groundwater observations are those recorded by the drilling foreman during light cable percussion drilling. Each time groundwater is encountered these records should comprise:

(a) The depth at which water is first encountered (known as a ‘water strike’).

(b) A description of the speed of inflow (e.g. slow, medium, fast). Boring is normally suspended following a water strike and groundwater levels observed to record the rise (if any) in groundwater levels. Ideally, monitoring should continue until the water level in the borehole stabilizes, but often the rise in water level is recorded for a fixed period (often 20 min) only, before boring recommences.

(c) The depth at which the groundwater inflow is sealed off by the temporary drilling casings.

Additionally the drilling foreman should record whether water was added to the borehole during drilling and should record the water level in the borehole at the start and end of the drilling shift (together with the corresponding depth of borehole and casing at that time). All groundwater details are recorded on the driller's daily record sheet, and should appear on the final borehole log.

When reviewing water levels recorded during boring the following points must be noted:

(a) In all but the most permeable soils, observing the water level rise for a short period following a water strike, may not allow sufficient time for the natural groundwater level to be apparent. In many cases the water level recorded following a water strike will be lower than the actual groundwater level in that stratum.
(b) A significant rise in borehole water level following a water strike may indicate the presence of a confined aquifer, particularly if the water strike occurs just after the borehole has passed from a low permeability stratum into a high permeability one. However, smaller rises in water levels may be observed, even in unconfined aquifers. This occurs when the driller has drilled a short distance below the water table before noticing the inflow. The drilling action will have removed some water from the borehole and when drilling stops the water level will rise up to the natural level.
(c) Because of the speed of boring, when drilling through soils of permeability from very low to moderate (such as clays, silts and silty sands) a water strike may not be noticed at all by the driller. The spoil from the borehole will be damp or moist, but there will not be time for free water to enter the borehole. There have been cases when groundwater inflow was not recorded in investigation borings through strata of silty sand, yet excavation works encountered groundwater and needed significant dewatering. This error could have been avoided by installing and monitoring observation wells, which allow time for equalization of water levels.
(d) The boring process will inevitably alter the water level in the borehole so it is not representative of that in the surrounding soil. The action of the drilling tool will remove water, and the driller may be deliberately adding water as part of the drilling process. The water level in the borehole at the end of the shift is likely to be very unreliable, being highly

influenced by the recent drilling activities. If drilling work is done on a day shift basis, the water level at the start of shift (next morning) may be more reliable as the borehole water level will have begun to equalize overnight. Even so, the start of shift level may still be unrepresentative, especially in low permeability soils, where longer periods may be necessary for full equalization.

Observations during rotary drilling are generally unsatisfactory, because during drilling the borehole is either kept topped up with water (if water-based flush fluids are used) or water is continually blown out of the borehole (if air-based flush fluids are used). It can be difficult to detect discrete groundwater inflows during drilling. As with boring methods, the water levels at the start of a shift tend to be more reliable than those observed at the end of the shift.

The principal drawback which affects all groundwater observations during trial pitting or boring is that, at best, they only give 'spot' readings of conditions at the time of investigation. As was discussed in Chapter 3, in general groundwater levels are not constant, but will vary with the seasons, with long terms trends (such as drought) or in response to external influences. Observations in trial pits and boreholes cannot give information on potential variations, and (depending on the time of year when they were drilled) may not indicate the highest groundwater level that can occur at a site. This type of information can only be obtained by the installation and monitoring of observation wells of some sort.

6.5.2 Observation wells

An observation well is an instrument installed in the ground at a specific location to allow measurement of the groundwater level or pore water pressure. When determining groundwater levels observation wells have a number of advantages over observations during boring:

1. Because they are long-term installations, appropriately designed observation wells can allow observation of equilibrium groundwater levels, even in very low permeability soils.
2. Observation wells can be installed as piezometers to record water levels or pore water pressures in a specific stratum.
3. Observation wells can be monitored for long periods of time, to observe the variations in groundwater level at a site.

The two most commonly used devices for monitoring groundwater levels in permeable soils are standpipes and standpipe piezometers.

The simplest from of observation well is the standpipe (see Fig. 6.3). This consists of a small diameter pipe, of which the bottom section (usually at least 1 m in length) is perforated or slotted, with the base plugged. The pipe

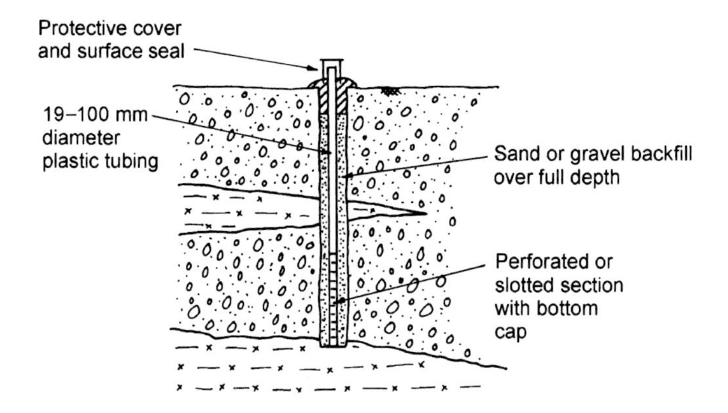

Figure 6.3 Typical standpipe installation.

is installed in the centre of a borehole and sand or gravel placed around the pipe, if necessary tamped into place. Backfilling should cease at a depth of about 0.5 m below ground level and the remainder of the hole sealed using puddled clay or bentonite/cement and capped off with concrete, to prevent surface or rainwater entering the borehole. It is advantageous to haunch the concrete to help to shed the surface water. Unless a special protective cover is required the pipe should project about 0.3 m above ground level and be provided with a suitable cap or threaded plug. In urban areas it is essential that the cover or capping arrangement is secure enough to resist vandalism. Some designs of covers (known as 'stop-cock' covers) can be installed flush with the ground surface. These covers are sometimes preferable in vandal-prone areas because they are unobtrusive and may not attract the attention of vandals.

Plastic tubing (such as PVC) is an ideal material for standpipe tubing. Typically supplied in 3 m or 5 m long pieces it can readily be sawn to desired length and joined using PVC couplings and solvent cement. The perforated lengths of pipe are usually supplied in 1.5 m lengths and are pre-drilled or pre-slotted. If necessary the plain pipe can be slotted on site using a hacksaw but it should be noted that the total area of perforations should be at least twice the cross-sectional area of the standpipe. The water level in a completed standpipe can be measured using a dipmeter (see Section 14.2). The preferred internal diameter for standpipe tubing is approximately 50 mm; this enables water samples to be taken and allows a small airline to be used to flush out the standpipe if it becomes blocked. Smaller diameter tubing is sometimes used, but the minimum acceptable internal diameter is usually 19 mm, because this is the smallest size down which many commercial dipmeters can pass.

A standpipe is simple and cheap to install, but it is only a basic instrument. The standpipe will respond to pore water pressures in water-bearing strata along its entire depth. This is acceptable if the standpipe is used in a simple unstratified unconfined aquifer, where the total head is constant with depth. However, if a standpipe is installed in a layered aquifer system water can enter from more than one water-bearing layer. If the groundwater levels are different in each layer (e.g. a main water table and a perched water table) the standpipe will show a 'hybrid' water level, between the two true water levels.

Standpipes are not suited to use in layered or complex groundwater regimes. In such cases it is necessary to use a 'piezometer' where the instrument is sealed into the ground so that it responds to groundwater levels and pore water pressures over a limited, defined, depth only. The most common type of piezometer is the standpipe piezometer.

Figure 6.4 shows typical construction details for standpipe piezometers. The aim is to produce a 'response zone' of sand or fine gravel at the level of the stratum in which the groundwater level is to be observed. Rigid PVC tubing is installed in the borehole in a similar way to a standpipe, with a 'piezometer tip' located in the centre of the response zone. Grout seals above and below the response zone ensure water can only reach the tip from the desired stratum. As with a standpipe, water level readings can be taken with a dipmeter.

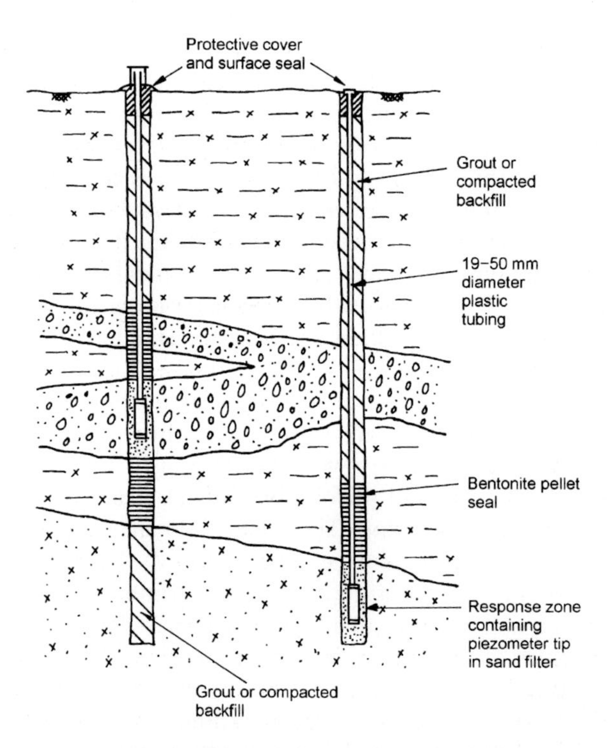

Figure 6.4 Typical standpipe piezometer installations. Two piezometers are shown, each with its response zone and piezometer tip in a different water-bearing stratum.

Installation of standpipe piezometers is more complex than for standpipes, and should be carried out with care. It is essential that the seals are effective, otherwise water may leak into the response zone from strata above or below. If clay backfill is used it must be adequately compacted (e.g. with the drill rods or shell) to reduce the risk of later settlement. Where grout is used to backfill parts of the boreholes it should be cement-bentonite grout of the appropriate consistency. A layer of bentonite pellets should be placed between the grout seal and the sand filter in the response zone, to avoid the sand becoming contaminated with grout (if pellets are not available, bentonite balls will have to be made up by hand). Once the lower bentonite seal is in place, and has had time to swell, it is good practice to flush out the dirty water in the borehole and replace it with clean water before installing the sand filter.

A 'generic' specification of the grading of sand filters is not possible. However, for guidance a filter consisting of a clean well-graded sand and gravel with only a small proportion of fine to medium sand, is suitable for soils with some clay or silt content. For a fine sand soil, the filter should consist of coarse sand or coarse sand and gravel with not more than a few per cent medium sand. Local material may have to be used, but it is essential that the filter material is free from clay and silt. Bentonite pellet and grout seals are installed above the sand filter. The tubing should be capped off at ground level with a secure cover or headworks.

The 'piezometer tip' typically consists of a porous plastic element or a porous ceramic element (sometimes known as a 'casagrande element'); tips are generally a 150–600 mm in length. It is good practice to soak the filter sand and ceramic element (if used) in water prior to installation – this helps avoid any air being trapped in the system, and speeds up the process of equilibration between the piezometer and the natural groundwater level.

It is possible to install two or more piezometer tips (each in its own response zone and separated by grout seals) in one borehole. If this is being contemplated, it is essential that it is carried out by experienced personnel and is carefully supervised. This is awkward work, in all but the largest boreholes and there is always the risk that the installation of the second tip and seals will affect the piezometer already installed. If possible, single piezometer installations in each borehole should be used, purely because the water level readings will be easier to interpret, with no worry of water leaking between response zones.

In many investigations for groundwater lowering projects piezometers will be installed in relatively permeable soils, where the water level inside the piezometer will respond rapidly to changes in the pore water pressure in the soil. However, piezometers in soils of moderate to very low permeability may respond slowly to changes in pore water pressure. This is because a finite volume of water must flow into or out of the piezometer to register the change in pressure. This leads to a 'time lag' between changes in pore water pressure in the soil and the registering of that change in the piezometer. The time lag is greater in soils of lower permeability, and is greater for piezometers where larger volume flows are needed to register pressure changes.

In a standpipe piezometer the prime factor controlling the equilibration rate is the internal diameter of the tubing; the smaller the diameter the shorter the time lag and the quicker the piezometer will respond to pressure changes. In soils of low to moderate permeability, it is normal to specify the internal diameter of the tubing as small as possible (19 mm is the lower practicable limit to allow monitoring by dipmeter). However, in permeable soils such as sand and gravels, the equilibration rate will tend to be rapid, and 50 mm diameter tubing could be used, allowing greater flexibility for sampling or flushing out of the piezometer.

A defining feature of observation wells is that (provided they are adequately protected from damage or vandalism) they can be used to observe groundwater levels long after the main ground investigation is complete. This can allow natural changes in groundwater levels to be determined. However, this requires the instruments to be monitored for extended periods, and the practicalities of this are sometimes overlooked. If readings are to be taken manually this will have to be included in the site investigation plan and associated costings. In remote or inaccessible sites it may be appropriate to use datalogging systems (see Section 14.5) to record groundwater levels, to reduce the cost associated with regular visits by personnel.

6.5.3 Other methods for determination of groundwater levels and pore water pressures

There may be occasions, especially in soils of low or very low permeability when the time lag is so great that the equilibration rate of a standpipe piezometer is too slow to give useful readings. In such cases, specialist instruments such as pneumatic piezometers or electronic vibrating wire pressure transducers could be used. These instruments are characterized by the very small volume of water which must flow into or out of the sensor in order to record a change in pressure; they have been used successfully to observe pore water pressure changes in silts, clays and laminated soils.

Such specialist instruments must be specified, installed and calibrated with care; instrument manufacturers can often provide useful advice. The instruments are installed in a discrete response zone in a similar way to a standpipe piezometer, but instead of an open pipe these instruments are connected to ground level by a cable or small bore tubing (Fig. 6.5). At ground level the cables can be terminated (to be read later using a portable read-out unit) or are connected to a datalogger system (see Section 14.5) to allow regular readings at pre-programmed intervals. The datalogger unit can be located some distance from the piezometer itself, being connected by cable or tubing. This

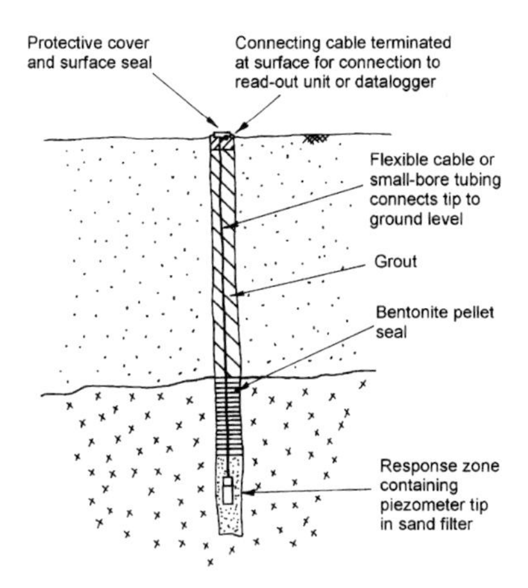

Figure 6.5 Specialist piezometers.

allows these instruments to be used where there is no permanent surface access for reading; with suitable grout seals, they have been used in boreholes beneath rivers and lagoons, with the datalogger located on the shore.

These instruments do not record total head, but record pore water pressure (above the level of the instrument). To interpret readings from the instruments correctly, the precise level at which the instrument was placed must be recorded during installation. Determination of total head or piezometric level from pore water pressure readings is described in Section 3.2.

Table 6.2 summarizes the advantages and disadvantages of methods of determining groundwater levels.

6.5.4 *Groundwater sampling and testing*

Knowledge of groundwater chemistry may be required for a variety of reasons (see Section 3.8). In particular groundwater chemistry can influence how discharge water can be disposed of, or may affect any potential side effects of dewatering. Groundwater samples should be obtained and tested as part of the site investigation.

Where groundwater is encountered during drilling and boring, obtaining a sample is relatively straightforward. The water sample should be taken using a clean sampling bailer as soon as possible after the seepage has begun. If water has been added during boring this should be bailed out prior to taking the sample.

Groundwater samples can also be taken from observation wells by use of a sampling pump. The water in the observation well has been exposed to the atmosphere and is unlikely to represent the true aquifer water chemistry.

Table 6.2 Advantages and disadvantages of methods of determining groundwater levels

Method	*Advantages*	*Disadvantages*
Observations in trial pits	Commonly carried out Allows seepage into excavation to be observed directly May allow perched water tables to be identified	Limited to shallow depth Standing water levels may be unrepresentative unless pit is left open long enough for equalization to occur
Observations during boring	Commonly carried out Normally sufficient to identify water inflows from major water bearing strata	True groundwater seepages and levels may be masked by drilling action and by addition and removal of water Seepages in soils of low to moderate permeability may not be identified Insufficient time is normally allowed for water levels to equalize to natural levels
Observations during drilling	Commonly carried out	Groundwater seepages and levels tend to be masked by the presence of flush medium
Observation wells – standpipes	Cheap and simple to install in boreholes Useful in simple unconfined aquifers Allow monitoring of groundwater levels in the long-term after boring is completed	Not appropriate for sites with confined aquifers, multiple aquifers or perched water tables May need to be purged or developed before use May need to be protected from vandalism and damage
Observation wells – standpipe piezometers	Relatively straightforward to install in boreholes Allow groundwater levels in specific strata to be observed Can be used on sites with complex groundwater conditions Allow monitoring of groundwater levels in the long-term after boring is completed	Closer supervision of installation is needed than with standpipes May need to be purged or developed before use May need to be protected from vandalism and damage
Specialist instruments – pneumatic and electronic piezometers	Can allow accurate pore water pressure measurements in low and very low permeability soils Can be read remotely Allow monitoring of groundwater levels in the long-term after boring is completed	Installation and calibration in boreholes can be complex; close supervision is advisable Readings may not be straightforward to interpret May need to be protected from vandalism and damage

Therefore, it is vital to fully 'purge' the well before taking a sample. Purging involves pumping the observation well at a fairly steady rate until at least three 'well volumes' of water have been removed (a 'well volume' is the volume of water originally contained inside the well liner). Specialist sampling pumps should be used in preference to airlifting, since the latter method may aerate the sample, increasing the risk of oxidation of trace metals and other substances. Obtaining a water sample during a pumping test is described in Appendix 6C.

In general, water samples should be of a least 1 litre volume. The bottles used for sampling should be clean with a good seal, and should be filled to the brim with water to avoid air bubbles in the sample. The sample bottle should be washed out three times (with the water being sampled) before taking the final sample.

Samples may degrade following sampling and should be tested as soon as possible after they are taken. Ideally, they should be refrigerated in the meantime. However, even then the sample may degrade in the bottle (e.g. by trace metals oxidizing and precipitating out of solution). Specialists may be able to advise on the addition of suitable preservatives to prevent this occurring. The choice of sample bottle (glass or plastic) should also be discussed with the laboratory since some test results can be influenced by the material of the sample bottle. Above all, it is vital to use an accredited, experienced laboratory for the chemical testing of water samples.

6.6 Determination of permeability

Permeability is an essential parameter to be determined for the design of groundwater lowering systems. There is a wide range of methods for determining permeability, some of which have been in use since the nineteenth century. The available methods can be classified into four main types.

(a) Laboratory testing (including particle size analysis and permeameter testing)
(b) Small-scale *in situ* tests (including borehole, piezometer and specialist tests)
(c) Large-scale *in situ* tests (including pumping tests and groundwater control trials)
(d) Other methods (including geophysics, visual assessment, and inverse numerical modelling).

While many test methods are available, obtaining realistic values of permeability is far from straightforward. The key problem is that, as was described in Section 3.2, soils and rocks are not homogenous isotropic masses. Permeability is likely to vary from place to place and to be different for different directions of measurement. As was stated by Preene *et al.* (1997) 'Even if it could be obtained, there is no single value of permeability in the ground waiting to be measured'.

If it is accepted that it can be difficult to determine meaningful values of permeability, that should still not deter those involved in site investigation and dewatering design from putting their best efforts into obtaining the most useful values practicable. The following sections outline the characteristics and limitations of the most commonly used methods of determining permeability. The final section in this chapter discusses the relative reliability of the various techniques. Selection of permeability for the design of groundwater lowering systems is dealt with in Chapter 7.

6.6.1 Visual assessment

Visual assessment of the permeability of a soil sample is the process of assessing the soil type or grading and, based on experience or published values (such as Table 3.1), estimating a very approximate range of permeability. This method is essential to allow corroboration of permeability test results. On *every* project the soil descriptions from borehole or trial pit logs should be reviewed to give a crude permeability range, against which later test results can be judged. Information gathered by the desk study, such as experience from nearby projects, can be useful in this regard.

For example a soil described as a medium sand might typically be expected to have a permeability of the order of 10^{-4} m/s; certainly such a soil would be unlikely to have a permeability greater than 10^{-3} m/s or less than 10^{-5} m/s. If permeability tests give results of 10^{-8} m/s there is clearly some discrepancy. Either the soil description is misleading or the test results are in error or unrepresentative. If this is recognized while testing is still going on there may be a chance to modify test types or procedures to get better results. Discrepancies of this magnitude are not rare, and visual assessment of permeability can often be a more useful guide to permeability than test results, especially if the latter are limited in scope and questionable in quality.

6.6.2 Inverse numerical modelling

This approach can only be used if extensive groundwater monitoring data (e.g. data from a number of observation wells over a long time period) are available. It also requires a thorough understanding of the geological structure and extent of the aquifers in the area. The method involves setting up a numerical model of the aquifer system in the vicinity of the site and running the model with a variety of permeability and other parameters, until an acceptable match is obtained between the model output and the groundwater monitoring data.

This approach is not straightforward and is carried out only rarely. Since a number of parameters, in addition to permeability, will be varied during the modelling, a 'non-unique' solution may result. In other words, a number of permeability values may give an acceptable fit with the data, depending on what other parameter values are used.

6.6.3 *Geophysics*

Geophysical methods have traditionally been used extensively in hydrogeological studies for the development of groundwater resources, but are used only rarely in investigations for construction projects. The methods available can be divided into two main types:

i Surface geophysical methods,
ii Downhole (or borehole) geophysical methods.

These methods do not generally give *direct* estimates of permeability values, but can allow: indirect estimation of permeability; identification of zones of relatively high or low permeability; or correlation of permeability values between different parts of a site (Macdonald *et al.* 1999).

Surface geophysical methods include seismic refraction methods, gravity surveying, electromagnetic surveying and the most widely used method, resistivity soundings (Barker 1986). These methods measure the variation in specific physical properties of the sub-surface environment, and apply theoretical and empirical correlations to infer the structure of the ground and groundwater regime. Typical applications include mapping the extent of gravel deposits within extensive clay strata, or locating buried channel features within drift deposits overlying bedrock.

Downhole geophysical methods involve surveying previously constructed boreholes. This is normally achieved by lowering various sensing devices (known as 'sondes') into the borehole. Some types of sonde investigate the properties of the aquifer material (this is known as formation logging) while others measure the properties of the water in the borehole (fluid logging). The various methods available are described in Beesley (1986) and BS7022 (1988). These methods are generally applicable in boreholes drilled into rock, when more permeable fissured or fractured zones can sometimes be identified.

Geophysical methods are specialized techniques. Their application and interpretation requires care. In all but the simplest of cases, specialist advice should be obtained from experienced geophysicists.

6.6.4 *Laboratory testing: particle size analysis*

There are a number of empirical methods available to allow the permeability of granular soils to be estimated from analyses of particle size distributions (PSDs) of samples. An American water works and sanitary engineer from New England, Allen Hazen (1892; 1900) was the first to propose an empirical correlation for the permeability of a sand from its PSD curve. Probably due to the great simplicity of his 'rule', Hazen is still widely used by many of today's geotechnical practitioners, often without due regard to the limitations that Hazen himself stated. His objective was to determine guidelines for suitable sand gradings for water supply filtration. He determined that the D_{10} particle size (called the 'effective grain size') and D_{60}/D_{10} (the 'uniformity coefficient') were both important factors.

Hazen included allowances for variations in the temperature of water. However, the temperature of the groundwater in the United Kingdom varies little between about 5 and 15°C; so Hazen's rule used to estimate permeability k may be stated as:

$$k = C(D_{10})^2 \tag{6.1}$$

where C is a calibration factor and D_{10} is the 10 per cent particle size taken from the particle size distribution curves (Fig. 6.6).

Hazen stated in his work that his rule was applicable over the range of D_{10} particle size from 0.1 mm to 3.0 mm and for soils having a uniformity coefficient less than five. He also stated that (when k is in m/s and D_{10} is in mm) his calibration factor C could vary between about 0.007 and 0.014. In practice, presumably for reasons of simplicity, C is normally taken to be 0.01. It cannot be stressed too strongly that, even within its range of application, Hazen's rule gives *approximate* permeability estimates only.

Since Hazen many others – particularly Slichter, Terzaghi, Kozeny and Rose (all reported in Loudon 1952) and Masch and Denny (reported in Trenter 1999) – have developed expressions for estimating permeability

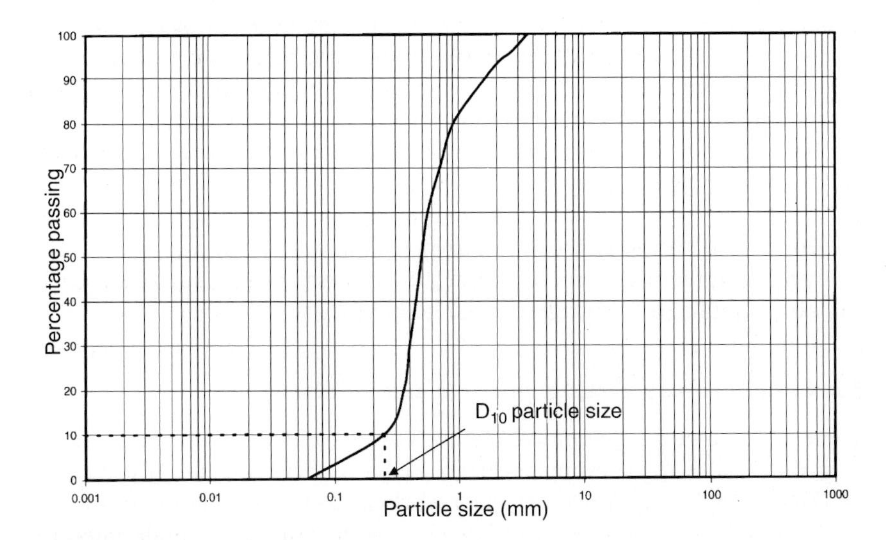

Figure 6.6 Application of Hazen's rule.

values from grain size distributions of sands. Unlike Hazen, who did not seek to address *in situ* soils, some have taken account of porosity, angularity of the grains and specific surface of the grains. None claim to be relevant to soils other than 'a wide range of sands'.

Loudon (1952) thoroughly reviewed various published formulae and supplemented his review with his own laboratory investigations. He concluded that the error in prediction using Hazen's rule could be of the order of plus or minus 200 per cent but that Kozeny's formula – which is similar to that of Terzaghi, though more complicated – was to be preferred to the various others. Loudon stated that an accuracy of about plus or minus 20 per cent can be expected from Kozeny's formula.

He also proposed that his own formula, based on Kozeny, should be used for reasons of simplicity.

$$\log_{10}(kS^2) = a + bn \tag{6.2}$$

where k is the permeability expressed in cm/s, n is porosity of the granular soil (a dimensionless ratio, expressed as a fraction not as a percentage), S is specific surface of grains (surface area per unit volume of grains) expressed in $(\text{cm}^2)/(\text{cm}^3)$, and a and b are calibration factors with values of 1.365 and 5.15 respectively.

Whilst the porosity of a sample can be determined in the laboratory, it is virtually impossible to determine the porosity of a sample *in situ*. This is a limitation on the usefulness of Loudon and other similar works and an explanation for the somewhat erratic results that they sometimes give. They take little or no account of density and heterogeneity of soils. Standard penetration test N values do give an indication of relative density of granular soils and so may afford some *tentative* indication of porosity. Refer to Appendix 6A for further information concerning Loudon's method.

In the 1950s in America the late Professor Byron Prugh researched and developed an empirical method for estimating permeability based on the use of particle size data together with *in situ* density field measurements. He checked his predictions against field measurements of permeability. Prugh's approach represents a return to the pragmatic co-ordination of academic and field observations.

Prugh plotted (Fig. 6.7) curves for various uniformity coefficients (D_{60}/D_{10}). The D_{50} grain sizes are plotted on the horizontal axis to a log scale. Permeability is plotted on the vertical axis, also to a log scale. Three separate sets of uniformity coefficient curves were compiled for:

(a) dense soils
(b) medium dense soils
(c) loose soils.

To use Prugh's curves first determine whether the soil sample is dense, medium dense or loose (based on standard penetration test N values from

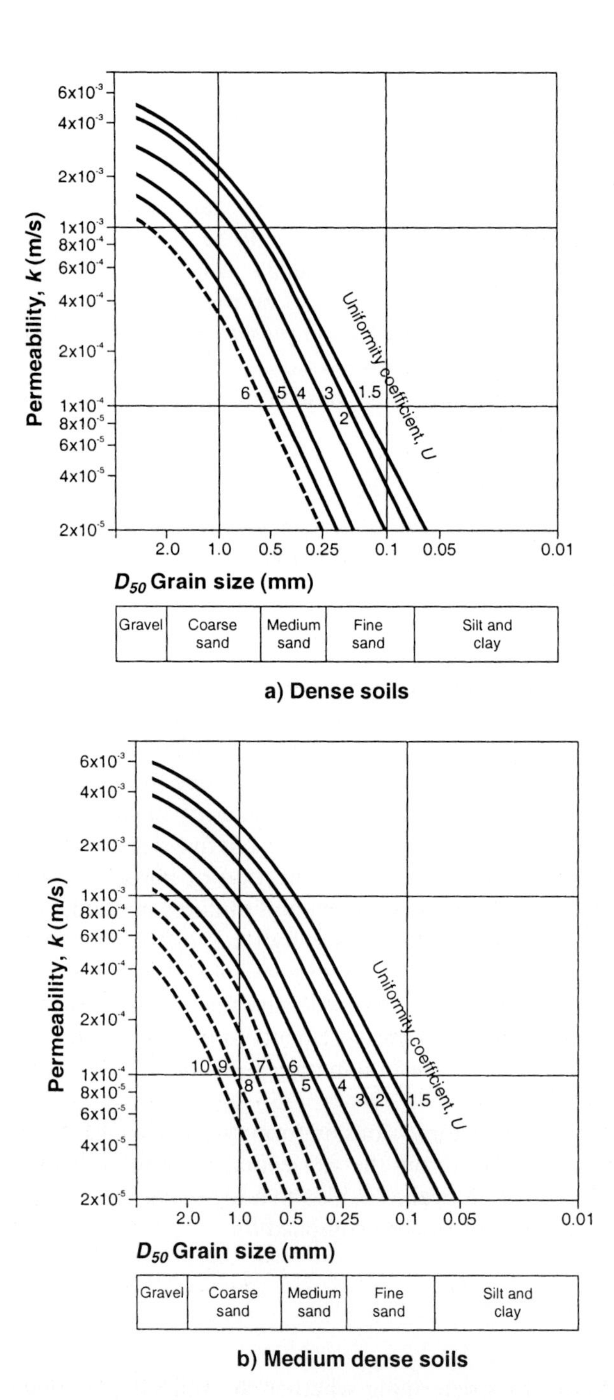

Figure 6.7 Prugh method of estimating permeability of soils (from Preene *et al.* 2000; reproduced by kind permission of CIRIA).

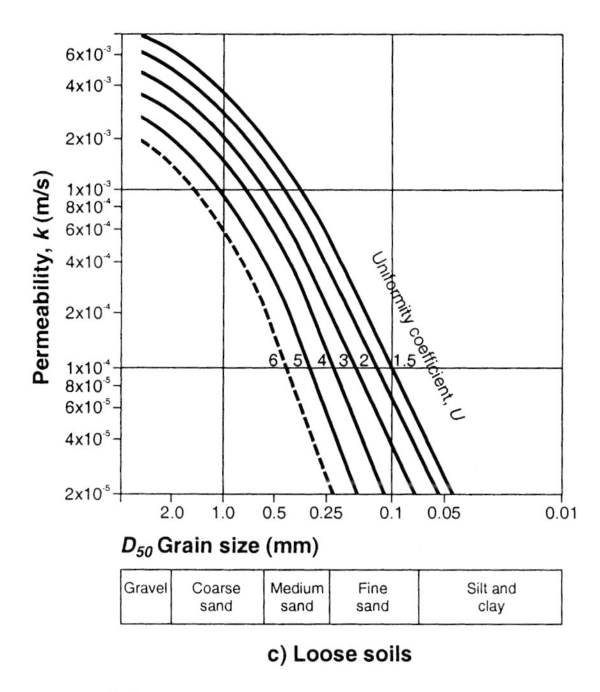

Figure 6.7 Continued.

borehole logs); then project upwards from the D_{50} grain size of the sample onto the appropriate uniformity coefficient curve; from the uniformity coefficient curve project horizontally to read off the permeability value.

Prugh's data indicate that as the uniformity coefficient increases (i.e. the sample becomes less and less a single-size material), the permeability decreases noticeably. The significance of the Prugh curves, apart from their usefulness, is that of helping greatly the understanding of the inter-relationship of various factors (other than D_{10} used in Hazen's rule) affecting soil permeability. His work has been published by Powers (1992) and in CIRIA Reports (Preene *et al.* 2000). Like others, Prugh did not claim his method to be relevant to soils other than 'a wide range of sands.'

Irrespective of which method is used to estimate permeability, these approaches all use data from samples recovered from boreholes, rather than tested *in situ*. This can lead to inaccuracies in permeability assessment, including:

1 Any soil structure or fabric present in the *in situ* soil will be destroyed during sampling and test specimen preparation. Permeability estimates

based on the PSD curve of the resulting homogenized sample are likely to be unrepresentative of the *in situ* permeability. If a clean sand deposit contains laminations of silt or clay, these will become mixed into the mass of the sample during preparation and the PSD curve will indicate a clayey or silty sand. Under-estimates of permeability will result if the Hazen or Prugh methods are applied to these samples.

2 The samples used for particle size testing may be unrepresentative. When bulk or disturbed samples are recovered from below the water level in a borehole there is a risk that finer particles will be washed from the sample. This is known as 'loss of fines'. Samples affected in this way will give over-estimates of permeability if the Hazen or Prugh methods are used. Loss of fines is particularly prevalent in samples taken from the drilling tools during light cable percussion boring. This can be minimized by placing the whole contents (water and soil) into a tank or tray and allowing the fines to settle before decanting clean water. Unfortunately, in practice this is rarely done. Loss of fines is usually less severe for tube samples such as SPT or U100 samples; these methods may give more representative samples in fine sands.

6.6.5 Laboratory testing: permeameter testing

There are a number of techniques for the direct determination of the permeability in the laboratory by inducing a flow of water through a soil sample – this approach is known as 'permeameter' testing. According to Head (1982) there are two main types of permeameter testing.

1 Constant head test. A flow is induced through the sample at a constant head. By measuring the flow rate, cross-sectional area of flow and induced head, the permeability can be calculated using Darcy's law. This method is only suitable for relatively permeable soils such as sands or gravels ($k > 1 \times 10^{-4}$ m/s); at lower permeabilities the flow rate is difficult to measure accurately.

2 Falling head test. An excess head of water is applied to the sample and the rate at which the head dissipates into the sample is monitored. Permeability is determined from the test results in a similar way to a falling head test in a borehole or observation well. These tests are suitable for soils of lower permeability ($k < 1 \times 10^{-4}$ m/s) when the rate of fall in head is easily measurable.

Tests may be carried out in: special permeameters; oedometer consolidation cells; Rowe consolidation cells; and triaxial cells. Methods of testing are described in BS1377 (1990) and in Head (1982).

While these tests are theoretically valid, in practice they are rarely used because of the difficulty of obtaining representative 'undisturbed' samples of the granular soils (silt, sand or gravel) of interest in dewatering design. Even if the sample is representative of the particle size distribution of the soil, the *in situ* density and hence void ratio of the soil is likely to be known only sketchily. This means that the *in situ* condition of the soil cannot be reproduced. Similarly, any soil fabric or layering in the sample will have a profound effect on the *in situ* permeability, but cannot be replicated in the laboratory.

The only time such tests should be considered is when the permeability of very low permeability soils needs to be determined during investigations of potential consolidation settlements. Even then results must be interpreted with care, as in clays with permeable fabric the size of the test sample may result in scale affects distorting the measured permeability (see Rowe 1972). Large (250 mm diameter) samples may give more representative results than the 76 mm diameter samples routinely tested, but such large samples are rarely available.

6.6.6 In situ *tests in boreholes: rising, falling and constant head tests*

This group of tests includes:

i Rising and falling head tests (collectively known as variable head tests),
ii Constant head tests.

These tests are carried out in the field on the soil *in situ*. They, therefore, avoid the problems of obtaining representative undisturbed samples that limit the usefulness of laboratory testing. Tests in boreholes are those carried out during interruptions in the drilling or boring process. When the test is complete drilling recommences – this allows several tests at different depths to be carried out in one borehole. These tests are distinct from tests carried out in observation wells following completion of the borehole, where tests can be carried out only at the fixed level of the response zone.

Execution of variable head tests is straightforward and requires only basic equipment. The borehole is advanced to the proposed depth of test, and the original groundwater level noted. The upper portion of the borehole is supported by temporary casing (which should exclude groundwater from those levels). The 'test section' of exposed soil is between the bottom of the casing and the base of the borehole.

For a falling head (or inflow) test (Fig. 6.8(a)) water is added to raise the water level in the borehole. Once the water has been added the water level in the borehole is recorded regularly to see how the level falls with time as water flows out of the borehole into the soil. The necessary equipment is a dipmeter, bucket, stopwatch and a supply of clean water (perhaps from a tank or bowser). It is essential that any water added is absolutely clean, otherwise any suspended solids in the water will clog the base of the borehole test section and significantly affect results. Particular attention should be given to the cleanliness of tanks, buckets, etc, so that the water does not become contaminated

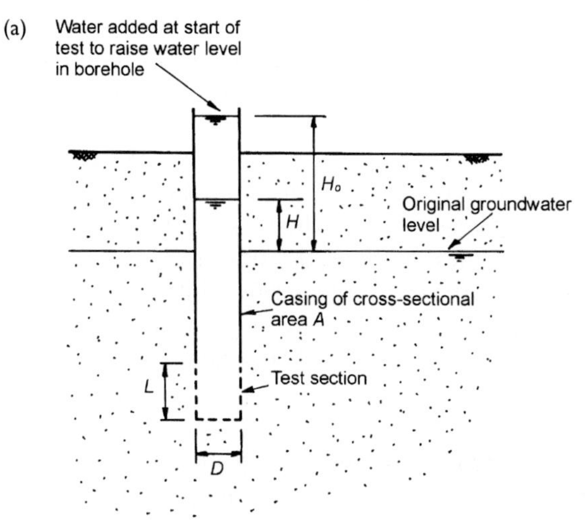

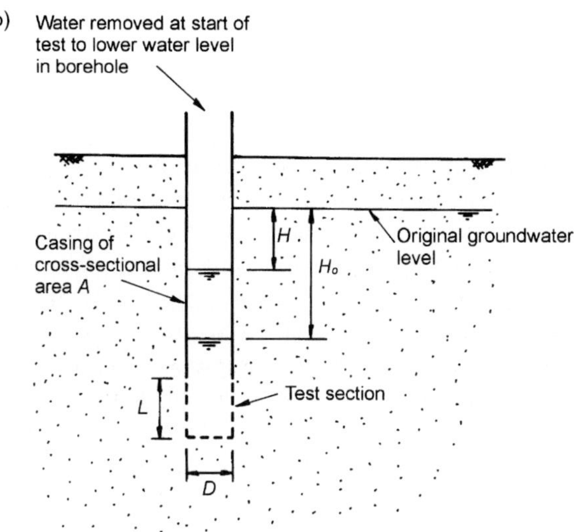

Figure 6.8 Variable head tests in boreholes. (a) Falling head (inflow) test, (b) rising head (outflow) test.

by those means. It can be difficult to carry out falling head tests in very permeable soils (greater than about 10^{-3} m/s) because water cannot be added quickly enough to raise the water level in the borehole. If the natural groundwater level is close to ground surface it may be necessary to extend the borehole casing above ground level to allow water to be added.

A rising head (or outflow) test (Fig. 6.8(b)) is the converse of a falling head test. It involves removing water from the borehole and observing the rate at which water rises in the borehole. The test does not need a water supply (which can be an advantage in remote locations), but does require a means of removing water rapidly from the borehole. The most obvious way to do this is using a bailer, which is adequate in soils of moderate permeability but it can be surprisingly difficult to significantly lower water levels if soils are highly permeable. Alternatives are to use airlift equipment or suction or submersible pumps.

For the relatively permeable soils of interest in groundwater lowering problems, variable head tests are analysed using the work of Hvorslev (1951) which is the basis of the methods given in BS5930 (1999). Hvorslev assumed that the effect of soil compressibility on the permeability of the soil was negligible during the test, and this is a tolerable assumption for most water-bearing soils. If *in situ* permeability tests are carried out in relatively compressible silts and clays different test procedures and analysis may be required (see Brand and Premchitt 1982).

For the Hvorslev analysis, permeability k is calculated using

$$k = \frac{A}{FT} \tag{6.3}$$

where A is the cross-sectional area of the borehole casing (at the water levels during the test), T is the basic time lag and F is a shape factor dependent on the geometry of the test section. T is determined graphically from a semi-logarithmic plot of H/H_0 versus elapsed time as shown on Fig. 6.9. H_0 is the excess head in the borehole at time $t=0$ and H is the head at time t (both H and H_0 are measured relative to the original groundwater level). Additional notes on the analysis of variable head tests are given in Appendix 6B.

Values of shape factor F for commonly occurring borehole test section geometries were prepared by Hvorslev (1951) and are shown in Fig. 6.10. Shape factors for other geometries are given in BS5930 (1999). The simplest test section is when the temporary casing is flush with the base of the borehole, allowing water to enter or leave the borehole through the base only. If soil will stand unsupported it may be possible to extend the borehole ahead of the casing to provide a longer test section. If the soil is not stable the borehole could be advanced to the test depth, and the test section be backfilled with filter sand or gravel as the casing is withdrawn to the top of the test section.

Constant head tests (Fig. 6.11) involve adding or removing water from a borehole at a known rate in order to maintain a constant head, which is recorded. Constant head tests are most often carried out as inflow tests, but outflow tests can also be carried out. The equipment required is rather more complex than for variable head tests, as some form of flow measurement (typically by the timed volumetric method) is required. In the simplest form of the test, appropriate to relatively permeable soils, the flow rate is adjusted until a suitable constant head is achieved, and the test allowed to continue

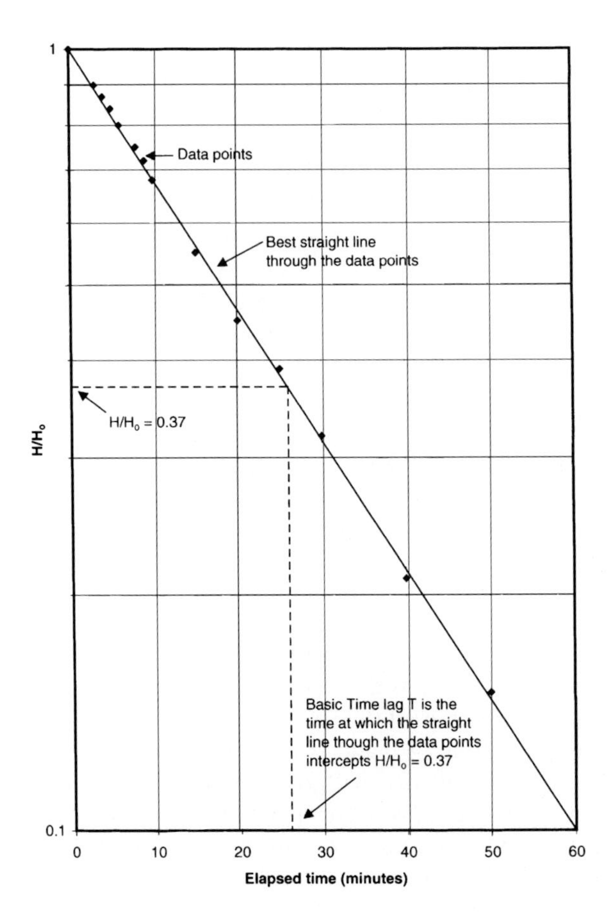

Figure 6.9 Analysis of variable head tests.

until a steady flow rate is established. A consistent supply of clean water is required for tests, and this can be a disadvantage in remote locations.

Permeability k is calculated from

$$k = \frac{q}{FH_c} \tag{6.4}$$

where q is the constant rate of flow, H_c is the constant head (measured relative to original groundwater level) and F is the shape factor (from Fig. 6.10).

Variable and constant head tests in boreholes have a number of limitations, and may be subject to a number of errors. When carrying out these tests (and when reviewing the results) it is essential that these factors are considered.

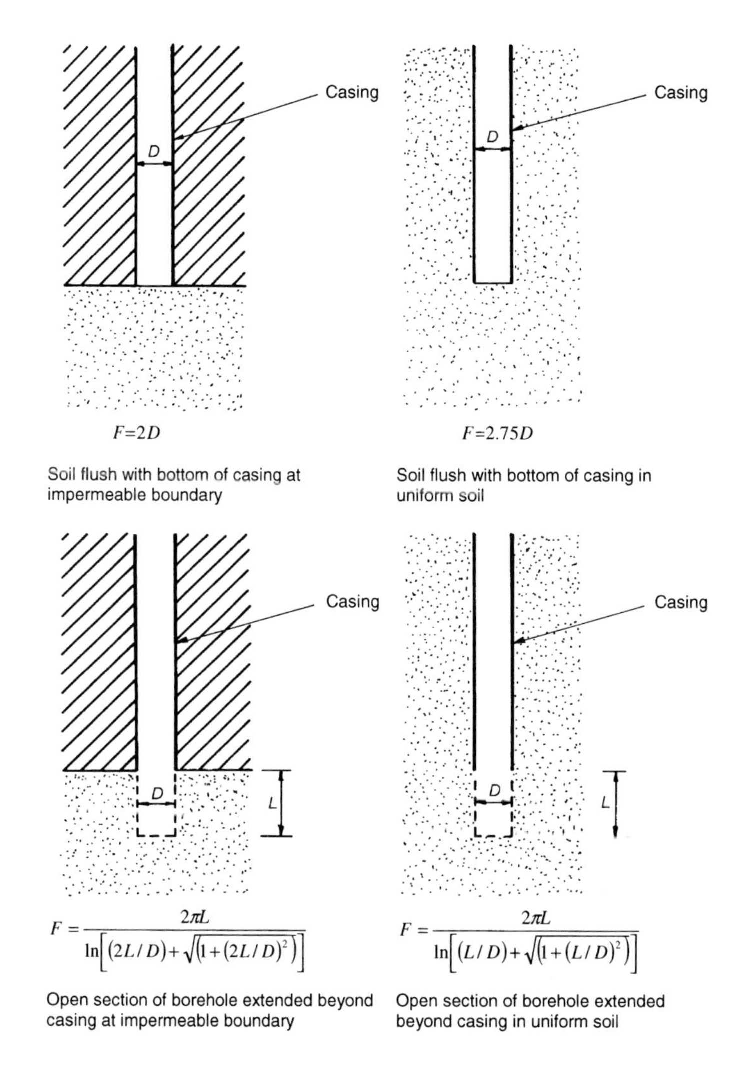

Figure 6.10 Shape factors for permeability tests in boreholes (after Hvorslev 1951).

1 Tests in boreholes only involve a relatively small volume of soil around the test section. If the soil is heterogeneous or has significant fabric, such tests may not be representative of the mass permeability of the soil. Large scale tests (such as pumping tests) may give better results.
2 Results of inflow tests (falling head and constant head tests) can be significantly affected by clogging or silting up of the test section as water is added. It is vital that only totally clean water is added but, even then,

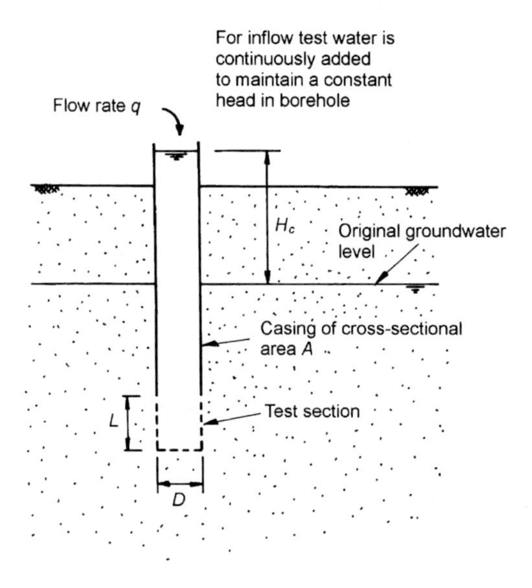

Figure 6.11 Constant head inflow test in boreholes.

silt already in suspension may block flow out of the borehole. It is not uncommon for inflow tests to under-estimate permeability by several orders of magnitude.

3 In loose granular soils outflow tests (rising head and constant head tests) may cause piping or boiling of soil at the base of the borehole. This could lead to over-estimates of permeability.

4 The drilling of the borehole may have disturbed the soil in the test section, changing the permeability. Potential effects include particle loosening, compaction, or smearing of silt and clay layers.

5 Reliable analysis of test results requires that the original groundwater level be known (this is discussed in Appendix 6B). However, if the natural groundwater level varies during the test (due to tidal or other influences) the test may be difficult to analyse. If significant groundwater level fluctuations are anticipated during a test of say one or two hours duration, tests in boreholes are unlikely to be useful.

Although these tests have a number of limitations they are inexpensive to execute, and are widely used. It is good practice to carry out both rising and falling head tests in the same borehole to allow results to be compared. In any event, results should be treated with caution until supported by permeability estimates from other sources.

6.6.7 In situ tests in boreholes in rock

The borehole testing techniques used in soil can also be applied to boreholes in rock. However, in practice, a different approach is often taken to the *in situ* testing of boreholes drilled through rock strata. This is because the flow of groundwater will be mostly along joints, fissures or other discontinuities. A borehole drilled through a stratum of rock may pass through relatively unfissured zones (which will be of low permeability) and through more fissured zones (of higher permeability). It is important that the level and extent of these zones are identified. Two of the most useful approaches are geophysical logging of boreholes and packer permeability testing.

Geophysical formation logging of unlined boreholes in rock can help identify the presence of more or less fissured zones. Fluid logging methods (including flowmeter and fluid conductivity and temperature logging) can be used to determine specific levels at which groundwater is entering the well (see Beesley 1986; BS7022 1988). The results from geophysical surveys can be used to specify the levels at which permeability testing should be carried out.

The packer test is one of the most common types of permeability test used in boreholes drilled in rock, provided the borehole is stable without casing. The method is a form of constant head test, carried out within a discrete test section isolated from the rest of the borehole by inflatable 'packers' (Fig. 6.12). Water is pumped into or out of the test section and the change in water pressure or level noted. Because discrete sections of borehole at various depths can be tested, the method can help identify any fissured permeable zones. The packer test was originally developed in the 1930s to assess the permeability of grouted rock beneath dam foundations, and is sometimes known as a Lugeon test after the French engineer who pioneered the method. Strictly speaking, a true Lugeon test is one particular form of packer test, carried out using specific equipment and injection pressures, and the term should not be used for packer tests in general.

The test method is described in Walthall (1990) and BS5930 (1999). In the double packer test inflatable packers are used to isolate a test section between two packers (Fig. 6.12). For a single packer test, the test section is between a packer and the base of the borehole. The most common type of packer test is an inflow test, where water is injected into the test section and the flow rate and head recorded. Outflow tests can also be carried out, although the test equipment is more complicated (Price and Williams 1993); nevertheless a number of studies (including Brassington and Walthall 1985) have concluded that outflow tests are preferable to inflow tests.

Packer tests are most suited to rocks of moderate to low permeability. If permeability is greater than 10^{-7} m/s friction losses in the pipework become significant and need to be included in calculations. When packer tests are carried out in highly permeable zones (greater than about 10^{-5} m/s) it is difficult to inject sufficient water to maintain the test pressure, and the

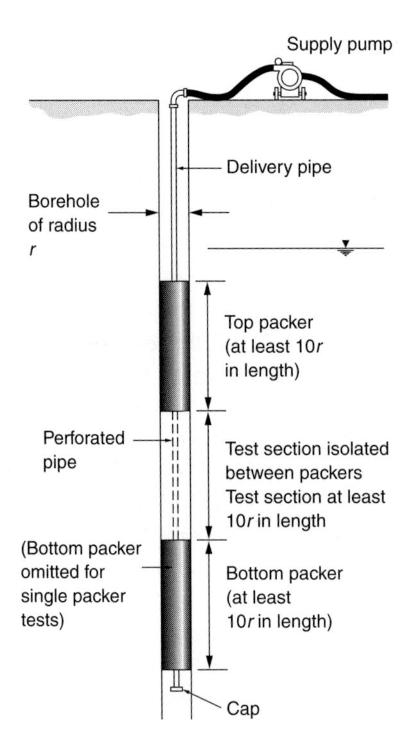

Figure 6.12 Packer test (from Preene *et al.* 2000; reproduced by kind permission of CIRIA).

test may have to be aborted. Analysis of packer tests is described in Clayton *et al.* (1995) and BS5930 (1999). Tests can be carried out at various depths in an unlined borehole, and may allow permeability depth profiles to be obtained. The following factors must be considered when carrying out packer tests (and when reviewing test results):

1 In most rocks, the overall permeability is dominated by flow through fissures. Measured permeability values will be affected if the drilling process has blocked or enlarged natural fissures. Walthall (1990) states that prior to a packer test the borehole should be cleaned out to remove all drilling debris, and also recommends that the borehole be developed by airlifting. Even without drilling-related effects, the mere presence of the borehole may lead to stress relief and stress re-distribution around the borehole, changing the local permeability.
2 As with any inflow test it is vital that the injected water is absolutely clean so that the risk of siltation of the test section is reduced.
3 Care must be taken to ensure that injection pressures are not so high as to cause hydraulic fracturing or uplift of the ground. Even without hydraulic fracturing, high water pressures may artificially dilate existing fissures.

4 Problems sometimes occur if packers do not form an effective seal with the borehole walls. This will lead to leakage from the test section and can give completely misleading results. It is good practice to select the test sections based on the drilling records of the test hole, and to try and locate packers in sections that are likely to give good seating for packers. If obtaining high quality test results is important it may be worthwhile carrying out a caliper survey prior to packer testing, and selecting packer setting depths on that basis.

6.6.8 In situ tests in observation wells

Variable and constant head tests can be carried out in observation wells (standpipes and standpipe piezometers) following completion of boring (Fig. 6.13). While these tests can only be carried out at the depth of the observation well response zone, they have the advantage that they can be executed (and repeated if necessary) without the time pressures after boring has been completed. Tests are analysed using the same methods as for tests in boreholes. For observation wells with cylindrical response zones the shape factor F derived by Dunn and Razouki (1975) can be used:

$$F = \frac{2.32\pi D(L/D)}{\ln\left[1.1(L/D) + \sqrt{(1 + (1.1L/D)^2)}\right]} \tag{6.5}$$

where L is the length of the borehole test section and D is the diameter of the borehole. Where L is large in relation to D the test will tend to determine the horizontal permeability k_h. In any event, if the horizontal permeability is

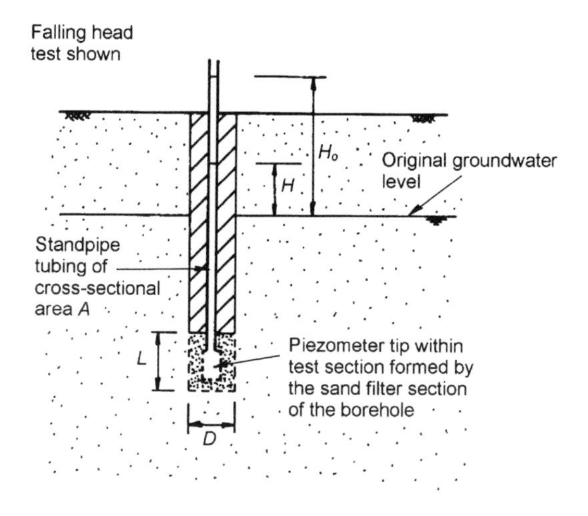

Figure 6.13 Variable head tests in observation wells.

much greater than the vertical, these tests will tend to measure k_h, whatever the ratio of L to D.

The limitations on variable and constant head tests in boreholes also apply to tests in observation wells. Additional problems may result from the nature of the observation well itself. If the standpipe or piezometer has not been installed to the highest standards it may be partially clogged. In such cases any tests will merely determine the permeability of the piezometer, rather than the permeability of the soil. To reduce this problem all observation wells should be purged or developed (by pumping or airlifting) prior to testing. Testing should not commence until the water level has recovered to its equilibrium level.

6.6.9 *Specialist in situ tests*

Specialist *in situ* tests are occasionally carried out as part of certain types of probing techniques. These include dissipation tests carried out in piezocone testing (Lunne *et al.* 1997) or *in situ* permeameters (Chandler *et al.* 1990). These tests are generally used only in fine-grained soils (such as silts and clays) of low or very low permeability, and are not generally used in investigations for groundwater lowering, except to investigate the permeability of aquitards, where the risk of settlement is of concern.

The most common of these specialist tests is the dissipation test, carried out as part of piezocone probing. When the piezocone is advanced during probing in low permeability soils an excess pore water pressure will build up at the cone, greater than the natural groundwater pressure, as a result of penetration. If penetration is stopped, the excess pore water pressure will dissipate at a rate controlled by the coefficient of consolidation with horizontal drainage c_h, which is related to the stiffness E'_0 and horizontal permeability k_h of the soil by

$$c_h = \frac{k_h E'_0}{\gamma_w} \qquad (6.6)$$

where γ_w is the unit weight of water. Analysis of the test involves determining the degree of dissipation U at a given time t, where

$$U = \frac{u_t - u_0}{u_i - u_0} \qquad (6.7)$$

and u_t = pore water pressure at time t; u_0 = equilibrium pore water pressure at the level of the test; and u_i = pore water pressure recorded at the start of the test. In general, the dissipation test should be continued at least until U is 0.50 or less. A typical dissipation test plot is shown in Fig. 6.14. To estimate c_h the time t_{50} for 0.5 degree of dissipation is determined and applied to the following formula:

$$c_h = \frac{T_{50}}{t_{50}} r_0^2 \qquad (6.8)$$

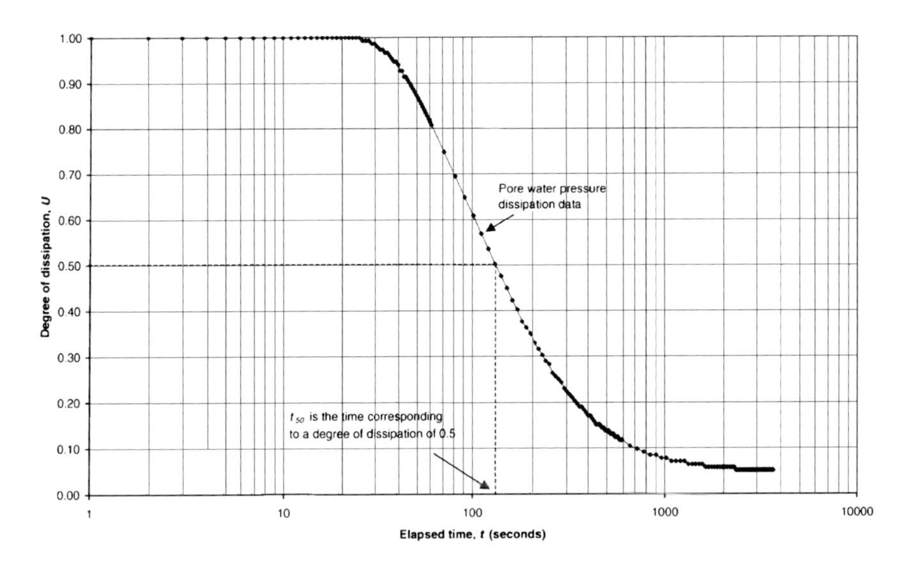

Figure 6.14 Analysis of pore water pressure data from piezocone dissipation test.

where the time factor T_{50} is obtained from published theoretical solutions and r_0 is the equivalent penetrometer radius. The theoretical solution to be used to estimate c_h should be chosen with care; it will be influenced by the strength characteristics of the soil, and by the location of the pore water pressure measuring point on the piezocone tip; Lunne *et al.* (1997) present a number of possible solutions that may be used to obtain T_{50} and r_0.

Once c_h has been determined, k_h can be estimated using equation (6.6), provided the soil stiffness is known. Determination of permeability from piezocone dissipation tests involves a lot of assumptions and uncertainties; values determined in this way should be treated as a general indicator only. The penetration of the piezocone itself may disturb the soil and may locally reduce the permeability around the cone. This can be particularly acute if the soil has a laminated structure, when the penetration of the cone through clay layers may smear clay over the more permeable sand layers. These factors may result in piezocone results underestimating the *in situ* permeability.

6.6.10 Pumping tests

The simplest form of pumping test involves controlled pumping from a well and monitoring the flow rate from the well and the drawdown in observation wells at varying radial distances away.

A correctly planned, executed and analysed pumping test is often the most reliable method of determining the mass permeability of water-bearing soils.

This is principally because the volume of soil through which flow of water is induced is significantly greater than in the cases of variable and constant head tests in boreholes and observation wells. Unfortunately, due to the relatively high cost of a pumping test compared to these methods, it is used less frequently than is desirable.

The essentials required for a pumping test (Fig. 6.15) are:

i A central water abstraction point, usually a single deep well – though it can be a ring of wellpoints or ejectors – which forms the 'test well'. Ideally this should penetrate the full thickness of the aquifer as this simplifies analysis of test results.
ii A series of observation wells (standpipes or standpipe piezometers depending on the aquifer type) installed in the aquifer at varying radial distances from the test well. These allow the depth to groundwater level to be measured (manually by dipmeter or using datalogging equipment)

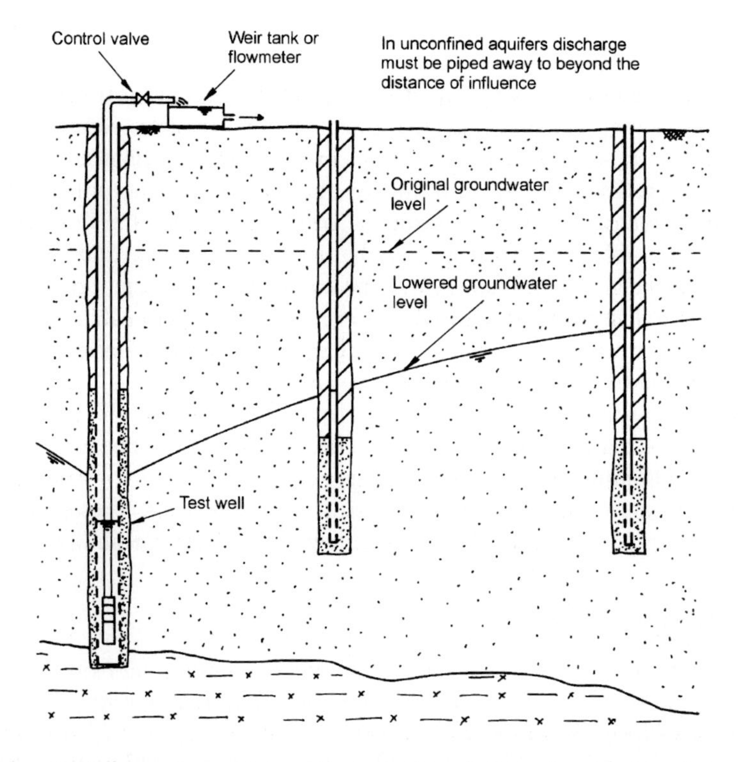

Figure 6.15 Components of a pumping test. For clarity, only two observation wells are shown but several are normally required.

to determine the drawdown at varying times after commencement of pumping and during recovery after cessation of pumping.

iii A means to determine the rate of pumping, such as a weir tank or flowmeter (see Section 14.3).

A pumping test typically consists of the following phases:

1 Pre-pumping monitoring,
2 Equipment test,
3 Step-drawdown test,
4 Constant rate pumping phase,
5 Recovery phase.

In addition to monitoring of groundwater levels and the discharge flow rate, a test is also a useful opportunity to obtain groundwater samples from chemical testing. Further details on the execution of pumping tests is given in Appendix 6C. Test methods are defined in BS5930 (1999) and BS6316 (1992). The test well itself may also provide additional information about ground conditions. Geophysical logging of the well may provide useful information (see Beesley 1986; BS7022 (1988)). In fissured rock aquifers geophysical logging can allow identification of fissured zones where groundwater is entering the well.

A pumping test is a relatively expensive way of determining permeability. The cost of a pumping test is seldom justified for a small project or for routine shallow excavations. However, for any large project or deep excavation, or where groundwater lowering is likely to have a major impact on the construction cost, one or more pumping tests should be carried out. It is essential that the pumping test is planned to provide suitable data for dewatering design. Issues to be considered include: the drawdown during the test (which in nearby observation wells should be at least 10 per cent of that in the proposed dewatering system); the number and position of observation wells (which should allow the drawdown pattern around the test well to be fully identified); and the design of the test well (which, ideally, should be of similar design to, and be installed by the same methods as, the proposed dewatering wells). To meet these aims a pumping test is generally carried out in the second or subsequent phases of ground investigation, when ground conditions have been determined to some degree. This allows the depth of the test well screen and observation well response zones to be selected on the basis of the data already gathered.

Analysis of the test results can provide information that is useful in a number of ways:

(a) Data from the step-drawdown test can be used to analyse the hydraulic performance of the well, to determine well losses and efficiency. This approach is widely used in the testing of water supply wells (see Clark

1977, for methods of analysis), but is less widely relevant to temporary works wells for groundwater lowering purposes. It can still be useful when trying to optimize the performance of deep well systems, and may allow comparison of the performance of wells drilled and developed by different methods.

(b) Data from the constant rate pumping phase can be used to estimate the permeability and storage coefficient of the volume of aquifer influenced by the test. The permeability estimated from the test results is used as an input parameter in the design methods described in Chapter 7. Permeability is estimated by conventional hydrogeological analyses (described below), which can also provide some information about the aquifer boundary conditions.

(c) Observations of the way drawdown reduces with distance from the test well can be used to construct a distance-drawdown plot, which can then be used to design groundwater lowering systems by the cumulative drawdown method (see Section 7.4). This method is interesting because the design does not need a permeability value, since that is implicit in the distance-drawdown plot.

Most pumping tests are analysed by 'non-steady state' techniques which are relatively flexible methods and can be applied to data even as the test is continuing. This allows data to be analysed almost in 'real time'. 'Steady-state' methods of analysis can be used, but may require much longer periods of pumping than is necessary with non-steady state methods. In general, analysis by non-steady state techniques is to be preferred.

There are a wide variety of methods of analysis which can be used to analyse the results of the constant rate pumping phase, many of which are usefully summarized in Kruseman and De Ridder (1990). Each of the methods is based on a particular set of assumptions about the aquifer system (unconfined, confined, leaky), the well (fully penetrating, partially penetrating) and the discharge flow rate (generally assumed to be constant). Methods suitable for analysis of data from the recovery phase are also available.

These methods should be viewed as a 'tool kit' providing a range of possible analysis methods. Provided the basic details of the aquifer and well are known, it is normally straightforward to select one (or more than one, if uncertainty exists over aquifer conditions) method appropriate to the case in hand. Commonly used methods of analysis fall into two main types:

i *Curve fitting methods*: These typically involve plotting on a log–log graph, for each observation well, drawdown against elapsed time. The data will generally form a characteristic shape. The data curve is then overlain with a theoretical 'type curve' and the relative positions of the two curves adjusted until the best match of the shape of the two curves is obtained. Once a match is achieved, permeability and storage coefficient can be determined by comparing values from each curve. These methods were developed from the work of Theis on simple confined aquifers, but variations are available for various other cases (see Kruseman and De Ridder 1990). The curve fitting process can be tedious, but in recent years computer programs for pumping test analysis have speeded up the process.

ii *Straight line methods*: This approach involves plotting sets of data so that characteristic straight lines are produced, allowing permeability and storage coefficient to be determined from the slope and position of the line. These methods are a special case of the Theis solution, and are based on the work of Cooper and Jacob (1946) and are often called the Cooper–Jacob methods. Two approaches are possible, and can be used on the same test. Time–drawdown diagrams involve plotting the drawdown data from one observation well against elapsed time since pumping began (Fig. 6.16); this process is repeated for all observation wells being analysed. Distance–drawdown diagrams plot the drawdown recorded (at a specific elapsed time) in all observation wells against the distance of each observation well from the test well (Fig. 6.17).

The Cooper–Jacob straight line method is the most commonly used method of analysis, mainly due to its relative simplicity. The original Cooper–Jacob method was based on horizontal flow to fully penetrating wells in confined aquifers, but can also be used in unconfined aquifers where the drawdown is a small proportion (less than 20 per cent) of the original aquifer saturated thickness.

For the time–drawdown data from a single observation well, aquifer permeability k and storage coefficient S are determined as follows. From the semi-log graph (Fig. 6.16) draw a straight line through the main portion of the data (the data will deviate from straight line at small times, and possibly at later times). From the graph obtain the slope of the straight line, expressed as Δs, which is the change in drawdown s per log cycle of time. Also determine t_0 the time at which the straight line intercepts the zero drawdown line – k and S are then obtained from:

$$k = \frac{2.3q}{4\pi \Delta s D} \tag{6.9}$$

$$S = \frac{2.25kDt_0}{r^2} \tag{6.10}$$

where q is the constant flow rate from the test well, D is the aquifer thickness, and r is the distance from the centre of the test well to the observation well. This process can be repeated for each observation well.

For the distance–drawdown approach data from all observation wells, at elapsed time t after pumping started, is plotted on the semi-log graph (Fig. 6.17). A straight line is drawn through the observation well data. If the test well has a very much larger drawdown than the observation wells this

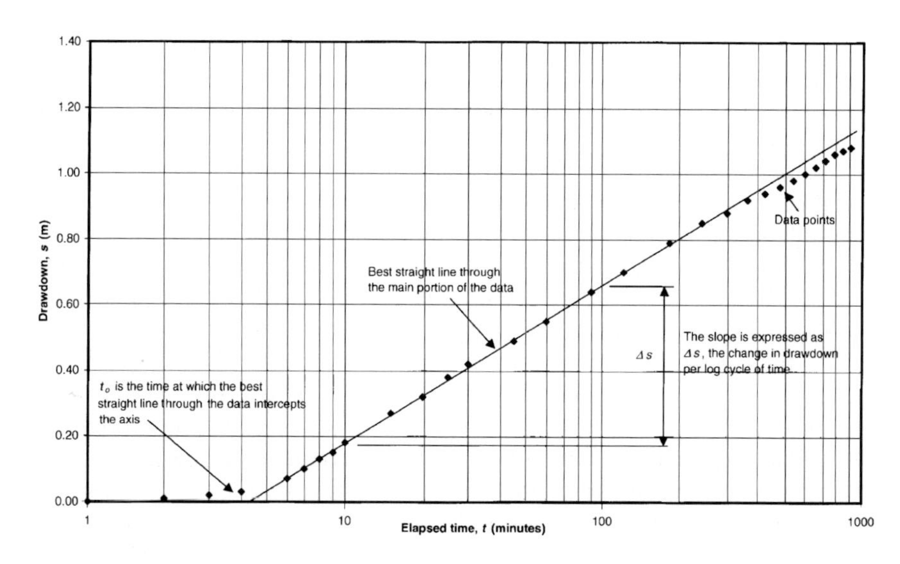

Figure 6.16 Analysis of pumping test data: Cooper–Jacob method for time–drawdown data.

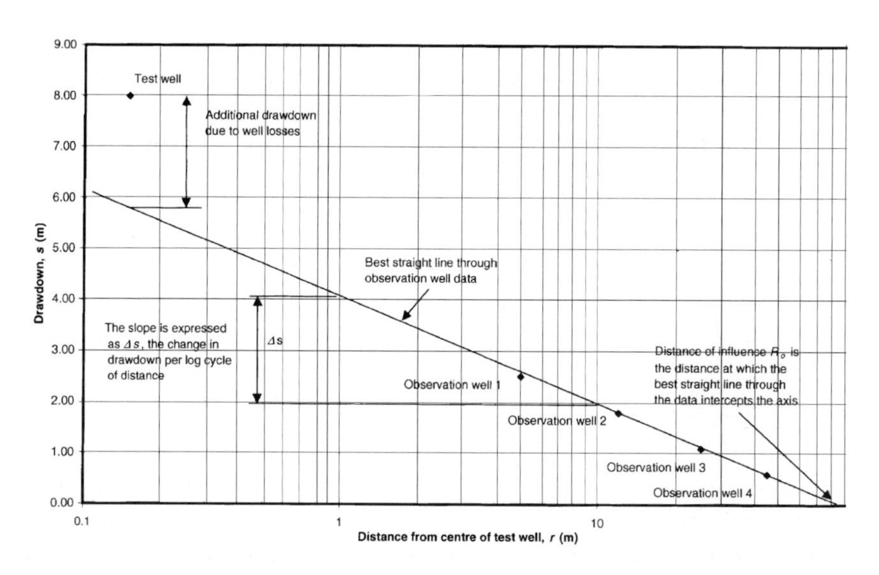

Figure 6.17 Analysis of pumping test data: Cooper–Jacob method for distance–drawdown data.

may be the result of well losses. In such cases the straight line should be based on the observation wells only, and should not include the test well drawdown. From the graph obtain the slope of the straight line, expressed as Δs, which is the change in drawdown s per log cycle of distance. Also determine R_0, the distance at which the straight line intercepts the zero drawdown line – k and S are then obtained from:

$$k = \frac{2.3q}{2\pi \Delta s D} \tag{6.11}$$

$$S = \frac{2.25kDt}{R_0^2} \tag{6.12}$$

The analysis can be repeated for various elapsed times. This approach also allows the distance of influence R_0 to be estimated, which can be a useful check on values used in later dewatering designs.

Care must be taken to ensure that consistent units are used in these equations. To obtain permeability k in m/s (its usual form in dewatering calculations) the following conventions are used: Δs, D, R_0 and r must be in metres; q must be in m^3/s (not l/s – this is a common cause of numerical errors); and t and t_0 must be in seconds (not in minutes, which is often the most convenient way to plot drawdown–time data).

Because the Cooper–Jacob method is only a special case of the more generic Theis solution, a check must be made to ensure the method is valid. The Cooper–Jacob method can be used without significant error provided $(r^2S)/(4kDt) \leqslant 0.05$. This means that the approach is valid provided t is sufficiently large and r is sufficiently small.

Comprehensive analysis of a pumping test may produce several values of permeability. Time–drawdown analysis will produce one result for each observation well, and distance–drawdown analysis can produce several values, depending on how many times are analysed. These permeability values are all likely to slightly differ, either due to uncertainties in analysis, or in response to changes in aquifer conditions across the zone affected by the test. This highlights that pumping test results may still need detailed interpretation; in complex cases it is prudent to obtain specialist advice.

Variations on the Cooper–Jacob method for certain other aquifer conditions are given in Kruseman and De Ridder (1990).

6.6.11 Groundwater control trials

Groundwater control trials are an extension of the ethos of pumping tests, in that they are a large-scale *in situ* test to determine the hydraulic properties of the ground. Further than that a carefully planned trial can provide other useful information.

Instead of a single test well, a groundwater control trial involves pumping from a line or ring of wells of some sort (wellpoints, deep wells or ejectors). Obviously, to specify a suitable trial the main dewatering design must be reasonably advanced, to allow selection of the appropriate pumping method, well spacing, well screen depth, etc. As with a pumping test, the trial is pumped continuously for a suitable period (typically one to four weeks), while monitoring discharge flow rate and water levels in observation wells. The results of such tests are normally analysed using the design methods of Chapter 7 to 'back-calculate' an equivalent soil permeability. Powrie and Roberts (1990) describe an example of a trial using ejector wells.

In addition to determining the permeability, trials allow opportunities to investigate other issues relevant to the proposed works:

i Because several wells have to be installed, the relative performance of proposed installation methods (e.g. drilling or jetting) can be assessed.
ii If the trial consists of a ring of wells, a trial excavation can be made inside the ring during the trial. This excavation can provide data on: ease of excavation of the soil; stability of dewatered excavations; and trafficability of plant across the dewatered soil. There have been cases where trial excavations have been combined with large-scale compaction tests to assess the suitability of the excavated material as backfill elsewhere on site.

Figure 6.18 shows a site where a 40 m by 40 m ring of wellpoints was installed in a silty sand and the area within the ring excavated during pumping. This trial not only demonstrated that the method was feasible at the site, but it allowed battered sides to be cut at various trial slopes and provided useful information on the handling characteristics of the dewatered soil.

Groundwater control trials are probably the most expensive form of *in situ* permeability test, but for large projects in difficult ground conditions they can reduce risks of problems during the main works, and the cost may be justified on that basis. If dewatering system is to be designed by the observational method (see Section 7.1) a trial can be a key part of the scheme.

6.6.12 Comparison of methods

The foregoing sections have described the methods commonly used to estimate permeability. Meaningful and representative values of permeability are essential for the design of groundwater lowering projects, but the designer may be presented with a wide range of values taken by different methods in the same aquifer. Some of the variations in reported values of permeability may result from heterogeneity of the aquifer. Unfortunately, limitations in the test methods may also introduce some 'apparent' variation (independent of the properties of the aquifer) in the results.

(a)

(b)

Figure 6.18 Wellpoint dewatering trial (courtesy of WJ Groundwater Limited). (a) A rectangular ring of wellpoints has been installed around the trial area. The wellpoints are pumped prior to commencement of trial excavation. (b) A trial excavation is made within the wellpoint ring. Trial batter slopes are formed to assess the stability of the soil following groundwater lowering.

Table 6.3 Tentative guide to the reliability of permeability estimates

Method	*Notes*	*Relative cost*	*Reliability*[1]
Visual assessment	For a soil or drift aquifer, can allow the order of magnitude of permeability to be estimated. Less applicable to fissured rock aquifers.	Very low	Good
Inverse numerical modelling	Only applicable if the hydrogeology of the aquifers being pumped is defined reasonably clearly, and if adequate monitoring data are available.	Low to moderate	Good
Geophysical methods	Can give indirect estimates of permeability and allow identification of zones of higher and lower permeability. Borehole geophysical methods are more applicable in rock aquifers to identify fissured horizons.	Moderate to high	Poor to good
Laboratory testing: particle size analysis	Can give reasonable results for samples of fairly uniform sand with low silt and clay content, provided loss of fines has not occurred. Loss of fines tends to be greater from bulk samples compared with tube samples; permeability estimated from bulk samples should be treated with caution. Gives very poor estimates in laminated or structured soils or where silt and clay content is significant.	Very low	Very poor to good
Laboratory testing: permeameter testing	Of little use in granular sand or gravels, as representative samples cannot be obtained without unacceptable disturbance. Can produce good results in clay and some silts where minimally disturbed samples can obtained. In soils with fabric and structure permeability estimates are affected by sample size; smaller samples may under-estimate permeability.	Low to moderate	Good
In situ tests in boreholes: falling head	Can test only a small volume of soil at the base of the borehole. May be influenced by soil disturbance due to boring. Prone to clogging of test section.	Very low	Very poor
In situ tests in boreholes: rising head	Can test only a small volume of soil at the base of the borehole. May be influenced by soil disturbance due to boring. Prone to clogging or loosening of test section. Better results are often obtained in coarse soils with little silt content.	Very low	Poor to moderate
In situ tests in boreholes: constant head	Can test only a small volume of soil at the base of the borehole. May be influenced by soil disturbance due to boring. Inflow tests prone to clogging of test section. Better results are often obtained in coarse soils with little silt content.	Low	Very poor to moderate

In situ tests in boreholes: packer tests	Normally carried out in unlined sections of boreholes in rock. Results may be dominated by the presence of fissures intercepting the test section. Inflow tests prone to clogging of test section. Outflow tests generally provide better results, but require more complex test equipment.	Low to moderate	Poor to good
In situ tests in observation wells: falling head	Highly dependent on the design of the observation well and any soil disturbance during installation. Can test only a small volume of soil near the test section. Prone to clogging of test section.	Very low	Very poor
In situ tests in observation wells: rising head	Highly dependent on the design of the observation well and any soil disturbance during installation. Can test only a small volume of soil near the test section. Better results are often obtained in coarse soils with little silt content.	Very low	Poor to moderate
In situ tests in observation wells: constant head	Highly dependent on the design of the observation well and any soil disturbance during installation. Can test only a small volume of soil near the test section. Inflow tests prone to clogging of test section. Better results are often obtained in coarse soils with little silt content.	Low	Very poor to moderate
Specialist *in situ* tests	Appropriate to silts, clays and some sands. Can test only a small volume of soil near the probe, but can be used to profile permeability with depth. If the tests involve penetrometers, disturbance or smear of soil layering may occur, affecting results.	Moderate	Moderate to good
Pumping tests	Appropriate to soils of moderate to high permeability such as sands or gravels. Rarely appropriate for clays and silts, unless pumped by ejectors and monitored by appropriate types of observation wells (such as pneumatic piezometers). Affects a large volume of soil and can provide good values of mass permeability. Results can be difficult to interpret if multiple aquifer systems are present beneath the site.	Moderate to high	Good to very good
Groundwater control trials	Appropriate for large-scale projects in difficult ground conditions or where the observational method is being used. Requires careful planning.	High to very high	Very good

Note: [1] When used in soil or rock conditions suitable to the method, and when analysed appropriately.

Permeability is a difficult parameter to determine accurately and, when selecting permeability values to be used in design, some uncertainty is unavoidable. During design this can be addressed by carrying out sensitivity analyses (see Chapter 7) to evaluate the effect of differing permeability on drawdown and flow rate.

When presented with permeability values from a range of methods, deciding on the range of permeability to be used in design calculations is made easier if the relative reliability of each method is known; some guidance is given in Table 6.3.

Appendix 6A: Estimation of permeability from laboratory data – Loudon method

Loudon (1952) proposed an empirical formula, based on earlier work by Kozeny to estimate the permeability of granular material:

$$\log_{10}(kS^2) = 1.365 + 5.15n \tag{6A-1}$$

where k is the permeability expressed in cm/s, n is porosity of the granular soil (a dimensionless ratio, expressed as a fraction not as a percentage) and S is specific surface of grains (surface area per unit volume of grains) expressed in (cm^2)/(cm^3). Note that the values in the formula are specific to these units. Care must be taken when applying this method not to introduce errors by using different units.

Estimation of specific surface S

According to Loudon, the specific surface S_i of spheres uniformly distributed in size between mesh sizes D_x and D_y of adjacent sieves is given by:

$$S_i = \frac{6}{\sqrt{D_x D_y}} \tag{6A-2}$$

which is accurate to within 2 per cent if (D_x/D_y) is not greater than two. The specific surfaces of spheres lying between given grain sizes are set out in Table 6A-1, based on data from Loudon (1952).

Of course, very few real soils have grains that are even approximately spherical. The actual specific surface S of a soil is influenced by the angularity factor f (also known as the coefficient of rugosity). The specific surface S of non-spherical grains is equivalent to the specific surface of spheres S_i multiplied by the angularity factor f.

The more angular the shape of the grains, the greater the specific surface will be. According to Loudon a visual estimate (using a hand lens) of the angularity factor seems quite good enough for the standard of accuracy generally needed. He stated f may be estimated roughly as follows:

Material type	*Angularity factor f*
Glass beads	1.0
Rounded sand	1.1
Sand of medium angularity	1·25
Angular sand	1·4

Thus in practice the range in angularity factor of natural soils is not great, being from about 1.1 to 1.4 – a full range variation for naturally occurring soils of about 25 per cent.

Loudon further indicated that very angular materials such as crushed marble, crushed quartzite or crushed basalt may have angularity factors of 1.5 to 1.7.

Once a particle size test has been carried out using British Standard sieve sizes the specific surface of a sand sample can be estimated by:

$$S = \left(x_1 S_1 + x_2 S_2 + \cdots + x_n S_n \right) \tag{6A-3}$$

where $x_1, \ldots, x_n$ are the fractions of the total mass (i.e. $x_1 + x_2 + \cdots x_n = 1$) retained on different sieves and $S_1, \ldots, S_n$ are the specific surfaces of spheres uniformly distributed within the corresponding sieved fractions, as listed in Table 6A-1.

Table 6A-1 Specific surface of spheres lying between given sieve sizes

British Standard sieve size	*Sieve size (mm)*	*Specific surface (cm^2/mm^3)*
7–14	2.4–1.2	35.2
14–25	1.2–0.6	70.6
25–52	0.6–0.3	143.0
52–100	0.3–0.15	284.0
100–200	0.15–0.075	558.0
7–10	2.4–1.9	29.6
10–14	1.9–1.2	41.9
14–18	1.2–0.8	59.2
18–25	0.8–0.6	83.8
25–36	0.6–0.45	119.0
36–52	0.45–0.3	170.0
52–72	0.3–0.2	242.0
72–100	0.2–0.15	336.0
100–150	0.15–0.1	476.0
150–200	0.1–0.075	675.0

Estimation of porosity *n*

The other relevant factor is *in situ* porosity n. The range of values for granular sands to which Loudon's method applies is from about 0.26 (densest state) to about 0.47 (loosest state), though mostly in the range 0.30–0.46 – quite a narrow range variation. Some values of porosity of typical soils are summarized in Table 6A-2.

Whilst the porosity of a sample can be determined in the laboratory, it is virtually impossible to determine the *in situ* porosity of a sample. The porosity should be estimated from published values (such as from Table 6A-2), appropriate to the soil description. This is a limitation on the usefulness of Loudon and other similar works and an explanation for the somewhat erratic results that they sometimes give.

Estimation of permeability

Once porosity n (as a fraction) has been estimated from Table 6A-2 and specific surface S (in cm^2/cm^3) has been calculated using equation (6A-3) and Table 6A-1, Loudon's formula can be used to calculate permeability k as:

$$k = \frac{10^{(1.365+5.15n)}}{S^2} \tag{6A-4}$$

Permeability is calculated in cm/s. To convert to m/s the k value from equation (6A-4) should be multiplied by 0.01.

Like many formulae for estimating permeability from particle size data, Loudon's formula is only valid for uniform sands between specific particle size ranges (in this case approximately between 1 mm and 0.075 mm). Loudon notes that the calculated permeability became unreliable when the soil contained more than 5 per cent of particles finer than 0.075 mm.

Table 6A-2 *In situ* porosity and void ratio of typical soils in natural state

Description	*Porosity* n	*Void ratio* e
Uniform sand, loose	0.46	0.85
Uniform sand, dense	0.34	0.51
Mixed-grained sand, loose	0.40	0.67
Mixed-grained sand, dense	0.30	0.43

Note: Based on Terzaghi *et al.* (1996).

Appendix 6B: Execution and analysis of variable head permeability tests in boreholes

General comments on performing permeability tests

Many factors may influence the results obtained from *in situ* permeability tests. Great care must be exercised to ensure that tests are made under controlled conditions; some tests are reliable only when performed according to specific conditions and these should be strictly adhered to. Some general guidance is given below:

(a) When making tests in boreholes, prior to the test the borehole should be carefully cleaned out to remove loose or disturbed material, the presence of which can introduce significant errors in the results obtained. Any sediment-laden water in the borehole should be flushed out and replaced with clean water.
(b) When carrying out inflow tests, the water used should be clean and attention should be given to cleanliness of tanks, buckets, pumps, etc., so that the water does not become contaminated by dirt in the equipment, since very small amounts of silt may cause serious errors in the results.
(c) The water level should be measured with a dipmeter relative to a stable datum such as the top of the borehole casing or a firmly positioned rod across the top of the borehole.
(d) In the early stages of rising and falling head tests, it is desirable to take readings of the water level at frequent intervals of about 30 seconds. To obtain reasonable accuracy in these circumstances, it is necessary to have two persons taking the readings: one to adjust the dipmeter to keep pace with the change in water level and to take the depth readings, and the other to call out the time intervals from a stop watch and record the water levels.
(e) The equations for calculating permeability are all based on the assumption that flow through the tested material is laminar. To be sure of laminar flow, velocity should not exceed 0.03 m/s. Therefore, the tests are only applicable when the flow rate (into or out of the borehole) divided by the area of test section is less than 0.03 m/s. The open area is taken as the peripheral area of the filter section around a well, or the intake area of the sides and base, in the case of an unsupported hole.
(f) Tests should, if possible, be made during dry weather. If a test has to be made while it is raining, measures must be taken to prevent surface run-off from entering the test borehole. If the borehole casing is close to ground level this may involve using sand bags to build a small dam around the top of the borehole.

Where the test is carried out in an unconfined aquifer and surface conditions are relatively pervious, rainfall during the test may produce a change in the groundwater level, which could affect the results obtained. In these circumstances, a careful note should be made of any showers immediately preceding or during a test. The groundwater level should be measured before and after the test.

Variable head inflow (falling head) test – procedure

Each test is carried out at a specific depth interval known as the 'test section' (Fig. 6B-1). The whole test section must be below the groundwater level. If possible, the groundwater level should be measured before making the test.

Prior to the test itself the borehole is advanced to the base of the test section. Where casing is not necessary to support the borehole, it must nevertheless be inserted to the top of the test section so that only the test section is exposed to the strata. The casing must be tight against the side of the borehole, otherwise the test water will leak around the casing. This test is suitable for testing at various depths during progress of boring, where the test section will stand unsupported.

If casing is required to support the sides of the hole, it must be taken down initially to the base of the test section. Fine gravel backfill is then placed in the borehole, whilst withdrawing the casing to expose the required length of unlined hole for the test; subsequently the gravel fill should be brought up with and always kept just above the bottom of the casing as this is withdrawn. It is good practice, prior to placing the gravel, to replace any water standing in the borehole with clean water; this helps avoid suspended silt becoming trapped in the test section, which would reduce permeability. This type of test can also be used at various depths during boring, but the gravel backfill will need to be drilled out following each test – this will slow down drilling progress.

It is best if these tests are carried out at the start of the day's work, as the original groundwater level will be less affected by drilling activities. If carried out during the day's drilling, the original water level recorded may not be the 'true' natural water level.

To start the test clean water is rapidly added to the borehole to raise the water level as high as possible in the casing; no more water is added during the test.

Once the water has ceased to be added, readings of the water level inside the borehole are taken at frequent time intervals, making a note of the time corresponding to each reading. Time $t=0$ is taken as the time of the first reading. No fixed time intervals can be specified, as the frequency of readings will depend on the rate of fall of water level. As a guide, an attempt should be made to take the readings so that each level differs from the previous reading by about equal increments. Since the water level will fall most rapidly at the start of the test, and then at a decreasing rate, readings will become less frequent as the test progresses. If the initial head of water in the

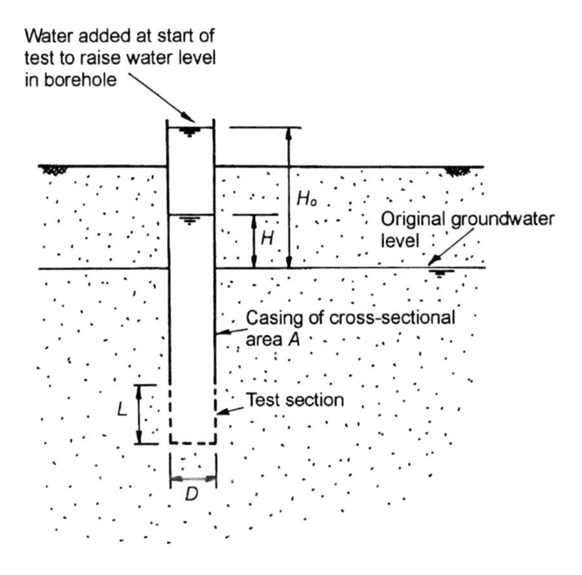

Figure 6B-1 General arrangement of variable head tests in boreholes.

borehole above original groundwater level is say 2 m, take readings so that the difference in successive readings is about 100 mm; if the head is smaller, the difference in level readings should be smaller, say 25–50 mm for initial head of 0.5–1 m. Ideally, readings should be continued until the head of water in the borehole above groundwater level is less than one fifth of the initial head above original groundwater level.

The following information should be recorded:

i Diameter of unlined borehole being tested and diameter of casing
ii Depth to the base of the borehole (bottom of test section)
iii Depth to bottom of casing (top of test section)
iv Depth to top of gravel backfill, if used
v Date and time that water level readings are started
vi Water level readings and time of each reading
vii Depth to the original groundwater level.

All depths and water level readings should be measured relative to a clearly defined datum, ideally ground level. If ground level is uneven or flooded the top of the casing can be used as a datum, provided the casing is secured so it will not slip or sink during the test.

Variable head outflow (rising head) test – procedure

This test should be used only in lined boreholes or in observation wells, provided baling or pumping-out is practicable. Each test is carried out at a specific depth interval known as the 'test section'. The whole test section must be below the groundwater level; ideally the borehole should penetrate below groundwater level by at least 10 times its diameter. If possible, the groundwater level should be measured before making the test.

Preparation of the borehole and test section is similar to that for an inflow test. At the start of the test water is rapidly bailed or pumped out of the borehole to just above the bottom of the casing. No more water is removed during the test.

Once the water has ceased to be removed, readings of the water level inside the borehole are taken at frequent time intervals, making a note of the time corresponding to each reading. Time $t=0$ is taken as the time of the first reading. Readings should be taken at the same general intervals as an inflow test and the same data recorded. Readings should be continued until the difference between the water level in the borehole and original groundwater level is less than one fifth of the initial difference in these levels.

It is best if these tests are carried out at the start of the day's work, as the original groundwater level will be less affected by drilling activities. If carried out during the day's drilling, the original water level recorded may not be the 'true' natural water level.

Variable head (falling and rising head) tests – calculation of results

The permeability k may be determined for a variable head test using the following formula:

$$k=\frac{A}{FT} \qquad (6B\text{-}1)$$

where A is the cross-sectional area of the borehole casing (at the water levels during the test), T is the basic time lag and F is a shape factor dependent on the geometry of the test section.

Determination of shape factor F

The geometry of the test should be compared with the analytical solutions of Hvorslev given in Fig. 6.10 and the appropriate shape factor calculated.

Determination of basic time lag T

(1) Where the original groundwater level is reliably known, plot values of H/H_0 on a logarithmic scale against corresponding values of elapsed time t on an arithmetic scale (Fig. 6B-2). H_0 is defined as the excess head of water (measured relative to the original water level) at the start of the test (the initial reading at $t=0$). H is the excess head of water at time t during the test. Draw the best fitting straight line through the experimental points. In some cases, the experimental points for values of H/H_0 near 1.0 may follow a

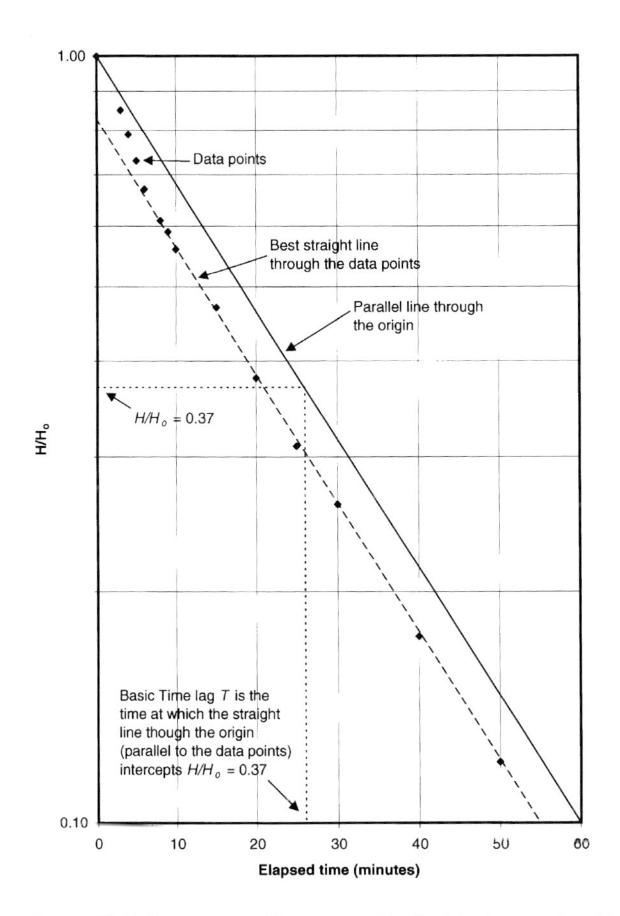

Figure 6B-2 Estimation of basic time lag (original water level known).

curve; these should be ignored and the straight line drawn through the remaining points; then draw a parallel straight line through the origin. The basic time lag T is obtained by reading off the value of t when $H/H_0 = 0.37$ using the straight line through the origin.

(2) If the original groundwater level is only approximately known, calculate values of H from an assumed groundwater level (making as accurate an estimate as is possible). Plot the resulting values of H/H_0 on a logarithmic scale against corresponding values of t on an arithmetic scale (Fig. 6B-3). If the slope of the line through these points decreases with increasing t (curve A), the assumed original groundwater level was too low. If the slope increases with increasing t, the assumed groundwater level was too high. By trial and error a groundwater level can be determined from which values of H/H_0 plot as a straight line against t, at least for the higher values of t.

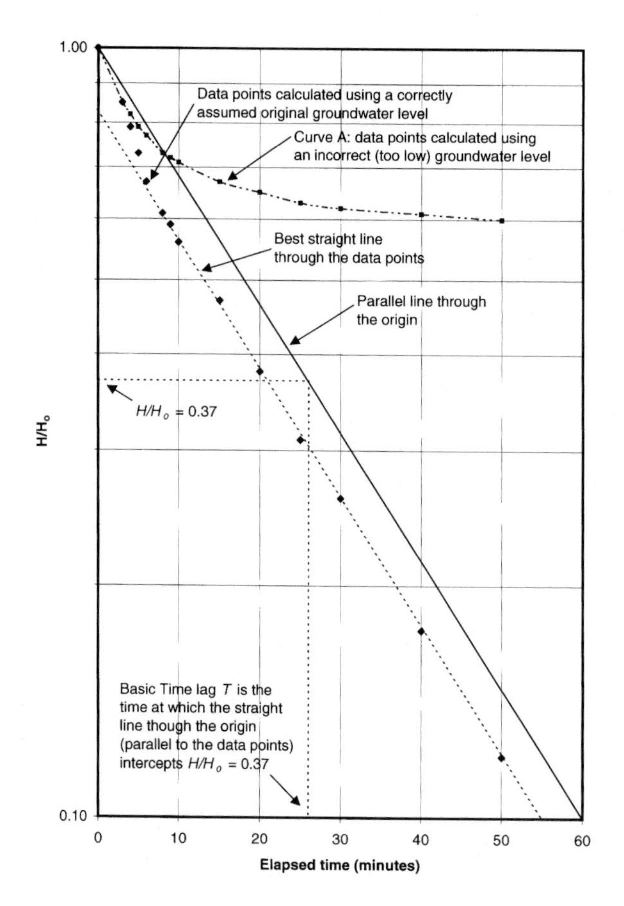

Figure 6B-3 Estimation of basic time lag (original water level unknown).

Experimental points for small values of t may follow a short curve, due to the flow settling down into equilibrium with the soil. Basic time lag T can then be determined in the same way as for step 1.

The values of A (in m^2), F (in metres) and T (in seconds) are applied into equation (6B-1) to produce the calculated permeability k in m/s.

Appendix 6C: Execution of well pumping tests

Planning of pumping tests

The simplest form of pumping test involves controlled pumping from a well – the 'test well' – and monitoring of the discharge flow rate from the well and the drawdown in observation wells at varying radial distances away (Fig. 6C-1). A pumping test is one of the more expensive *in situ* tests, and requires careful planning to ensure that time, effort and money are not wasted.

A pumping test consists of a number of phases:

1. Pre-pumping monitoring: This involves monitoring natural groundwater levels in the observation wells (and test well, if possible) for a period from a few days to several weeks prior to commencing pumping. The aim is to determine any natural or artificial variations in groundwater level that may affect the drawdowns observed during pumping.
2. Equipment test: This is a short period of pumping (typically 15–120 min) to allow correct operation to be ascertained of the pumps, flowmeters, dataloggers and to check for leaks in the discharge pipework. It is normal to take some crude readings of flow rate and drawdown in the well to allow selection of flow rate for subsequent phases.
3. Step-drawdown test: This is a period of continuous pumping typically lasting 4–8 h, during which the flow rate is increased in a series of

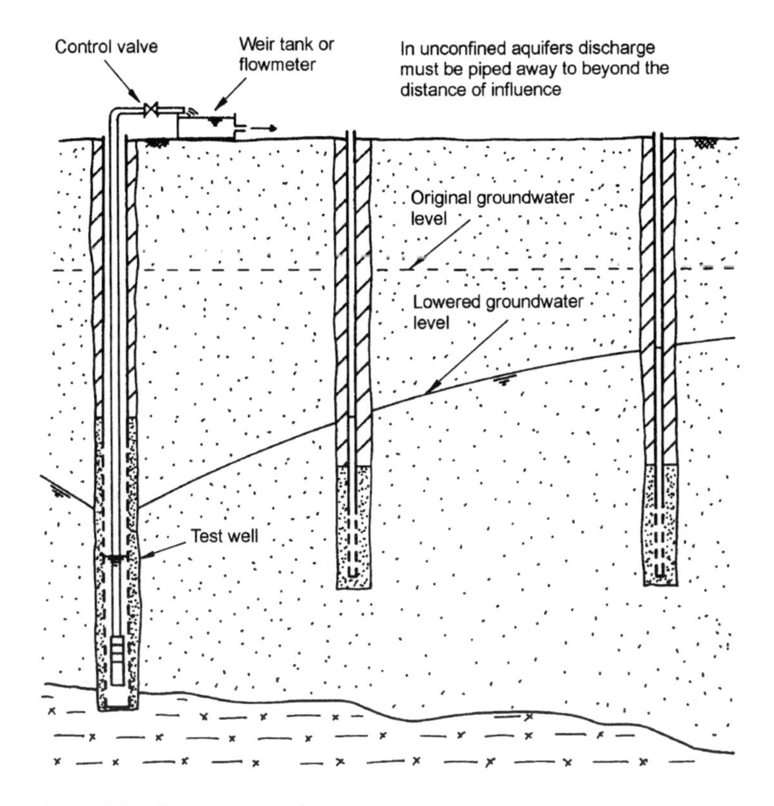

Figure 6C-1 Components of a pumping test.

steps. Each step is of equal duration (normally 60 or 100 min) and an appropriately designed test should have four or five steps at roughly equal intervals of flow rate. Water levels are normally allowed to recover for at least twelve hours before the constant rate pumping phase can be started.

4. Constant rate pumping phase: This is often considered to be the main part of the test and involves pumping the well at a constant flow rate (chosen following the step-drawdown test). The constant rate test normally lasts between one and seven days, although test durations of up to twenty eight days are not unknown.
5. Recovery phase: Water levels in the observation wells and test well are monitored as they recover following cessation of pumping. This phase often lasts between one and three days.

Guidance on the execution of pumping tests is given in BS6316 (1992) and Kruseman and De Ridder (1990).

Pumping tests as part of groundwater lowering investigations need to be designed to meet certain criteria consistent with providing high quality data to the dewatering designer.

(a) The aim should be to achieve drawdowns (measured in piezometers within 10–20 m of the test well) of at least 10 per cent of the required drawdown in the proposed groundwater lowering scheme.
(b) If the site is subject to natural groundwater level variations (such as tidal effects) the test should aim to achieve drawdowns significantly greater than the background variations.
(c) It is essential that pumping is absolutely continuous during the initial hours of pumping. If pumping is interrupted due to whatever cause, such as a pump or power failure, it is essential that the test is suspended until groundwater levels to return to equilibrium, or those levels recorded prior to commencement of any pumping. Only then should pumping recommence. Once the test pumping has been going on for 24 h or more, occasional short interruptions (for example for daily checks of generator oil levels) are permissible. These interruptions should be as short as possible (ideally no longer than a few minutes); any longer interruptions may require the test to be suspended and re-started later. Ideally, pumping should continue until steady-state drawdown conditions are established in observation wells near the test well.
(d) Most of the simpler methods used to analyse pumping test data assume the well fully penetrates the aquifer. Ideally, the test well should be fully penetrating. However, provided the screened length is of the order of 70 per cent to 80 per cent of the aquifer thickness the accuracy of the assessment of permeability is generally acceptable. Where the amount of penetration is less the flow will not be predominantly horizontal, and more special corrections may need to be applied to the drawdown data.

(e) Sufficient observation wells should be installed, in suitable locations, to allow groundwater flow patterns to be identified. Ideally, there should be at least three lines of observation wells radiating out from the test well, with each line spaced 120 degrees apart. These need not penetrate the full depth of the aquifer but must be deeper than the expected drawdown (an observation well that becomes dry is not acceptable!). Each line should consist of three or more observation wells. Depending upon the permeability of the aquifer, the distance from the test well to the nearest observation well should be in the range of 2–3 m (in a low to moderate permeability aquifer) or to 5–10 m (in a high permeability aquifer) from the test well. More distant observation wells should located within the anticipated distance of influence of the test well.

General comments on performing pumping tests

The test well should be designed, installed and developed according to the guidelines of Chapter 10 and observation wells should ideally be consistent with the good practice outlined in Section 6.5. The most common form of test well is a deep well pumped by a borehole electro-submersible pump. Suction pumps can be used, but only in highly permeable soils with a high water table, otherwise, as drawdown occurs, the pump may become starved of water, giving a varying flow rate. Whatever the method of pumping it is desirable to measure the water inside the test well during pumping. This will require the installation of a dip tube in the well so that the dipmeter tape does not become entwined around the pump riser pipe due to the swirl of water near the pump intake.

The principal parameters to be measured during a test are water levels in the test well and observation wells, and the discharge flow rate from the well. All readings need to be referenced to the time elapsed since the start of pumping (or recovery) for that test phase. The frequency of monitoring must be matched to the rate of change of the measured parameter. During a test the flow rate should remain approximately constant, but the water levels will fall rapidly immediately following the start of pumping, and will fall slower as pumping continues. Measurements of water levels must be taken very frequently at the start of the test, and become less frequent as time passes. A commonly used schedule of timings for readings is given in BS6316 (1992). This is summarized below:

- Readings to be taken immediately before the discharge flow rate is started, changed or stopped;
- Readings to be taken every minute for the first 10 min of the test phase;
- Readings to be taken every 5 min thereafter up to an elapsed time of 60 min;

- Readings to be taken every 15 min thereafter up to an elapsed time of 4 h;
- Readings to be taken every 30 min thereafter up to an elapsed time of 8 h;
- Readings to be taken every 60 min thereafter up to an elapsed time of 18 h;
- Readings to be taken every 2 h thereafter up to an elapsed time of 48 h;
- Readings to be taken every 4 h thereafter up to an elapsed time of 96 h;
- Readings to be taken every 8 h thereafter up to an elapsed time of 188 hours;
- Readings to be taken every 12 h thereafter, unless other influences warrant more frequent measurement.

These timings are for guidance only, and should be applied with judgement. The key point is to ensure that sufficient readings are taken to allow the rate of change of water levels to be clearly identified. If water levels are only changing slowly, it may be possible to increase the intervals between readings without reducing the quality of the data. This would avoid generating unnecessary readings and can reduce the work involved in data analysis.

Water levels can be measured by manual dipping with a dipmeter (see Section 14.2). However, at the start of the test readings need to be taken very frequently. If more than a few observation wells are to be monitored manually, several people may be needed to take the frequent readings at the start of the test, creating potential problems of co-ordination between personnel. There is also the problem of actually gathering enough capable people together in the first place. This can lead to poor recording of the precise timing of measurements taken during the first few minutes of pumping, leading to difficulties in subsequent analysis. If taking manual water level readings with limited numbers of observers, it is best to concentrate initial monitoring on points nearest the test well. As time passes, and the time period between readings increases, more observation wells can be included in the monitoring.

The use of electronic datalogging equipment, linked to pressure transducers in the observation wells (see Section 14.5), may help to overcome such staffing problems. Once programmed to take readings at the appropriate intervals such equipment not only reduces the personnel requirements, but will also produce the test measurements in electronic form, allowing rapid analysis using spreadsheet programs.

If a test is carried out in an aquifer with significant tidal response, the monitoring intervals later in the test (after several days pumping) must be selected with care. Standard tables of monitoring intervals (such as in BS6316 (1992)) allow monitoring frequencies of one reading every few hours. If such large time gaps between readings are permitted, data will be very difficult to interpret because the tidal responses will not be accurately recorded. For tidal pumping tests the monitoring intervals for the first few hours of the test should be as per published guidelines, but the remainder of the test should be monitored at 15–30 minute intervals to ensure the tidal fluctuations are fully resolved. This requirement means that a datalogging system is almost essential for pumping tests on tidal sites.

The pump discharge can be measured by means of a tank or gauge box fitted with a V-notch or rectangular weir (see Section 14.3) or by installing an integrating flowmeter into the discharge pipeline. Even if flowmeters are used in preference to weir tanks, it is good practice to include a settlement tank in the discharge line so that it can be visually checked for suspended solids in the discharge water. In general, once adjusted at the start of the test phase, the flow rate should not vary significantly. Immediately following commencement of pumping, flow rate does not need to be monitored as frequently as water levels for the first hour or so of pumping. The outlet of the discharge should be well away from the test area to avoid the return seepage affecting the drawdown levels.

It is normal to take water samples from the pumped discharge during the test. Obtaining a sample is relatively straightforward if it is possible to fill a sample bottle directly at the discharge tank. However, when taking groundwater samples from the discharge flow the following factors should be considered.

(a) Try to minimize the exposure of the sample to the atmosphere. Try and obtain it directly at the point where the pump discharges into the tank. Totally fill the bottle and try and avoid leaving any air inside when it is sealed. If the pump discharge is 'cascading' before the sampling point the water will become aerated and oxidation may occur. The discharge arrangements should be arranged so the sample can be obtained before aeration occurs.
(b) Samples may degrade between sampling and testing. The samples should be tested as soon as possible after they are taken and, ideally, should be refrigerated in the meantime. The bottles used for sampling should be clean with a good seal. However, the sample may degrade while in the bottle (for example by trace metals oxidizing and precipitating out of solution). Specialists may be able to advise on the addition of suitable preservatives to prevent this occurring. The choice of sample bottle (glass or plastic) should also be discussed with the laboratory since some test results can be influenced by the material of the sample bottle.

The wellhead chemistry (see Section 3.8) can be determined using probes or sensors immersed in the flowing discharge water. These can be portable meters, recorded manually, or may be linked to datalogging systems.

Execution of pumping tests

Before any pumping is carried out, water levels in observation wells must be recorded regularly over a period of days. This is to try and determine whether any natural (or artificially induced) variations in water level are occurring. There are no generic guidelines on the duration and frequency of background monitoring. On sites where it is anticipated that variations will be small (e.g. remote inland sites) monitoring three times daily for three to five days is the minimum acceptable. On coastal sites with a significant tidal response it has been necessary to use dataloggers to record water levels every 15 min, 24 h per day for up to thirty days. If in doubt, it is best to do more than the minimum monitoring. The time to find out about background variations is prior to test pumping, not during the test itself.

The equipment test should be used to determine the most suitable pump discharge rate for subsequent phases of the test, so that at the end of pumping the water in the borehole is not drawn down to the pump intake. It should also be used to ensure that flow measuring devices function and that discharge pipework is not blocked and does not leak – to check in general whether the test is ready to proceed. Ideally, these preparations should be tried out at least a day before test pumping is commenced so that the test well and observation wells can be left to ensure that the groundwater level is realistically re-established.

The following general information should be recorded prior to testing:

i Elevation of ground surface at the test well and at each observation well.
ii Elevation of reference datum for water levels at each well (the top of the well casing is often used as a datum).
iii Depth of the well screen in the test well and the depth of response zones in all observation wells.
iv Distances from the centre of the test well to all observation wells.

A step-drawdown test (if carried out) has the aim of investigating the performance of the well at increasing flow rates (see Clark 1977). The well is pumped in a number of steps (ideally four, or an absolute minimum of three); the flow rate in each step is constant, with the rate for each step greater than the last. Increments for flow rate should be roughly equal. For example a four step test might be designed as:

Step 1: one quarter of maximum well yield
Step 2: one half of maximum well yield
Step 3: three quarters of maximum well yield
Step 4: maximum well yield (estimated from the equipment test).

Step-drawdown pumping test results can be analysed to provide information about well efficiency. This is a function of friction head losses through the filter pack and the well screen. It can also be affected by the techniques used to bore and develop the test well.

When starting any pumping (step-drawdown or constant rate) or recovery phase it is important that the monitoring team has been adequately briefed and is ready. Checks like ensuring all the dipmeters work and that all observers have a pen or pencil and paper to record readings are obvious, but can be very embarrassing if overlooked! If datalogging equipment is being used it should be checked in advance for battery function and for accuracy of clock setting. Above all, one person should be responsible for deciding when to start pumping, and he or she should resist being pressurised to start before everyone is ready. It is not necessary to start a test on the hour of local time, but it makes recording and analysis easier if the test starts on, say, a multiple of 10 min past the hour.

The time when the pump is started should be recorded. The control valve of the test pump should be adjusted to achieve the desired flow rate as quickly as possible after the start of pumping (or the step in a step-drawdown test). Once the flow rate has been set the valve should not be further adjusted as this will affect the drawdown and complicate analysis of results. At the start of the test phase, when readings are being taken frequently, it is important that readings in the test well and observation wells are taken at as nearly the same instant as possible. If all the wells are close together one of the observers may be able to make a visible or audible signal to the others that it is time for a reading. Otherwise each observer will need to have a clock of some sort, and the clocks will all need to be synchronized. If there are few observers and many wells it is best to abandon a rigid schedule of monitoring and just take readings as rapidly as possible, but record the precise time each reading was taken.

Monitoring during the test phases normally comprises:

i Water levels in the test well and observation well, with the time of each reading recorded. Readings are normally taken at specified intervals (very frequently at the start of the test, less frequently as pumping proceeds). Guidance is given in BS6316 (1992).
ii Pump discharge flow rate, recorded at the same time as the water levels (during the first hour or so of pumping, it is acceptable to record the flow rate less frequently than water levels, especially if the number of observers is limited).
iii Clarity of discharge water (as a check on any sand or silt in the discharge water).
iv Discharge water temperature and chemistry (by taking samples for testing or using portable meters). It is normal to take a water sample during the constant rate phase immediately after pumping commences, again after a few hours, and just before pumping ceases.

Upon cessation of pumping, monitoring of water levels in piezometers should be continued until full recovery of water level is approached. During the initial recovery period, readings should be at frequent intervals, at similar intervals to the start of pumping. As the rate of water level rise slows down the time intervals between readings may be extended progressively.

Plotting of test results

All water level and flow rate data gathered during the test should be plotted in graphical form while the test is proceeding. Even if this is done on site in rough form on a graph pad it will help identify any anomalies or inconsistencies. These may be due to occasional human error. If readings are plotted on site as the test is in progress and an anomaly shows up a further check reading can be made immediately. If the second reading confirms the earlier reading it is possible that some fault has developed with the dipmeter, datalogger transducer or observation well and so immediate remedial measures can be implemented.

Plotting of the test results also allows the aquifer drawdown response to be observed almost in 'real time'. This can be a useful guide when deciding whether a pumping phase needs to be extended or whether, pumping can be stopped early.

The most useful method of plotting data is to use the Cooper–Jacob straight line method described in Section 6.6. Drawdown is plotted on the vertical axis (linear scale) against elapsed time on the horizontal logarithmic scale (Fig. 6C-2). The first few readings curve upward and, in theory, then form a straight line. Obviously there will be times when the data will not conform precisely to the theory, but this form of presenting the data is nevertheless very useful for viewing trends in the drawdown data.

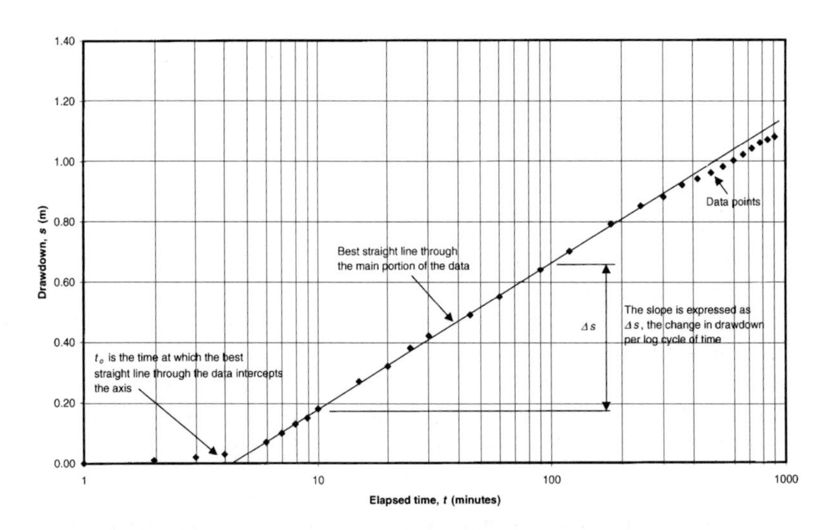

Figure 6C-2 Cooper–Jacob straight line method of plotting data.

References

Barker, R. D. (1986). Surface geophysical techniques. *Groundwater, Occurrence, Development and Protection* (Brandon, T. W., ed.). Institution of Water Engineers and Scientists, Water Practice Manual No. 5, London, pp 271–314.

Beesley, K. (1986). Downhole geophysics. *Groundwater, Occurrence, Development and Protection* (Brandon, T. W., ed.). Institution of Water Engineers and Scientists, Water Practice Manual No. 5, London, pp 315–352.

Binns, A. (1998). Rotary coring in soils and soft rocks for geotechnical engineering. *Proceedings of the Institution of Civil Engineers, Geotechnical Engineering*, **131**, 63–74.

Brand, E. W. and Premchitt, J. (1982). Response characteristics of cylindrical piezometers. *Géotechnique*, **32**(3), 203–216.

Brassington, F. C. (1986). The inter-relationship between changes in groundwater conditions and civil engineering construction. *Groundwater in Engineering Geology* (Cripps, J. C., Bell, F. G. and Culshaw, M. G., eds). Geological Society Engineering Geology Special Publication No. 3, London, pp 47–50.

Brassington, F. C. and Walthall, S. (1985). Field techniques using borehole packers in hydrogeological investigations. *Quarterly Journal of Engineering Geology*, **18**, 181–193.

BS1377. (1990). *Methods of Testing of Soils for Civil Engineering Purposes*. British Standards Institution, London.

BS5930. (1999). *Code of Practice for Site Investigation*. British Standards Institution, London.

BS6316 (1992). *Code of Practice for Test Pumping of Water Wells*. British Standards Institution, London.

BS7022. (1988). *Geophysical Logging of Boreholes for Hydrogeological Purposes*. British Standards Institution, London.

Card, G. B. and Roche, D. P. (1989). The use of continuous dynamic probing in ground investigation. *Penetration Testing in the UK*. Thomas Telford, London.

Chandler, R. J., Leroueil, S. and Trenter, N. A. (1990). Measurements of the permeability of London Clay using a self-boring permeameter. *Géotechnique*, **40**, (1), 113–124.

Clark, L. (1977). The analysis and planning of step drawdown tests. *Quarterly Journal of Engineering Geology*, **10**, 125–143.

Clayton, C. R. I., Matthews, M. C. and Simons, N. E. (1995). *Site Investigation*, 2nd edition. Blackwell, London.

Cooper, H. H. and Jacob, C. E. (1946). A generalised graphical method for evaluating formation constants and summarising well field history. *Transactions of the American Geophysical Union*, **27**, 526–534.

Dumbleton, M. G. and West, G. (1976). *Preliminary Sources of Information for Site Investigations in Britain*. Transport and Road Research Laboratory, LR403, Crowthorne.

Dunn, C. S. and Razouki, S. S. (1975). Interpretation of in situ permeability results on anisotropic deposits. *US Transportation Board, Transportation Research Record*, **532**, 43–48.

Hazen, A. (1892). Some physical properties of sands and gravels with special reference to their use in filtration. *24th Annual Report*. Massachusetts State Board of Health, p 539.

Hazen, A. (1900). *The Filtration of Public Water Supplies*. Wiley, New York.

Head, K. H. (1982). *Manual of Soil Laboratory Testing, Volume 2: Permeability, Shear Strength and Compressibility Tests*. Pentech Press, London.

Hvorslev, M. J. (1951). *Time Lag and Soil Permeability in Groundwater Observations*. Waterways Experimental Station, Corps of Engineers, Bulletin No. 36, Vicksburg, Mississippi.

Institution of Civil Engineers. (1991). *Inadequate Site Investigation*. Thomas Telford, London.

Irvine, D. J. and Smith, R. J. H. (1992). *Trenching Practice*. Construction Industry Research and Information Association, CIRIA Report 97, London.

Kruseman, G. P., and De Ridder, N. A. (1990). *Analysis and Evaluation of Pumping Test Data*. International Institute for Land Reclamation and Improvement, Publication 47, 2nd edition, Wageningen, The Netherlands.

Loudon, A. G. (1952). The computation of permeability from simple soil tests. *Géotechnique*, **3**, 165–183.

Lunne, T., Robertson, P. K. and Powell, J. J. M. (1997). *Cone Penetration Testing in Geotechnical Practice*. Blackie, London.

MacDonald, A. M., Burleigh, J., and Burgess, W. G. (1999). Estimating transmissivity from surface resistivity soundings: an example from the Thames Gravels. *Quarterly Journal of Engineering Geology*, **32**, 199–205.

McCann, D. M., Eddleston, M., Fenning, P. J. and Reeves, G. M., (eds) (1997). *Modern Geophysics in Engineering Geology*. Geological Society Engineering Geology Special Publication No. 12, London.

Meigh, A. C. (1987). *Cone Penetration Testing: Methods and Interpretation*. CIRIA Ground Engineering Report, Butterworths, London.

Powers, J. P. (1992). *Construction Dewatering: New Methods and Applications*, 2nd edition. Wiley, New York.

Powrie, W. and Roberts, T. O. L. (1990). Field trial of an ejector well dewatering system at Conwy, North Wales. *Quarterly Journal of Engineering Geology*, **23**, 169–185.

Preene, M., Roberts, T. O. L., Powrie, W. and Dyer, M. R. (2000). *Groundwater Control – Design and Practice*. Construction Industry Research and Information Association, CIRIA Report C515, London.

Price, M. and Williams, A. T. (1993). A pumped double-packer system for use in aquifer evaluation and groundwater sampling. *Proceedings of the Institution of Civil Engineers, Water, Maritime and Energy*, **101**, April, 85–92.

Rowe, P. W. (1972). The relevance of soil fabric to site investigation practice. *Géotechnique*, **22**(2), 195–300.

Site Investigation Steering Group. (1993). *Site Investigation in Construction, Volume 1: Without Investigation Ground is a Hazard*. Thomas Telford, London.

Terzaghi, K., Peck, R. B. and Mesri, G. (1996). *Soil Mechanics in Engineering Practice*, 3rd edition. Wiley, New York.

Trenter, N A. (1999). A note on the estimation of permeability of granular soils. *Quarterly Journal of Engineering Geology*, **32**, 383–388.

Walthall, S. (1990). Packer testing in geotechnical engineering. *Field Testing in Engineering Geology* (Bell, F. G., Culshaw, M. G., Cripps, J. C. and Coffey, J. R., eds). Geological Society Engineering Geology Special Publication No. 6, London, pp 345–350.

Chapter 7

Design of groundwater lowering systems

7.0 Introduction

The philosophy and basic methods for the design of groundwater lowering systems are outlined in this chapter. The main emphasis of this chapter, and indeed of most of this whole book, is to deal with field proven, simplistic but practical, methods applicable to many common situations, while providing advice on the approach to more complex problems.

The uncertainty inherent in any ground engineering process, requires a 'questioning' or 'testing' approach be adopted in design, where nothing is taken for granted. Sensitivity or parametric analyses can be used to assess how the design could cope with differing conditions. Alternatively, the observational method might be used to vary the design based on records taken during construction. Both these approaches are outlined, and the vital importance of developing a realistic conceptual model is stressed. The important effect that geometry and geological structure has on design is also described.

The basic designer's 'tool kit' – the formulae and concepts used in routine designs – are presented and their application discussed. Methods for estimation of steady-state discharge flow rate, and for selection of well yield and spacing are described in detail. Other design issues (such as time to achieve drawdown) are also outlined. The basic tenets of groundwater modelling are discussed in relation to more complex problems. Several simple design examples are presented in an appendix.

7.1 Design approach

There are two main philosophical views of the design of groundwater lowering or dewatering systems:

1 That the design process is essentially a seepage calculation problem, which can be assessed using application of groundwater and hydrogeological theory. This approach implies that the major problem is estimating the

discharge flow rate, and that selection and design of equipment is a secondary matter, carried out once the flow rate has been determined. We could call this the 'theoretical' approach. Alternatively;

2 That the design process must concentrate on selecting the appropriate type of well, well spacing and pump size for the ground conditions. Direct estimation of the flow rate is a less important issue (and indeed may never actually be calculated precisely). This can be thought of as the 'empirical' approach using case history experience.

The authors consider both the theoretical and empirical, have their advantages and disadvantages depending on ground conditions and the nature of the project.

For example, there are a number of cases which are sufficiently common that, once site investigation has confirmed there are no unusual complications, they can be designed purely empirically – almost by rule of thumb, based on the established capabilities of standard dewatering equipment. For example, shallow trench excavations in homogenous sand deposits of moderate permeability can almost always be dewatered by wellpoint systems with a spacing of 1–2 m between wellpoints. This has been practically proven over many decades. The empirical method is not applicable in more complex (or less clearly identified) geometries and ground conditions. If applied in such circumstances it can lead to considerable difficulties.

The theoretical approach can be used in cases whether or not there is empirical experience of the case in hand. The approach may involve fairly simple calculations or analyses, or may require more complex numerical modelling. The results of the calculations or modelling are used to specify the number and type of wells, pumps, etc, that will be required. Problems often arise when the theoretical approach does not take into account the limitations and advantages of the various dewatering techniques. If these issues are not considered, impractical or uneconomic system designs may result.

The best design approaches incorporate elements of both the theoretical and empirical methods. The theoretical method requires a 'conceptual model' of the ground and groundwater regime to be developed, following which calculations are carried out. Simple and fairly basic calculations are perfectly acceptable, and may be preferred in many cases, provided they are compared with an empirical approach. The empirical method should be used as a 'sanity check' to ensure that the proposed groundwater lowering system is realistic and practicable. For example, if the output of a theoretical design recommends a single stage of wellpoints to achieve a drawdown of 8 m this is clearly not going to work. More subtly, if a wellpoint spacing of, say, 15 m was recommended, this should be looked into more closely, since this is outside the normal range of wellpoint spacings – there may be problems with the conceptual model, methods of analysis, or selection of dewatering method.

This chapter presents the methods that can be used in combined theoretical and empirical approaches. The main emphasis is to deal with field proven practical means pertinent to the majority of groundwater lowering projects. The various methods presented form a 'tool kit' of techniques which, if selected with care, can deal with a wide range of real problems.

One defining feature of the design of any geotechnical process (including groundwater lowering) is that there will be some uncertainties in the ground. These uncertainties may result from the site investigation being of limited or inappropriate scope. Alternatively, even following a comprehensive site investigation, the sheer variability and complexity of the revealed ground conditions may give rise to uncertainty in design.

Uncertainty will affect the way the design is progressed. In principle there are two basic approaches to design:

i *Pre-defined designs*. This might be thought of as the 'traditional' design process, whereby one geological profile and set of parameters is selected and used to produce a single set of design predictions. The design is implemented with fairly basic monitoring, limited to checking that the design is 'effective' in the gross sense. Design and construction methods are not reviewed unless the original design is 'ineffective' (e.g. not achieving the target drawdown in a dewatering design). In such cases corrective action (alternative design and construction methods) would be taken.

ii *The observational method*. This contrasting design process begins with more than one geological profile and set of parameters and more than one option for the required dewatering system. These might range from 'most probable' to 'most unfavourable', perhaps with a number of intervening conditions as well. More than one set of design predictions are produced, together with measurable 'trigger values' to allow the detailed effectiveness of the design to be observed during construction. The data taken during construction are continually reviewed, to allow design and construction methods to be altered incrementally to match the behaviour of the ground and groundwater.

7.1.1 Pre-defined designs for groundwater lowering

The pre-defined approach to dewatering design is to select a geological profile and important parameters (primarily permeability), and then apply these to an appropriate method of seepage analysis (such as one of those presented in Section 7.4). The result of this calculation is the 'estimated' or 'design' discharge flow rate and drawdown distribution. Such an approach sometimes has an air of certainty or finality, especially if carried out by civil engineers more used to relatively consistent and reliable materials like steel and concrete.

In reality, few successful groundwater lowering systems are designed in such a regimented way. As was discussed in Chapter 3, permeability and aquifer boundaries can have a dramatic effect on dewatering systems; the difficulty of accurately determining permeability was described in Section 6.6. These factors mean that it is unrealistic to carry out a single seepage analysis and to then expect the dewatering system to have a high likelihood of performing adequately.

The best groundwater lowering systems are flexible and robust in nature, able to cope with ground conditions slightly different from those anticipated with few, if any, minor modifications. Such systems are also easy to modify or upgrade if ground conditions are substantially different from those expected.

The pre-defined approach can still be used, but normally more than one set of calculations is carried out as part of 'sensitivity' or 'parametric' analyses.

A sensitivity analysis is a set of repeated calculations, using varying values of a single parameter in each calculation. Typically for dewatering designs permeability is a key parameter, so seepage calculations are carried out for a range of possible values. The question being addressed is 'can the groundwater lowering system, as designed, cope with the range of discharge flow rates corresponding to the possible range of permeability?' If the answer is yes, all is well. If the answer is no, but the system could be easily (and quickly) modified to handle flow rates at either extreme end of the range, then the system may still be acceptably robust. If the answer is no, and the system cannot be readily modified, there is a risk that the system will not be able to handle all the possible flow rates. In such cases it may be prudent to try and develop an alternative, more flexible, design and test that against the results of the sensitivity analysis.

Parametric analyses are a rather broader version of sensitivity calculations. The question addressed is 'what parameters are influential to the design?' A number of parameters or conditions are systematically varied in calculations to investigate which have the greatest effect on the design. This may allow the design to be varied, or may prompt additional site investigation work to refine estimates of certain parameters or clarify key issues. In addition to permeability, aquifer boundary conditions can have an important effect on dewatering designs. A parametric study might look at the effect of a close source of recharge affecting the distance of influence, or the depth of the aquifer being deeper than expected. Again, the design process should include an assessment of the ability of the proposed dewatering system to cope with such possible conditions.

7.1.2 The observational method for groundwater lowering

The observational method allows a rational approach to construction where there is uncertainty over ground conditions or over the most suitable and economic dewatering or ground treatment options. The method uses construction observations to gather information about the behaviour of the ground and groundwater, and then modifies methods in response. The method is much more than carrying out parametric or sensitivity analyses. It should be a continuous and deliberately planned and managed process of design, construction control, monitoring and review that allows previously defined modifications to be incorporated into the construction process when necessary. The object is to achieve a robust process which provides for economic construction without compromising safety. The background, methods and case histories of the observational method are described in CIRIA Report C185 (Nicholson *et al.* 1999). Two main variants are normally identified:

(a) *Ab initio*: applied from inception of the project.
(b) *Best way out*: applied during construction to allow progress when unexpected problems occur on site. It is often applied when a 'pre-defined' design has proved unsatisfactory, and modifications are required.

The observational method can be applied to many groundwater lowering systems. This is because they can be easily modified (by the addition of extra wells or by using pumps of different capacity) and because easily observable parameters (such as drawdown and discharge flow rate) can be used to interpret how the system is performing. Examples of the observational method applied to groundwater lowering systems are given in Roberts and Preene (1994), Nicholson *et al.* (1999) and Preene *et al.* (2000).

The best way out application of the observational method is when a pre-defined method does not work and an alternative must be developed. It has been used when dewatering systems have had to be uprated or modified when the pre-defined design fails to achieve the design aims. Effectively the initial system (which was not installed with the observational method in mind) is monitored and used as a large-scale pumping test or trial to allow remedial measures to be selected.

The *ab initio* approach has been widely applied to larger projects, where ground conditions are known to be complex, or where the project design is not going to be finalized until construction is well advanced. The number of wells and pumping capacity can be optimized, based on the data gathered during the groundwater lowering works themselves. Optimization of the systems should be carefully considered, to avoid the temptation to use the bare minimum number of wells. A truly optimized system will have adequate standby plant, alarm systems, plus some additional wells (over and above the minimum) as an allowance against loss of wells or performance due to construction damage, clogging or biofouling and so on.

One possible reason why the *ab initio* application of the observational method has mainly been applied to larger projects is the need for clear management of the design and construction process. This is necessary to allow construction feedback to be obtained and linked into the ongoing design process in a timely manner. If smaller contracts are managed on this basis there is no reason why the benefits of the observational method could not be applied to a wider range of projects.

The methods presented in the remainder of this chapter are based largely on pre-defined methods of design, but applied to give robust and flexible systems. The observational method will not be considered further, but it is an approach that the dewatering designer should be familiar with, since it is another part of the 'tool kit' to be applied when appropriate.

7.2 Development of conceptual model

The essential first step in the design process is the development of a conceptual model of the ground and groundwater conditions. Until the conceptual model is developed, it is not possible to make rational decisions about which of the methods in the designer's 'tool kit' is suitable for the case in hand. In order to develop a conceptual model the designer needs to be familiar with the concepts of groundwater flow, aquifers, boundary conditions and so on (see Chapter 3).

If a conceptual model is a poor match for actual conditions, then much of the subsequent design work may be of dubious value. Even if the design work is diligently and carefully carried out, if it is based on an inappropriate conceptual model the design may be going in totally the wrong direction – and this is likely to lead to poorly performing dewatering systems. Developing the conceptual model at an early stage also forces the designer to review the available data. If the designer has insufficient reliable data to formulate a model this could be a sign that further site investigation is needed.

The conceptual groundwater model depends on a number of factors that are dictated by ground conditions, and are beyond the designer's control:

i Aquifer type(s) and properties. Aquifers (the water bearing strata in which groundwater levels are to be lowered) may be classified as confined or unconfined. These types behave in quite different ways; the aquifer type(s) must be identified. It is essential that the likely permeability range of each aquifer be identified. If transient analyses are to be carried out the storage coefficient must also be estimated.
ii Aquifer depth and thickness. These dimensions must be estimated, and a judgement taken as to whether they are effectively constant across the area affected by the dewatering, or whether the aquifer thickness varies significantly.
iii Presence of aquitards and aquicludes. Very low permeability silt or clay layers may act as barriers to vertical groundwater flow. The presence of such strata may necessitate well screens above and below the aquiclude or aquitard.
iv Distance of influence and aquifer boundaries. Is all the pumped water likely to be derived purely from storage in the aquifer? If not are there

any nearby sources of recharge water? The presence of any barrier boundaries can also influence groundwater flow.

v Initial groundwater level and pore water pressure profile. The initial groundwater level determines the amount of drawdown required. Complications may arise if the groundwater level slopes across the site, or if groundwater levels vary with tidal or seasonal influences.

vi Presence of compressible strata. If present in significant thickness this indicates that potentially damaging consolidation settlements may occur.

There are also some factors that are not directly related to ground conditions. Some of these can be controlled by the designer.

vii Geometry of the works. The depth and size of the excavation will have a direct influence on the groundwater lowering requirements. The target lowered groundwater level is normally set a short distance (say 0.5 to 1 m) below excavation formation level. The drawdown needed is taken as the vertical distance from the original groundwater level to the target lowered groundwater level.

viii Groundwater lowering technique. Different dewatering methods may interact with the groundwater regime in quite different ways.

ix Period for which groundwater lowering is required. The time for which pumping is to continue may influence the choice of technique.

x Depth of wells. In very deep aquifers it may be more economic to install wells which do not penetrate to the base of the aquifer. Instead shallower, partially penetrating, wells may be more suitable.

xi Environmental constraints. The location or nature of the site and its surroundings may give rise to limits on discharge rates or drawdown being imposed by the client or environmental regulator. These limits are normally imposed to reduce the risk of detrimental side effects from groundwater lowering (see Chapter 13). Occasionally these constraints can severely limit the options available to the dewatering designer.

If all, or at least most, of these questions can be answered a conceptual model can normally be developed. The conceptual model need not be particularly complex, and may simply be a list of the expected conditions. Figure 7.1 shows an example of a simple pro-forma to record key data. Conceptual models are outlined in each of the design examples given in Section 7.8.

Any consideration of groundwater flow in general, or of the equations presented later in this chapter highlights that aquifer permeability is a critically important parameter. Chapter 6 has described the plethora of techniques available to estimate permeability, from the simple to the very complex. When assessing permeability values to be used in calculation, the designer should not visualize a single permeability value to be used in the conceptual model, but rather a range of realistic values to be used in sensitivity analyses. The range

PMC DEWATERING	
Site:	Eastern resewerage, phase 1
Location:	Anytown
Prepared by:	MP
Basic Information	
Aquifer type and properties:	Unconfined aquifer. Generally described as fine to medium sand. PSD results indicate k range of 1 to 5 x 10^{-4} m/s. Falling head tests give lower results.
Initial groundwater level:	1.5 mbgl at northern end, varying steadily to 1.8 mbgl at southern end. No long-term monitoring carried out, so background or seasonal variations unknown.
Depth to base of aquifer:	Not proved. Deepest borehole penetrated to 18 m depth and was still in sand.
Aquitards/Aquicludes present:	Some thin clay layers at 3 to 5 m depth in southern section; could result in overbleed seepage.
Recharge/Barrier boundary:	None apparent
Excavation depth and geometry:	Trench works, mainly 2.5 to 3.5 m deep. Trench width less than 1.5 m.
Maximum drawdown:	Maximum drawdown is to 0.5 m below deepest dig = 4 m depth. This implies a maximum drawdown of 2.5 m.
Dewatering period:	4 weeks per 100 m section.
Compressible strata:	None indicated in site investigation data.
Possible dewatering technique:	Wellpoints parallel to trench. Wellpoints 6 m deep, so will be partially penetrating. Risk of overbleed in south.
Sketch:	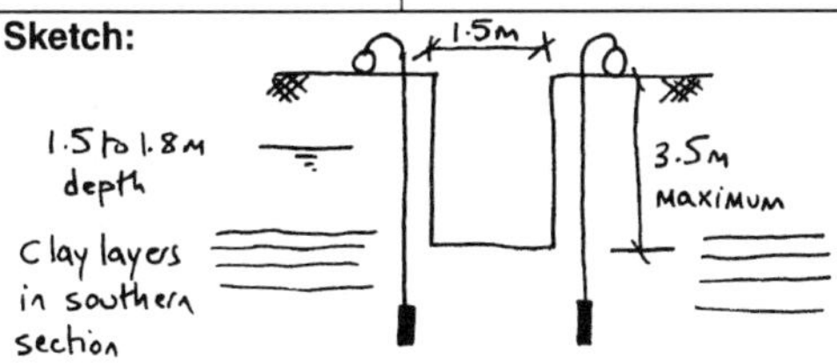
Other Notes	

Figure 7.1 Example of simple conceptual model for groundwater lowering system.

of permeability values may represent uncertainty due to natural variations in permeability, or limitations in the permeability test methods or results.

It is difficult to give simple, useful guidelines on selection of realistic permeability ranges. There will always be some reliance on judgement and experience, but the following advice is relevant:

- Be aware that different methods of assessing permeability produce results of greater or lesser reliability. Table 6.3 provides some guidance. Consider the relative merits of each method when assessing permeability from the available results.
- Always compare the permeability results with 'typical' values of permeability from published correlations with soil types (such as Table 3.1) or, even better, from experience at nearby sites. This approach is vital in excluding unrealistically high or low permeability results. For example, few experienced engineers would expect a slightly silty sandy gravel to have a permeability of 1×10^{-8} m/s, yet falling head tests in such soils often produce results of that order.
- If permeability estimates from various differing techniques produce broadly similar results, in agreement with typical values for those soil types, then the design range of permeability could be assessed from the full range of data. If there are large discrepancies in the data from various methods, some of the data may have to be excluded from the assessment process. Again Table 6.3 may be of help in assessing the reliability of data.
- Always consider the important aspects of the design when selecting the permeability range to be used in calculations. Mistakes have been made when designers have focussed too much on estimating the highest likely permeability, to ensure a pumping capacity sufficient for the maximum possible flow rate. This is not always the most appropriate approach. It is true that in high permeability aquifers, the total flow rate may be critical, and assessing permeability at the upper end of the possible range may be a robust approach. But alternatively, in soils of low to moderate permeability, a critical case in design may be if the permeability is at the lower end of the possible range when yields may be very low, necessitating unfeasibly large numbers of wells.

Once the conceptual model exists, an initial view must be taken of the dewatering method to be used, and the likely geometry of the system.

7.3 Selection of method and geometry

There are a number of groundwater lowering methods available (see Section 5.4), and part of the design process is to select an appropriate technique which will satisfy the various constraints on the project in hand.

A useful starting point when selecting a technique is shown in Fig. 5.6. Knowing the required drawdown and estimated soil permeability from the conceptual model the appropriate method can be chosen. Where more than one method is feasible, the choice between them may be made on cost grounds, local availability of equipment, or expertise of those carrying out the works.

7.3.1 Equivalent wells and slots

In practice, seldom can the required drawdown for an excavation be achieved by a single well; most dewatering systems rely on several wells acting in concert. The rare exceptions being, in a high permeability stratum, the use of a large central sump pumping installation with radial drains (see Section 8.4 for a case history). The collector well system (see Section 11.4) is a variation on this concept, but is more generally used for agricultural or land drainage purposes than for civil engineering construction.

For dewatering purposes the established principle is to install an array of water abstraction points, generally sited immediately adjacent to the area of the proposed excavation. This preferred location of wells (at the perimeter of the excavation) is based on practical experience. If wells are sited within the area to be excavated they are greatly at risk since maintaining wellheads, riser pipes, power cables and discharge mains is rarely compatible with the activities of heavy excavation plant. Where possible wells are best located immediately outside the excavation area or within the batter of excavations. Wells located in slope batters, whilst presenting an initial excavation and access impediment, are inherently more secure (in comparison with wells in the main excavation area) as work progresses deeper.

Very large or wide excavations often cannot be dealt with adequately or economically by wells sited only around the perimeter. As highlighted above the practicalities of maintaining and/or reinstalling damaged wells in the middle of a 'live' area of excavation are far from trivial. In many such cases unpumped relief wells (see Section 11.3) could be used inside the excavation area. These act as vertical drainage paths and their usefulness is not affected by progressive shortening as the excavation is deepened. The disadvantage of relief wells is their continuous water discharge; it must be disposed of by an effective drain and sump system. The disposal problem is readily controlled where the rates of flow are small (i.e. the underlying stratum is of low permeability, say less than 10^{-5} m/s).

In general, the multiple well systems used for groundwater lowering can be classified as either 'linear' systems (for installation alongside trench excavations) or 'ring' systems (for installation around circular or rectangular excavations). Figure 7.2 shows definitions of these geometries.

This leads to the practical question of 'how can we model the flow to such systems?' It would be very tedious to consider the flow to each individual well, and the complex interaction between them. A useful approach is to consider

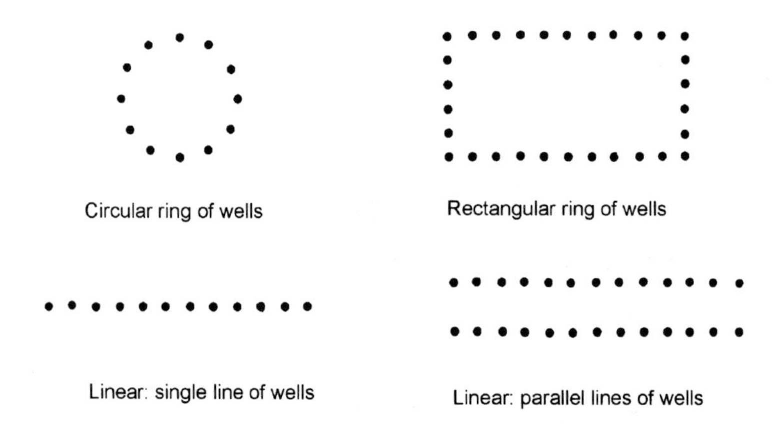

Figure 7.2 Plan layouts of groundwater control systems: linear and ring arrays.

the groups of wells as large 'equivalent wells' or 'equivalent slots', thereby allowing simple and accessible formulae to be used to estimate flow rate.

We can define an equivalent well as a groundwater lowering system where, on a gross scale, flow of groundwater to the system is radial. Radial flow implies that flow lines converge to the well from a distant diffuse source of water. The concept of the 'zone of influence' of a well (introduced in Section 3.4) describes the area of the aquifer affected by pumping from a well (Fig. 7.3(a)). The radius of influence R_0 is a theoretical concept describing the zone of influence. R_0 is defined as the distance from the centre of the well to the edge of an idealized circular zone of influence.

An equivalent slot is a line of closely-spaced wells forming a groundwater lowering system. If the line of wells is very long flow of groundwater to the system is plane (although there will be some radial flow to the ends of the slot). Under plane conditions, flow lines do not converge, but are parallel, resulting in quite different flow conditions to radial flow. The edge of the theoretical zone of influence will be parallel to the line of wells (Fig. 7.3(b)). The distance of influence L_0 is the distance from the line of wells to the idealised edge of the zone of influence.

Of course the equivalent well and slot concepts are approximations, introduced purely to make the estimation of discharge flow rate more amenable to simple solutions. Nevertheless, there is considerable justification for using these simplifications in appropriate conditions. The equivalent well concept was proposed by Forcheimer (1886). He based his work on that of Dupuit and analysed radial flow towards a group of wells. By means of correlation with field data, he demonstrated the acceptability of the concept for most practical purposes; provided that the wells are spaced in a regular pattern. The equivalent well concept was later endorsed by Weber (1928), based on extensive field data. The equivalent slot approach,

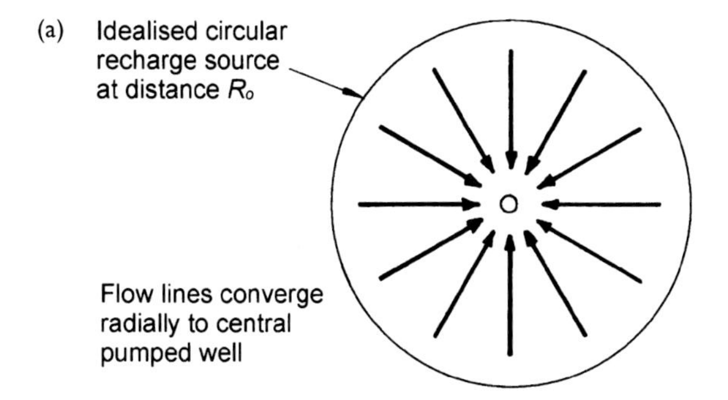

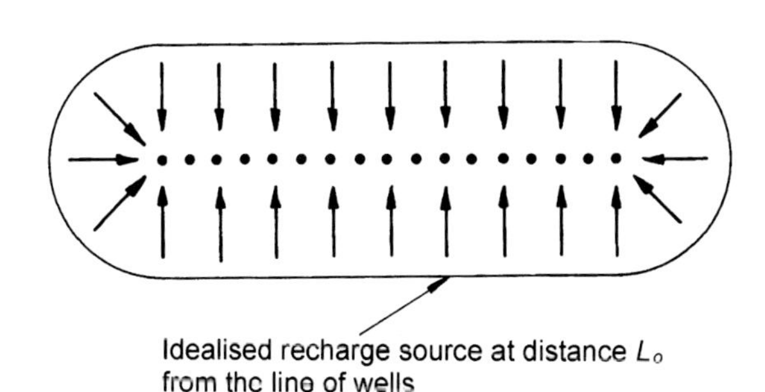

Figure 7.3 Zone of influence. (a) Radial flow, (b) plane flow.

where a long line of closely-spaced wells is treated as one continuous water abstraction slot, is implicit in the work of Chapman (1959) who studied flow to wellpoint systems. The equivalent well and slot simplification is an established practical method, used in the design sections of Powers (1992) and CIRIA Report C515 (Preene *et al.* 2000).

For radial flow to rings of wells this approach requires estimation of the equivalent radius r_e of the system (Fig. 7.4). For a circular ring of wells r_e is simply the radius of the ring. For a rectangular ring of wells of plan dimensions a by b, the equivalent radius can be estimated by assuming a well of equal perimeter:

$$r_e = \frac{(a+b)}{\pi} \tag{7.1}$$

or equal area:

$$r_e = \sqrt{\frac{ab}{\pi}} \tag{7.2}$$

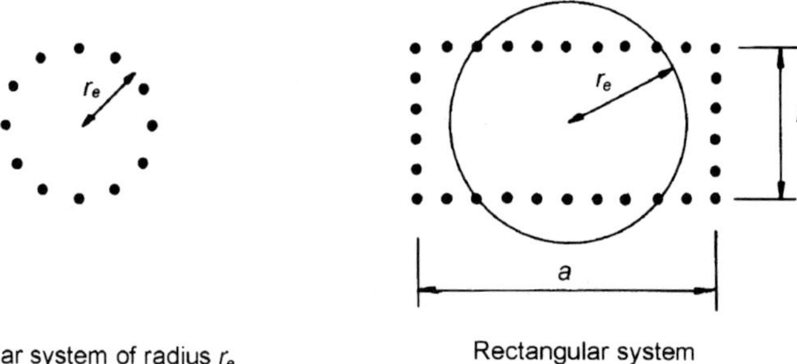

Figure 7.4 Equivalent radius of arrays of wells.

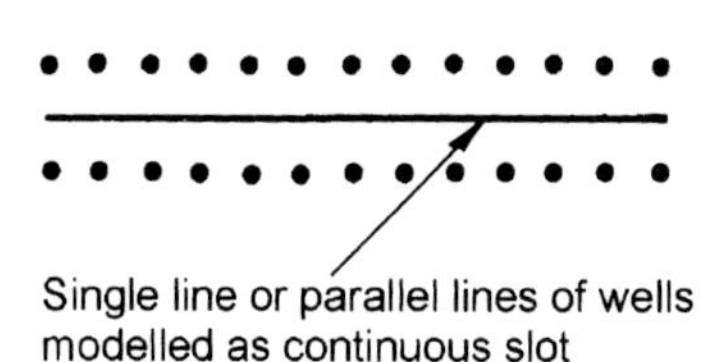

Figure 7.5 Equivalent slots.

In practice both formula give similar results, provided the ring of wells is not very long and narrow (i.e. provided a is not very much greater than b). The estimate of r_e determined in this way can be used in the well flow equations presented in Section 7.4.

Long narrow systems consisting of lines of closely-spaced wells (where a is much greater than b), or where the distance of influence is small, are likely to operate in conditions of plane flow (as opposed to radial flow). These systems may be better simplified to equivalent slots (Fig. 7.5). In addition to plane flow to the sides there will be a component of radial flow at the end of the line of wells (Fig. 7.3). For relatively long systems the radial flow component is likely to be relatively minor, and is sometimes neglected. For shorter systems the radial flow to the ends may be a significant proportion of the total discharge, and should be incorporated in calculations.

7.3.2 Geological structure, well depth and underdrainage

Geological layering and structure may have a controlling effect on the geometry of groundwater lowering systems, in particular the well depth and the level of well screens. There are situations where it may be possible to use the geological structure to advantage to enhance the performance of the

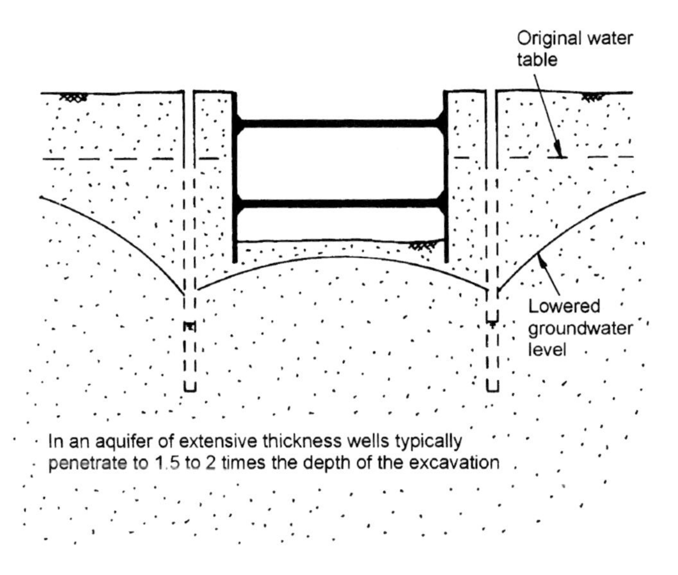

Figure 7.6 Wells in aquifers extending to great depth.

dewatering system. The potential to do this should have been identified in the conceptual model. Some options are discussed below.

In a homogenous permeable aquifer the wells must penetrate to sufficient depth to achieve the required drawdown. As a rule of thumb, widely spaced wells should penetrate to one and a half to two times the depth of the excavation, to ensure that the wells have adequate 'wetted screen length' (see Section 7.5) even after drawdown. In an aquifer that extends to great depth the wells may not need to penetrate to the base of the aquifer, but may be designed to be 'partially penetrating' with a depth controlled by the need for adequate wetted screen length (Fig. 7.6).

If the aquifer does not extend to great depth below excavation formation level, and is underlain by an aquiclude or aquitard, the wells will have to fully penetrate the aquifer. In practice obtaining the required drawdown for excavation can be very problematic if the residual aquifer thickness below the excavation is much less than around one-third of the original saturated thickness (Fig. 7.7). Obtaining the final part of the drawdown is difficult because the presence of the low permeability layer restricts the wetted screen length of each well. This reduces the yield of each well and its corresponding ability to generate further drawdown. There are two possible approaches in this case. One is to install the wells at much closer spacings than normal to obtain adequate wetted screen length by having a large number of wells. This approach can be economic with the wellpoint or ejector method, but less so with deep wells, due to the greater cost per well. The

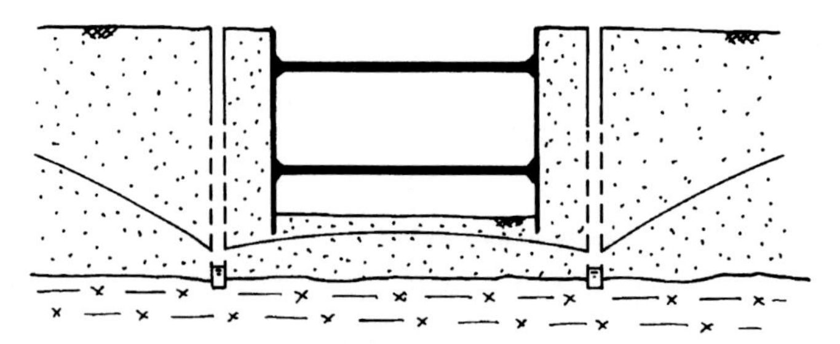

Figure 7.7 Wells in aquifers of limited thickness.

second approach is to install an economic number of wells and to then try and manage the residual 'overbleed seepage' (see Section 4.4) by protecting the faces of the excavation with sand bags, geotextile filters or other erosion prevention measures (Fig. 9.13(b)). This allows water to enter the excavation without causing damaging loss of fines; the seepage water must then be removed by sump pumping.

If the aquifer is not homogenous, but consists of layers of greater and lesser permeability, well depth and screen level must be dictated by the layering. The basic requirement is to abstract water *directly* from the most permeable layers (or more strictly the layers of the highest transmissivity) in preference to intermediate or lower permeability layers. This will ensure that well yields are maximized because the permeable layers will readily feed water to the well screens.

If the most permeable layer (such as a gravel stratum) is beneath layers of moderate permeability (such as a silty sand) the most efficient system will involve wells pumping directly from the gravel. This will promote downward drainage of water from the sand, to the gravel, and thence to the wells (Fig. 7.8). This important case is known as the 'underdrainage' approach, and is widely used where ground conditions allow. It is nearly always the best option, even if the wells have to be slightly deeper than first thought in order to intercept the permeable layer.

In contrast, where a permeable layer overlies a less permeable layer it is likely that a dewatering system abstracting purely from the lower layer may not achieve the required groundwater lowering. The rate of recharge to the more permeable stratum will exceed the rate at which water can be abstracted from the underlying lower permeability layer. Hence, if an excavation has to penetrate into the underlying stratum of lesser permeability, it will be necessary to provide two abstraction systems. One system must pump from the

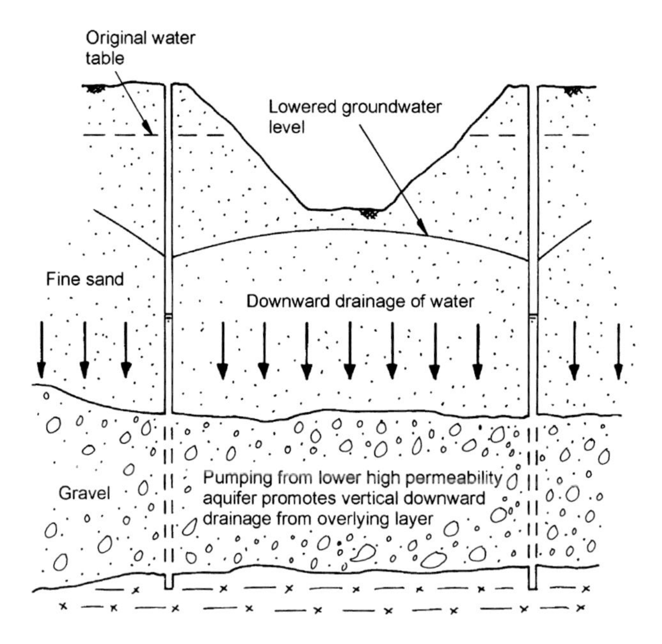

Figure 7.8 Pumping from a layered aquifer system using the underdrainage principle.

overlying more permeable layer. This allows the second pumping installation, screened in the underlying and less permeable stratum, to operate without being overwhelmed by seepage from the overlying layer.

7.4 Estimation of steady-state discharge flow rate

Two key unknowns to be determined during design are the steady-state discharge flow rate and the yield, number and design of wells necessary to achieve that flow rate. Some commonly used methods to estimate discharge flow rate are described in this section. Estimation of well yields is described in Section 7.5.

7.4.1 Steady-state well and slot formulae

This section presents simple formulae that can be used to estimate the steady-state discharge flow rate from systems treated as equivalent wells or slots. The emphasis here is on simple methods, to be used when the conceptual model indicates conditions not too different to the idealizations and assumptions discussed below. For more complex cases, or where conditions differ dramatically from the simple conditions discussed here, analysis by numerical modelling may be appropriate (see Section 7.7).

The simple formulae for radial flow to wells are generally based on the work of Dupuit (1863), which used certain simplifications and assumptions about the aquifer properties and geometry (Fig. 7.3).

- The aquifer extends horizontally with uniform thickness in all directions without encountering intermediate recharge or barrier boundaries within the radius of influence.
- Darcy's law is valid everywhere in the aquifer.
- The aquifer is isotropic and homogeneous – thus the permeability is the same at all locations and in all directions.
- Water is released from storage instantly when the head is reduced.
- The pumping well is frictionless and fully penetrates the aquifer.
- The pumping well is very small in diameter compared to the radius of influence, which is an infinite source of water forming a cylindrical boundary to the aquifer at distance R_0.

In reality, several of these assumptions are unlikely to be fully satisfied. For example, soils are usually stratified and generally exhibit horizontal permeabilities in excess of those in the vertical direction; often by more than one order of magnitude. Similarly in rock the permeability may be dominated by fissure flow and may vary greatly from point to point.

Dupuit made a further, important, assumption. This was that the groundwater flow to the well was horizontal. This is a valid assumption for fully penetrating wells in confined aquifers, but is invalid (at least close to the pumping well) in unconfined aquifers or if the well is only partially penetrating. Dupuit's analysis was purely for the radial flow case, but Muskat (1935) did analogous studies for plane flow to slots, using similar, idealized, assumptions.

Nevertheless, despite the idealizations and simplifications inherent in the formula, experience has demonstrated that the Dupuit-based formulae can be successfully used to estimate the steady-state pumping requirements for relatively short-term dewatering purposes. These methods are used in the design sections of Mansur and Kaufman (1962) and Powers (1992).

The empirical evidence that the Dupuit methods give reasonable estimates of flow rate, is supported by a number of theoretical studies. Hantush (1964) stated that 'The Dupuit–Forcheimer well discharge formulae, despite the shortcomings of some of the assumptions, predict the well discharges within a high degree of accuracy commensurate with experimental errors'. The assumptions have a more significant effect on the accuracy of the lowered groundwater level profile around a well, but even then it is generally accepted that the Dupuit approach can predict drawdowns to acceptable accuracy at distances from the well of more than one and a half times the aquifer thickness.

The commonly used formulae for estimation of the steady-state discharge flow rate are listed in Table 7.1, together with diagrams of the idealized

Table 7.1 Simple formulae for estimation of steady-state flow rate

Case	*Schematic diagram*	*Formula for steady-state flow rate Q*		*Notes*
Radial flow to wells Fully penetrating well, confined aquifer, circular source at distance R_0 (Theim equation)	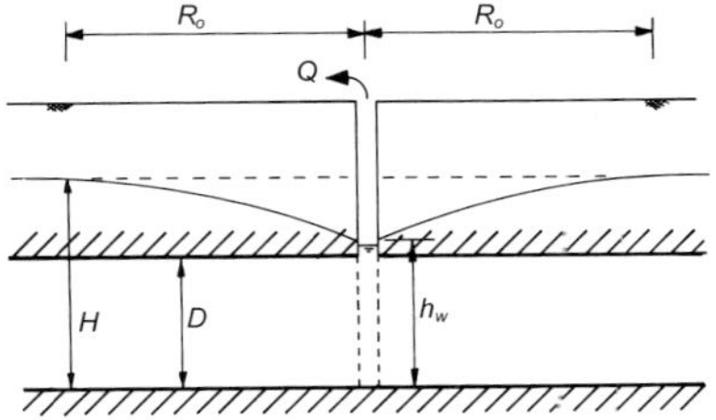	$Q=\dfrac{2\pi kD(H-h_w)}{\ln[R_0/r_e]}$	(7.3)	k = soil permeability; D = thickness of confined aquifer; H = initial piezometric level in aquifer; h_w = lowered water level in equivalent well; r_e = equivalent radius of well; R_0 = radius of influence.
Fully penetrating well, confined aquifer, line source at distance L_0 (method of images)	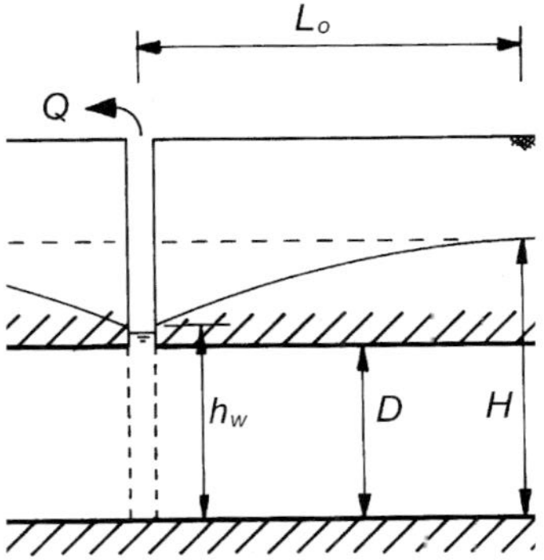	$Q=\dfrac{2\pi kD(H-h_w)}{\ln[2L_0/r_e]}$	(7.4)	k = soil permeability; D = thickness of confined aquifer; H = initial piezometric level in aquifer; h_w = lowered water level in equivalent well; r_e = equivalent radius of well; L_0 = distance to line source.

Table 7.1 Continued

Case	*Schematic diagram*	*Formula for steady-state flow rate Q*	*Notes*
Fully penetrating well, unconfined aquifer, circular source at distance R_0 (Dupuit–Forcheimer equation)	R_o R_o Q H h_w	$$Q=\frac{\pi k(H^2-h^2_w)}{\ln[R_0/r_e]} \quad (7.5)$$	k = soil permeability; H = initial water table level in aquifer; h_w = lowered water level in equivalent well; r_e = equivalent radius of well; R_0 = radius of influence
Fully penetrating well, unconfined aquifer, line source at distance L_0 (method of images)	L_o Q h_w H	$$Q=\frac{\pi k(H^2-h^2_w)}{\ln[2L_0/r_e]} \quad (7.6)$$	k = soil permeability; H = initial water table level in aquifer; h_w = lowered water level in equivalent well; r_e = equivalent radius of well; L_0 = distance to line source.

Partially penetrating well, confined aquifer	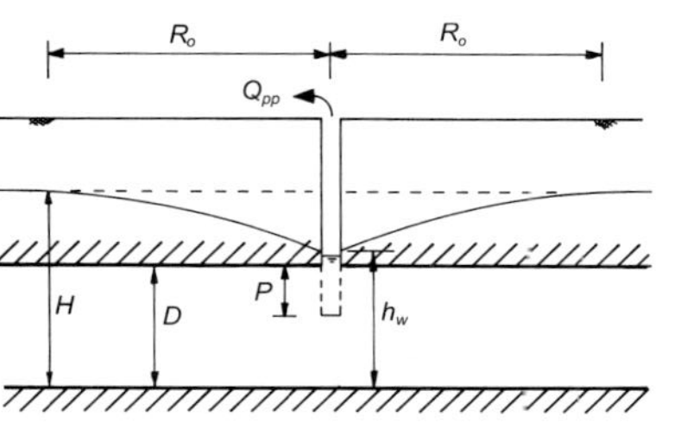 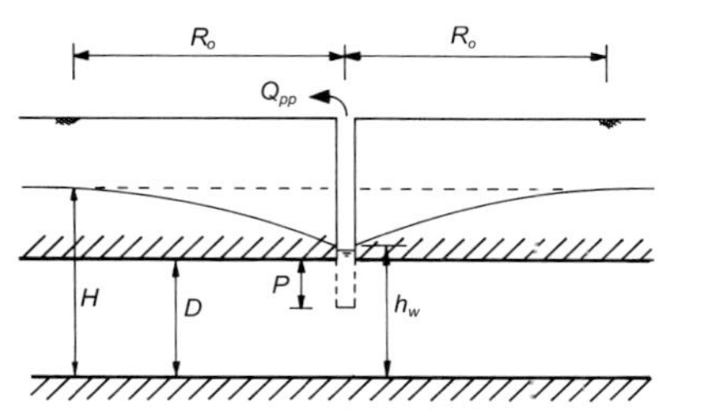	$Q_{pp} = BQ_{fp}$ (7.7)	Q_{pp} = flow rate from partially penetrating well; Q_{fp} = flow rate from fully penetrating well; B = partial penetration factor for radial flow (obtained from Figure 7.9(a)).
Partially penetrating well, unconfined aquifer	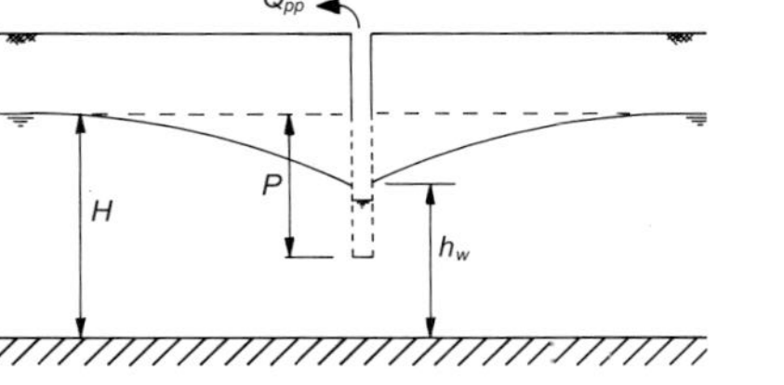	$Q_{pp} = BQ_{fp}$ (7.7)	Q_{pp} = flow rate from partially penetrating well; Q_{fp} = flow rate from fully penetrating well; B = partial penetration factor for radial flow (obtained from Figure 7.9(b)).
Plane flow to slots Fully penetrating slots, confined aquifer, flow from line sources on both sides of slot		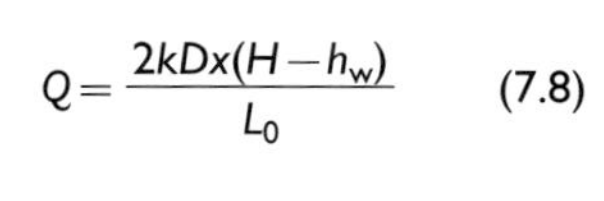 $Q = \dfrac{2kDx(H - h_w)}{L_0}$ (7.8)	x = linear length of slot; k = soil permeability; D = thickness of confined aquifer; H = initial piezometric level in aquifer; h_w = lowered water level in equivalent slot; L_0 = distance of influence;

Table 7.1 Continued

Case	*Schematic diagram*	*Formula for steady-state flow rate Q*	*Notes*
Partially penetrating slots, confined aquifer, flow from line sources on both sides of slot	L_o; L_o; Q_{pp}; H; D; P; h_w	$Q_{pp}=\dfrac{2kDx(H-h_w)}{(L_0+\lambda D)}$ (7.9)	x = linear length of slot; k = soil permeability; D = thickness of confined aquifer; H = initial piezometric level in aquifer; h_w = lowered water level in equivalent slot; L_0 = distance of influence; λ = partial penetration factor (obtained from Figure 7.9(c)).
Fully penetrating slots, unconfined aquifer, flow from line sources on both sides of slot	L_o; L_o; Q; H; h_w	$Q=\dfrac{kx(H^2-h^2_w)}{L_0}$ (7.10)	x = linear length of slot; k = soil permeability; H = initial water table level in aquifer; h_w = lowered water level in equivalent slot; L_0 = distance of influence.

Partially penetrating slots, unconfined aquifer, flow from line sources on both sides of slot	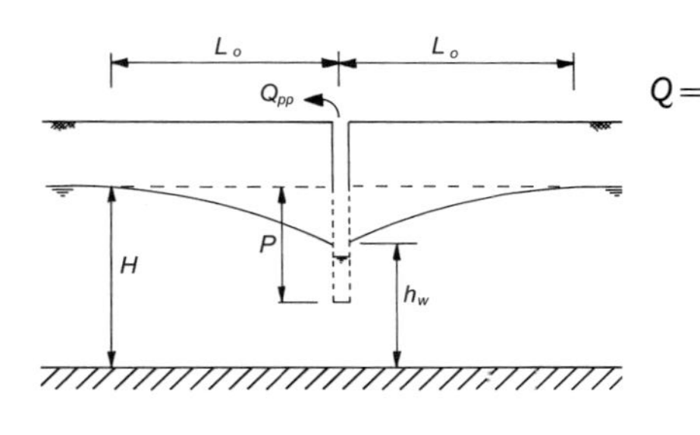 	$Q=\left[0.73+0.23(P/H)\right]\dfrac{kx(H^2-h^2_w)}{L_0}$	(7.11)	x = linear length of slot; k = soil permeability; H = initial water table level in aquifer; h_w = lowered water level in equivalent slot; L_0 = distance of influence; P = depth of penetration of slot below original water table.
Plane and radial flow Rectangular systems, confined aquifer		$Q=kD(H-h_w)G$	(7.12)	k = soil permeability; D = thickness of confined aquifer; H = initial piezometric level in aquifer; h_w = lowered water level in equivalent slot; G = geometry shape factor (obtained from Fig. 7.10).

geometry. The formulae are categorized by whether the aquifer is confined or unconfined, whether flow is radial or plane and whether the well or slot is fully or partially penetrating. All these conditions must be clarified during development of the conceptual model before the formulae can be applied.

A significant qualification on the use of these formulae is that the results will only be as valid (or invalid!) as the parameters used in them. Previous sections have discussed the selection of permeability values for design purposes, and the need for sensitivity and parametric analyses. A similar

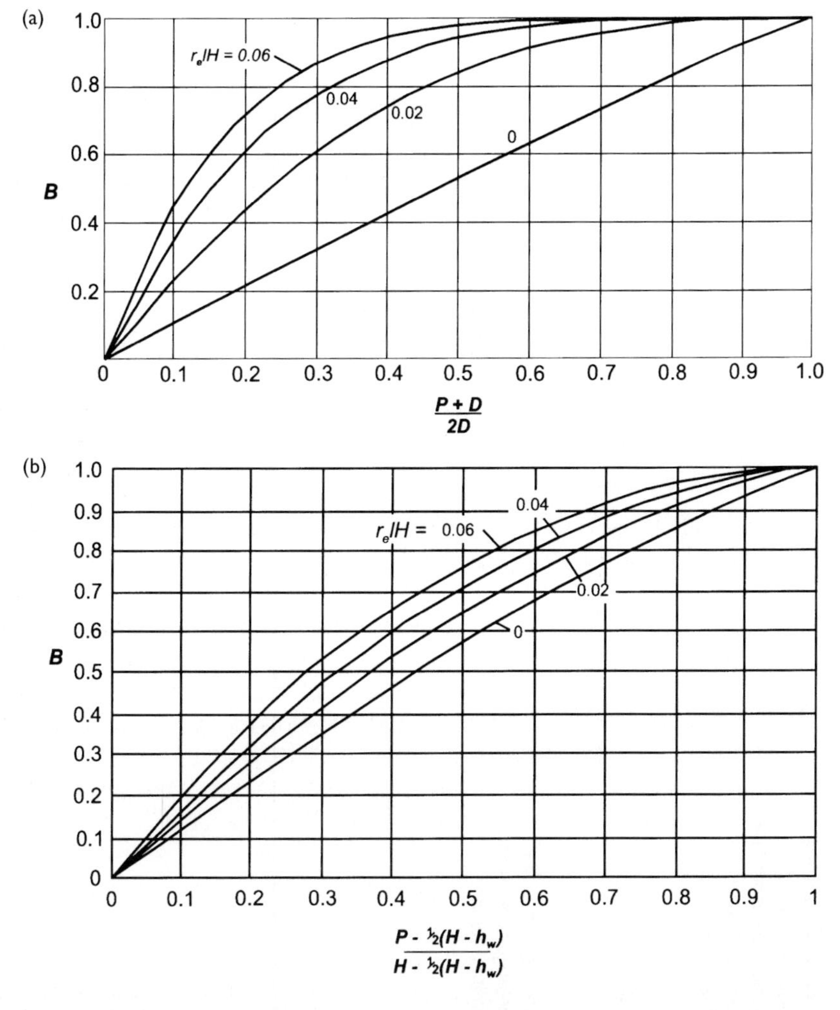

Figure 7.9 Partial penetration factors for wells and slots (after Mansur and Kaufman 1962). (a) Radial flow to wells in confined aquifers, (b) radial flow to wells in unconfined aquifers, (c) plane flow to slots in confined aquifers.

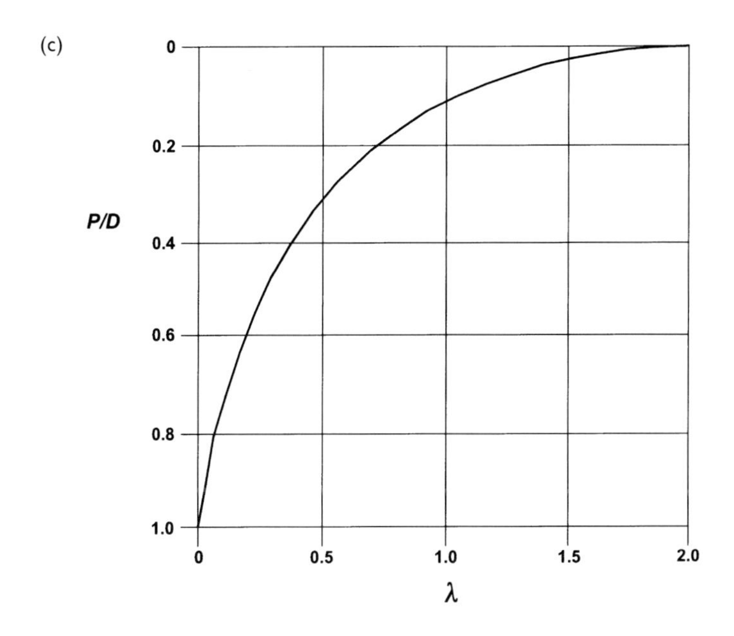

Figure 7.9 Continued.

approach should be applied when using the steady-state formula. Additionally, other parameters should be selected with care, including:

(a) Equivalent radius (r_e) of system. For radial flow cases this can be estimated from equation (7.1) or (7.2).
(b) Radius of influence (R_0) for radial flow cases. The radial flow cases assume a circular recharge boundary at radius R_0. This is a theoretical concept representing the complex behaviour of real aquifers (see Section 3.4); the distance of influence is not a constant on a site, but is initially zero and increases with time. However, the simplification of an empirical R_0 value is a useful one. The most reliable way of determining R_0 is from pumping test analyses presented as a Cooper–Jacob straight-line plot of distance–drawdown data (see Section 6.6). If no pumping test data are available approximate values of R_0 (in metres) can be obtained from Sichardt's formula (which is actually based on earlier work by Weber)

$$R_0 = 3000(H - h_w)\sqrt{k} \tag{7.13}$$

where $(H - h_w)$ is the drawdown (in metres) and k is the soil permeability (in m/s). This formula needs to be modified when used to analyse large

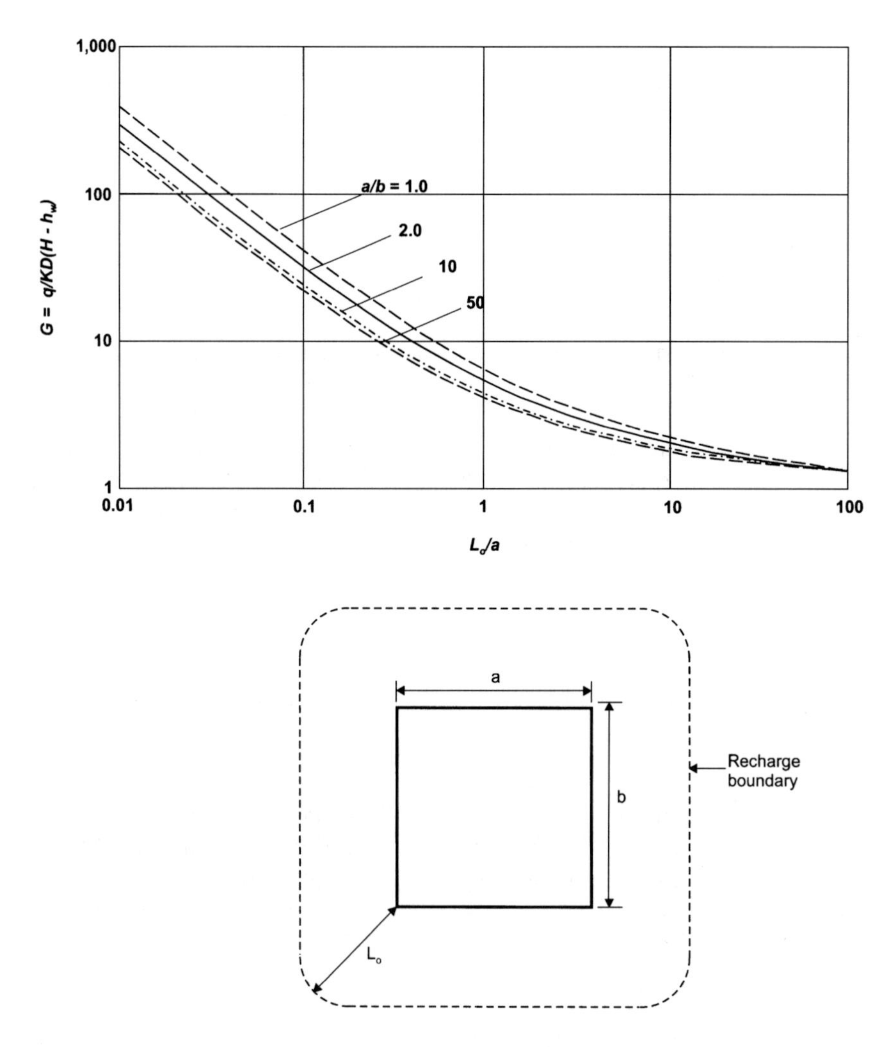

Figure 7.10 Shape factor for confined flow to rectangular equivalent wells (from Powrie and Preene 1992, with permission).

equivalent wells. Dupuit assumed that the radius of the well was small in comparison to the radius of influence, but often the radius r_e may be large in comparison to R_0. In such cases the following equation can be used.

$$R_0 = r_e + 3000(H - h_w)\sqrt{k} \quad (7.14)$$

When estimating R_0, it is important to review the calculated distance of influence to avoid using wildly unrealistic values. In the authors' experience values of less than around 30 m or more than 5,000 m are rare and should be viewed with caution. It may be appropriate to carry out sensitivity analyses using a range of distance of influence values to see the effect on calculated flow rates. For the radial flow case R_0 appears in a log-term so small errors do not have a significant effect on calculated flow rates, but the possibility of gross error exists if a very large or very small R_0 is used.

(c) Distance of influence (L_0) for plane flow cases. L_0 (in metres) can be estimated from Sichardt's formula, but a different calibration factor must be used.

$$L_0 = 1750(H - h_w)\sqrt{k} \tag{7.15}$$

where k is in m/s and $(H - h_w)$ is in metres. The distance of influence appears as a linear term in the plane flow equations – the estimated flow rate is inversely proportional to L_0. The distance of influence must be chosen with care, and sensitivity analyses are strongly recommended.

(d) Lowered water level (h_w) inside the equivalent well or slot. The equivalent well or slot method requires that the lowered water level (inside the well or the slot) used in equations is the groundwater level in the excavation area itself. Obviously, the water level in each individual well will be lower (perhaps considerably so), but this drawdown would not be representative of the drawdown in the equivalent well or slot.

7.4.2 Cumulative drawdown analysis – theoretical method

The formulae described in the previous section are used to analyse systems of closely-spaced wells, modelled as equivalent wells or slots. Such an approach is less satisfactory if the wells are widely-spaced; in those cases a cumulative drawdown (or superposition) method may be more suitable.

This method takes the advantage of the mathematical property of superposition applied to drawdowns in confined aquifers. In essence, the total (or cumulative) drawdown at a given point in the aquifer, resulting from the action of several pumped wells, is obtained by adding together (or superimposing) the drawdown from each well taken individually (Fig. 7.11). This approach is theoretically correct in confined aquifers, but is invalid in unconfined aquifers where the changes in saturated thickness that occur during drawdown complicate the interaction of drawdowns.

Expressed mathematically, the superposition principle means that the cumulative drawdown $(H - h)$ at a given point as a result of n wells pumping from a confined aquifer is the sum of the drawdown contribution from each well.

$$(H - h) = \sum_{i=1}^{n} (H - h)_i \tag{7.16}$$

Established mathematical expressions for the drawdown from an individual well can be applied to equation (7.16) to estimate the drawdown at a given point. For example, using the method of Theis (1935) in a homogeneous and

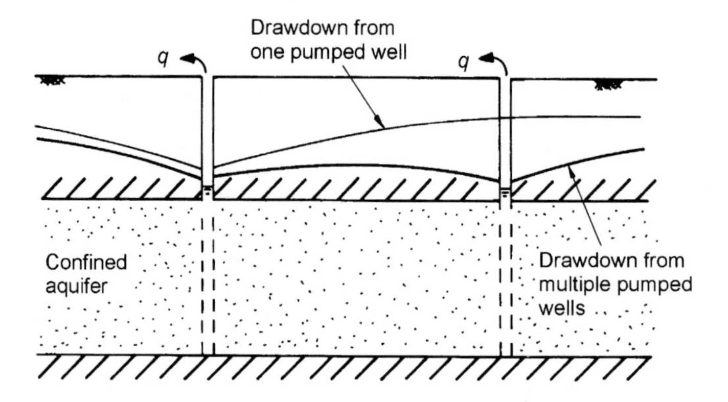

Figure 7.11 Superposition of drawdown from multiple wells.

isotropic confined aquifer of permeability k, thickness D and storage coefficient S, the cumulative drawdown from n fully penetrating wells, each pumped at a constant rate q_i, at time t after pumping commenced is:

$$(H-h)=\sum_{i=1}^{n}\frac{q_i}{4\pi kD}W(u_i) \tag{7.17}$$

where $W(u)$ is the Theis well function, values of which are tabulated in Kruseman and De Ridder (1990), $u=(r^2S)/(4kDt)$ and r is the distance from each well to the point under consideration. For values of u less than about 0.05 the simplification of Cooper and Jacob (1946) can be applied, giving:

$$(H-h)=\sum_{i=1}^{n}\frac{q_i}{4\pi kD}\left\{-0.5772-\ln\left[\frac{r_i^2S}{4kDt}\right]\right\} \tag{7.18}$$

In many aquifers the condition of $u<0.05$ is satisfied after only a few hours pumping, which means that equation (7.18) can generally be used for the analysis of groundwater control systems in confined aquifers.

Knowing the target drawdown $(H-h)$ in the excavation area, these equations can be solved to determine the number, location and yield of wells necessary to achieve the required drawdown. This also allows the total discharge flow rate (the sum of flow from all the wells) to be determined. This method is most suitable for systems of relatively widely-spaced wells. It is mainly used for deep wells, and occasionally for ejector systems; it is rarely used for wellpoint systems.

The following points should be considered when applying the method:

i The method has been reliably applied to estimation of drawdown within the area of excavation, away from the pumped wells themselves. Estimating the cumulative drawdown inside each well is more difficult because well losses may not be accurately known. If large well losses occur, the method is less reliable because the drawdown contribution becomes uncertain.

ii Application of the method requires that the aquifer parameters and well yields be estimated. In practice, the most reliable way to obtain suitable estimates is from analysis of a pumping test. If pumping test data are not available the estimated cumulative drawdowns should be treated with caution, unless there is a high degree of confidence in the parameter values used in calculations. If a pumping test has been carried out the graphical cumulative drawdown method (described in the subsequent section) may be a more appropriate method of analysis.

iii It may be possible to obtain the required drawdowns in the proposed excavation using a few wells pumped at high flow rates, or a larger number of wells of lower yield. Similarly, varying the well locations around (or within) the excavation may produce significantly different drawdowns in the area of interest. In years gone by investigating the effect of the various options was a tedious process. However, since the advent of widely-available personal computers it is possible to write routines or macros for spreadsheet programmes to evaluate equation (7.18), allowing many options to be rapidly considered. When evaluating the various options, it is vital that realistic well yields are used (see Section 7.5), otherwise too many or too few wells will be specified.

iv Equations (7.17) and (7.18) include a term for the time since pumping began, so each cumulative drawdown calculation is for a discrete time t. The time used in calculation will depend on the construction programme. If the programme shows that a two week period is available for drawdown (between installation of the dewatering system and commencement of excavation below original groundwater level) then that case should be analysed. However, in reality there may be problems with installation of a few of the wells and pumps so all n wells will not be pumping for the full two week period. It may be prudent to design on the basis of obtaining the target drawdown in a rather shorter time.

v The assumptions inherent in equations (7.17) and (7.18) (isotropic confined aquifer, fully penetrating wells, constant flow rate from each well) obviously will not apply in all cases. Provided that the basic aquifer conditions are confined or leaky it may be possible to use equation (7.17) for other conditions by substituting an alternative expression in place of the Theis well function $W(u)$. Kruseman and De Ridder (1990) give well functions for a number of cases, including leaky aquifers, anisotropic permeability, partially penetrating wells and variable pumping rates.

The cumulative drawdown method assumes that individual wells do not interfere significantly with each other's yield. For wells at wide spacings (greater than around 20 m) in confined aquifers (where the aquifer thickness does not change with drawdown), interference is usually low. In such cases the observed drawdowns are likely to be close to those predicted directly from the cumulative drawdown method. However, in general observed drawdowns will be slightly less than predicted. It is not unusual for observed drawdowns to be between 80 and 95 per cent of the calculated values, when applied using reliable parameters derived from pumping tests. To allow for this, the total well yield (or the number of wells to be installed) should be increased (by dividing by an empirical superposition factor J of 0.8–0.95). For example the total system flow rate Q is determined from the sum of the individual flow rates q_i from n wells.

$$Q=\frac{1}{J}\sum_{i=1}^{n} q_i \tag{7.19}$$

The cumulative drawdown method is invalid in unconfined aquifers (or confined aquifers where the drawdown is so large that local unconfined conditions develop). This is because the saturated thickness decreases at drawdown increases, making each additional well less effective compared with the initial wells. Although the method is theoretically invalid in unconfined conditions, where drawdowns are small (less than 20 per cent of the initial saturated aquifer thickness) the method has been successfully applied using an empirical superposition factor J of 0.8–0.95. For greater drawdowns in unconfined aquifers the cumulative drawdown method has been applied using empirical superposition factors of 0.6–0.8.

7.4.3 Cumulative drawdown analysis – graphical method

If distance–drawdown data are available describing the aquifer response to the pumping of a single well, a graphical cumulative drawdown method can be used. This approach is based on the Cooper–Jacob straight line method of pumping test analysis (see Section 6.6) which uses equation (7.18) expressed as:

$$(H-h)=\sum_{i=1}^{n} \frac{q_i}{2\pi kD}\ln\left(\frac{R_0}{r_i}\right) \tag{7.20}$$

where all terms are as described previously, apart from R_0 which is the radius of influence at time t. The equation is evaluated graphically, and is used to obtain the total drawdown *(H−h)* at the selected location, resulting from a given array of wells, without the need to evaluate the aquifer parameters.

The method is described in detail by Preene and Roberts (1994), and involves the following steps:

1 Determine the target drawdown level in critical points of the excavation. Typically, critical points where drawdown is checked include the centre and corners of the excavation. Normally the target drawdown is a short distance (0.5–1 m) below excavation formation level.

2 From the pumping test data, construct a drawdown–distance plot on semi-logarithmic axes. Drawdowns recorded in observation wells at a given time after pumping commenced are plotted, and a best straight line drawn through the data (Fig. 7.12(a)). For short duration pumping tests the data used are normally from the end of the test. The drawdown in the pumped well is normally ignored as it may be affected by well losses.
3 Convert each drawdown data point to specific drawdown by dividing by the discharge flow rate recorded during the test. A straight line is then drawn through the observation well data to obtain the design

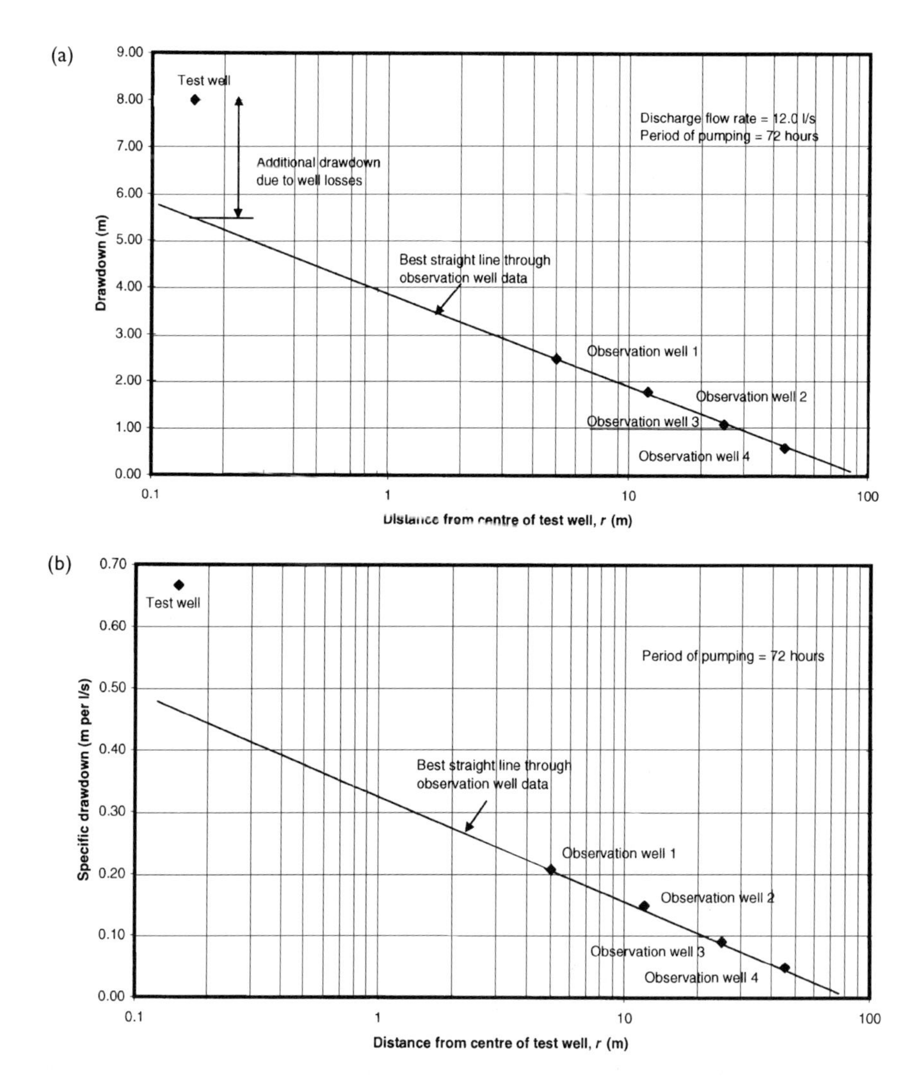

Figure 7.12 Cumulative drawdown analysis: graphical method. (a) Distance–drawdown plot, (b) specific–drawdown plot.

specific drawdown plot (Fig. 7.12(b)); this plot shows the drawdown which results from a well pumped at a unit flow rate.

4 Draw a plan of the excavation and groundwater lowering system, marking on the well locations and the points where drawdown is to be checked. Measure and record the distances from each well to each drawdown checking point.
5 Estimate the yield of each well in the system. This may be based on the pumping test results (based on step–drawdown data) or may involve the guidelines of Section 7.5.
6 At each drawdown checking location, calculate the drawdown which will result from the assumed set of well locations and yields. This is done using the specific drawdown plot (Fig. 7.12(b)). The drawdown contribution $(H-h)_i$ from each well is calculated by reading the specific drawdown at the appropriate distance and then multiplying by the assumed well yield. The total calculated drawdown $(H-h)$ is the sum of the contribution from each well, multiplied by an empirical superposition factor J.

$$(H-h) = J\sum_{i=1}^{n}(H-h)_i \tag{7.21}$$

J is normally taken to be between 0.8 and 0.95 in confined aquifers. As with the theoretical cumulative drawdown method, the graphical method has been applied in unconfined aquifers where drawdowns are less than 20 per cent of the initial saturated aquifer thickness. An example calculation in a confined aquifer (from Preene and Roberts 1994) is shown in Fig. 7.13; in that case J was back-calculated to be 0.92.
7 The calculated drawdown at each checking location is compared with the target drawdowns from step 1. If the drawdown is insufficient, the calculation is repeated after having either: changed well locations; increased the number of wells; or increased individual well yields. It is vital that the well yields assumed are achievable in the field. If the assumed yields are too large the system will not achieve its target drawdowns.

While the graphical method is most commonly used where site investigation pumping test data are available, the technique can also be used with the observational method. In this approach one of the first wells in the groundwater lowering system is pumped on its own in a crude form of pumping test. Drawdowns are observed in the other dewatering wells (which are unpumped at that time); these data allow distance–drawdown plots to be produced. The cumulative drawdown calculations are then used to help with the decision making progress to decide whether to install additional wells. As each new well is installed and pumped, further drawdown data are collected and the predicted drawdowns compared with the actual. In these cases it is often found that, as each additional well becomes operational, the empirical superposition factor J reduces further, as interference between wells increases.

Estimation of drawdown at Well 8 due to pumping on four other wells

Well No.	Discharge flow rate per well (l/s)	Distance to Well 8 (m)	Specific drawdown (m per l/s)	Calculated drawdown (m)
1	8.5	82	0.079	0.67
2	8.5	100	0.072	0.60
6	11.0	50	0.082	0.91
7	11.0	20	0.103	1.13
Total flow rate	**39.0**		**Total drawdown at Well 8**	**3.31**

Actual drawdown recorded at Well 8 after 44 hours = 3.06 m

Therefore, drawdown achieved is 3.06/3.31 = 92 per cent of calculated cumulative drawdown

Figure 7.13 Case history of cumulative drawdown calculation (data from Preene and Roberts 1994, with permission).

7.4.4 Storage release and uprating of pumping capacity

The steady-state discharge flow rates calculated using the methods presented above assume that 'dewatered' conditions have developed, and that a zone of influence of drawdown exists in the aquifer around the groundwater lowering system. To reach this condition, water must be released from storage in the aquifer within the zone of influence (see Section 3.4). This means that, during the initial period of pumping, before the steady-state is approached, an additional volume of water must be pumped.

In confined aquifers the quantity of water from storage release is small and is only significant during the first few hours of pumping. Its effect on the necessary pumping capacity is often neglected. In contrast, in unconfined aquifers the water from storage may be significant, and may persist for several weeks or more, dependent on the aquifer permeability and the pumping rate. Powers (1992) suggests that storage release is likely to be a major issue in permeable unconfined aquifers where the proposed discharge rate is more than 60–70 l/s.

Storage release means that either:

i A system designed with a capacity equal to the steady-state flow rate will take longer than anticipated to achieve the target drawdown or;
ii the design system flow rate should be increased above the steady-state estimate to deal with water from storage and to ensure that drawdown is achieved within a reasonable time period.

The release of water from storage has the same effect as reducing the distance of influence used in calculation. If the 'long-term' distance of influence is used in design, drawdown may only be achieved slowly. Alternatively, if the designer uses a 'short-term' distance of influence in design, drawdown may be achieved rapidly, but the system may be over-designed in the long term. This is only likely to be an issue in high permeability unconfined aquifers.

If the distance of influence used in design is taken from analysis of short-term pumping test data (see Section 6.6) this will include storage release, and calculated flow rates are likely to achieve drawdowns fairly rapidly. This also applies to systems designed by the graphical cumulative drawdown method, where pumping test data are used directly.

If no pumping test data are available and the distance of influence is estimated from empirical formulae such as equations (7.13)–(7.15), then judgement must be used where distances of influence of several hundred metres are predicted in high permeability soils. Systems with pumping capacities designed on that basis will be able to cope with steady-state inflows, once the zone of influence has developed. However, they may be overwhelmed by storage release during the early stages of pumping, and drawdown may take a long time to be achieved, leading to delays in the construction programme. Designers sometimes overcome this problem by using rather smaller distances of influence which predict higher flow rates; this helps ensure that drawdown is achieved in reasonable time.

If the distance of influence used in the design has been estimated from equations (7.13)–(7.15), the following equations can be used to crudely estimate the time t, which would be required for this distance of influence to develop. The equations are for radial flow (from Cooper and Jacob 1946)

$$R_0 = \sqrt{\frac{2.25kDt}{S}} \tag{7.22}$$

and for plane flow (from Powrie and Preene 1994)

$$L_0 = \sqrt{\frac{12kDt}{S}} \tag{7.23}$$

where D is the aquifer thickness, k is the permeability and S is the storage coefficient. Strictly, equations (7.22) and (7.23) are only valid in confined aquifers, but can be used in unconfined aquifers where the drawdown is not a large proportion of the original saturated thickness. Typical values of specific yield (approximately equal to S in an unconfined aquifer) are given in Table 3.2.

If the design distance of influence will take a long time (more than a few days) to develop, it is possible that the system should be designed assuming a smaller R_0 or L_0 – (values typically used in high permeability soils are in the range 200–500 m). Changing the design in this way would increase the required pumping capacity of the system. This would allow the water released from storage to be handled by the system, and would result in the drawdown within the excavation area being achieved within a reasonable length of time.

7.4.5 Other methods

Occasionally other methods are used to estimate steady-state flow rates.

Flow net analyses are sometimes used to model flow patterns not amenable to simplification as equivalent wells or slots. A flow net is a graphical representation of a given two-dimensional groundwater flow problem and its associated boundary conditions. Flow nets are one of the common forms of output of numerical groundwater models (see Section 7.7). However, as described by Cedergren (1989), hand sketching of flow nets can be used to obtain solutions to certain flow problems, considered either in plan or cross-section, for isotropic or anisotropic conditions. Typical problems where flow nets are used include seepage into excavations or cofferdams where the presence of partial cut-off walls alters the groundwater flow paths (see Williams and Waite 1993).

Very rarely, physical models or electrical resistance or resistance–capacitance analogues (see Rushton and Redshaw 1979) are used to analyse groundwater flows. In the past they were used more commonly to analyse complex problems, but in recent times advances in numerical modelling methods have made these techniques largely obsolete. A rare recent application of the use of electrical analogues is described by Knight *et al.* (1996).

7.5 Specification of well yield and spacing

Having determined the total required pumping rate, the next step is to determine the yield and spacing of wells (be they wellpoints, deep wells or ejectors).

7.5.1 Well yield

Each well must be able to yield sufficient water so that all the wells in concert can achieve the required flow rate, and hence the required drawdown.

Water enters a well where the well screen penetrates below the lowered water table in an unconfined aquifer, or where it penetrates a saturated confined aquifer. This depth of penetration is known as the 'wetted screen length' and is an important factor in the selection of well depths to achieve adequate yield (Fig. 7.14).

In theoretical terms, the yield q into a well can be described (from Darcy's law) by:

$$q = 2\pi r l_w k i \qquad (7.24)$$

where, r = the radius of the well borehole (not the diameter of the well screen) – this assumes the well filter media are of significantly greater permeability than the aquifer; l_w = wetted screen length below the lowered

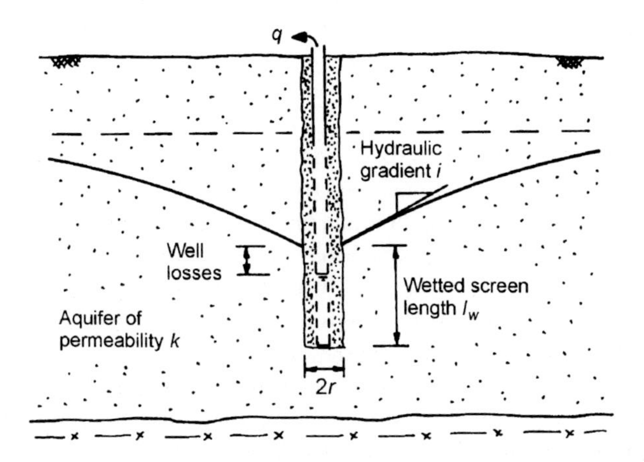

Figure 7.14 Wetted screen length of wells.

groundwater level; k = aquifer permeability; i = hydraulic gradient at entry to the well.

The designer has no control over aquifer permeability, but can vary the length and diameter of the well within limits determined by the geology of the site and the availability and cost of well drilling equipment. Experience suggests that any well will have a maximum well yield, beyond which the well will 'run dry', that is, the water level in the well will reach the pump inlet or suction level, preventing further increases in flow rate.

In 1928 Sichardt published a paper entitled *Drainage capacity of well-points and its relation to the lowering of the groundwater level*, which examined the yield of pumped wells based on the records of numerous groundwater lowering projects. He determined empirically that the maximum well yield is limited by a maximum hydraulic gradient i_{max} which can be generated in the aquifer at the face of a well. The Sichardt limiting gradient is generally taken as relating the aquifer permeability k (in m/s) to the maximum hydraulic gradient at the face of the well by:

$$i_{max} = \frac{1}{15\sqrt{k}} \tag{7.25}$$

This is the theoretical maximum amount of water that a *well* can yield – in very high permeability aquifers the potential well yield may be so large that the actual flow rate is controlled by the *pump* rather than the well.

The Sichardt gradient is probably reasonable when used to estimate the maximum well yield in a relatively high permeability aquifer (k greater than about 1×10^{-4} m/s). However, work by Preene and Powrie (1993), who analysed well yields in a large number of dewatering systems in fine-grained soils, has suggested that equation (7.25) may not be appropriate for lower permeability aquifers. The Sichardt gradient may overestimate hydraulic gradients (and hence well yields) in soils of permeability less than about 1×10^{-4} m/s. The work of Preene and Powrie indicated that in lower permeability soils hydraulic gradients were generally less than ten, and that an average of six was not unreasonable. These two approaches are combined in Fig. 7.15 to give the maximum well yield per unit wetted screen length for wells of various diameters of bored hole. Yield per unit length for other diameters can be calculated using equation (7.24) and a limiting hydraulic gradient appropriate to the aquifer permeability.

Figure 7.15 can be used to design deep well or ejector systems. Wellpoint systems (or systems of ejectors with short screens at very close spacings) are analysed rather differently, as will be described later.

For wells at wide spacings in an aquifer that extends to some depth below the excavation, once a well diameter is assumed, Fig. 7.15 can be used to determine the minimum wetted screen length (below the lowered water level) needed to obtain the total flow rate from the deep well system. This would then allow the number of wells and wetted depth per well to be estimated. For example, if 120 m of wetted screen length was estimated, this could be equivalent to twelve wells with 10 m wetted screen each, or eight

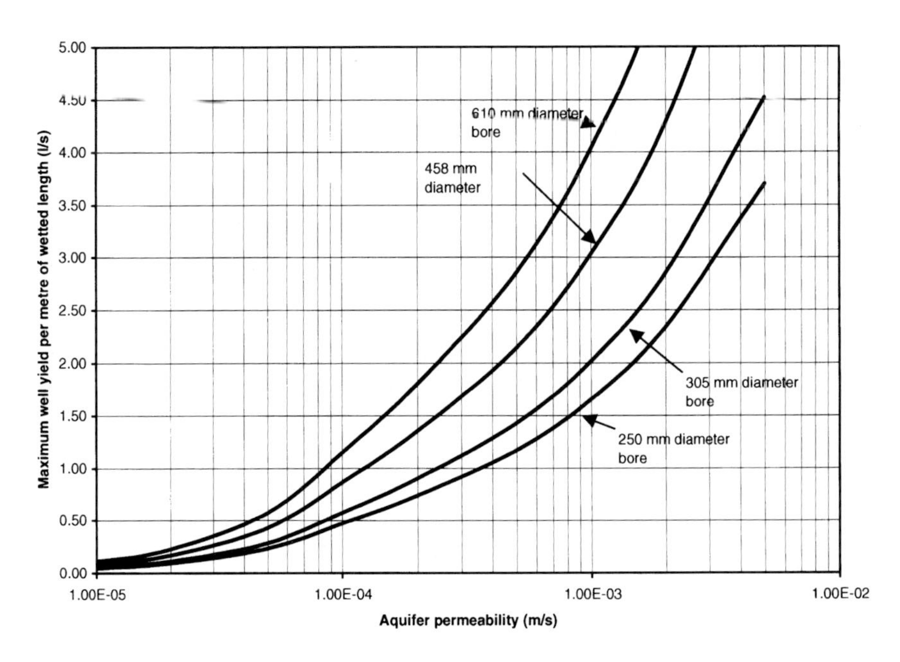

Figure 7.15 Maximum yield per unit wetted length of wells.

wells with 15 m wetted screen each, and so forth. Obviously, a check then needs to be made that the originally assumed well diameter is large enough to accommodate well screens and pumping capacity to produce the design yield. If necessary the well diameter must be increased. Guidance on the size of deep wells and ejector wells required to produce given discharge rates per well are given in Chapters 10 and 11 respectively.

Estimation of well yield is a point in design where judgement and experience can be vital, so the designer should consider the following:

1 The wetted screen length l_w will be rather less than the penetration of the well below the 'general' lowered water level. This is because of the additional drawdown around each well due to each individual cone of depression (Fig. 7.14). The difference between the wetted depth and the excavation drawdown level can be estimated from pumping test results. If no pumping test data are available, l_w must be estimated based on engineering judgement.
2 The yields calculated by the methods described above are theoretical maximums and may not be achieved in practice. Well yields may be reduced by use of inappropriate filter material or poor development. The method of drilling will also influence yield, with jetted boreholes generally being more efficient than rotary or cable tool percussion drilled holes. Experience also suggests that two competent drillers using the same methods and equipment on holes a few metres apart can produce wells with wildly differing yields. There is no conclusive explanation for this phenomenon, but it is likely to be related to the precise way each driller uses the bits, casing and drilling fluids and how that affects a thin layer of aquifer just outside the well.
3 There are certain rules of thumb about well depth, built up over many years of practice, and results of calculations should be compared with these to check for gross errors. If the aquifer extends for some depth below the base of the excavation being dewatered, the wells should be between 1.5 and 2 times the depth of the excavation. Wells significantly shallower than this are unlikely to be effective unless they are at very close spacing (analogous to a wellpoint system).
4 If the geology does not consist of one aquifer that extends to great depth, this may affect well depth and will restrict the flexibility of the designer in specifying l_w. If the aquifer is relatively thin screen lengths will be limited and a greater number of wells will be required. If there is a deeper permeable stratum that, if pumped, could act to underdrain the soils above, it may be worth deepening the wells (beyond the minimum required) in order to intercept the deep layer (see Section 7.3).
5 It is always prudent to allow for a few extra wells in the system over and above the theoretically calculated number. Typically for small systems at least one extra well is provided, or for larger systems the number of

wells may be increased by around 20 per cent. This allows for some margin for error in design or ground conditions but also means that the system will be able to achieve the desired drawdown if one or two wells are non-operational due to maintenance or pump failure.

In a similar manner, the potential yield of a wellpoint or an ejector installed in a jetted hole can be assessed using equation (7.24) and a limiting hydraulic gradient appropriate to the aquifer permeability. Figure 7.16 shows the theoretical maximum yield of 0.7 m long wellpoint screens. Yield from longer or shorter screens can be estimated pro-rata. The figure shows that even in very high permeability soils a conventional wellpoint is unlikely to yield more than around 1 l/s (special installations of larger diameter and longer screen length may yield more, but such applications are rare). A maximum possible wellpoint yield of 1 l/s is a useful practical figure for the designer to remember.

Figure 7.16 is based on a disposable wellpoint installed in a 200 mm diameter sand filter, and a self-jetting wellpoint installed in a jetted hole of 100–150 mm diameter. When dealing with jetted wells the diameter of the jetted hole may be uncertain, requiring further judgement to be used. In theory, a wellpoint system could be installed in an analogous manner to a deep well system, by determining the necessary total wetted screen length, the corresponding number of wellpoints and then adding additional wells as a

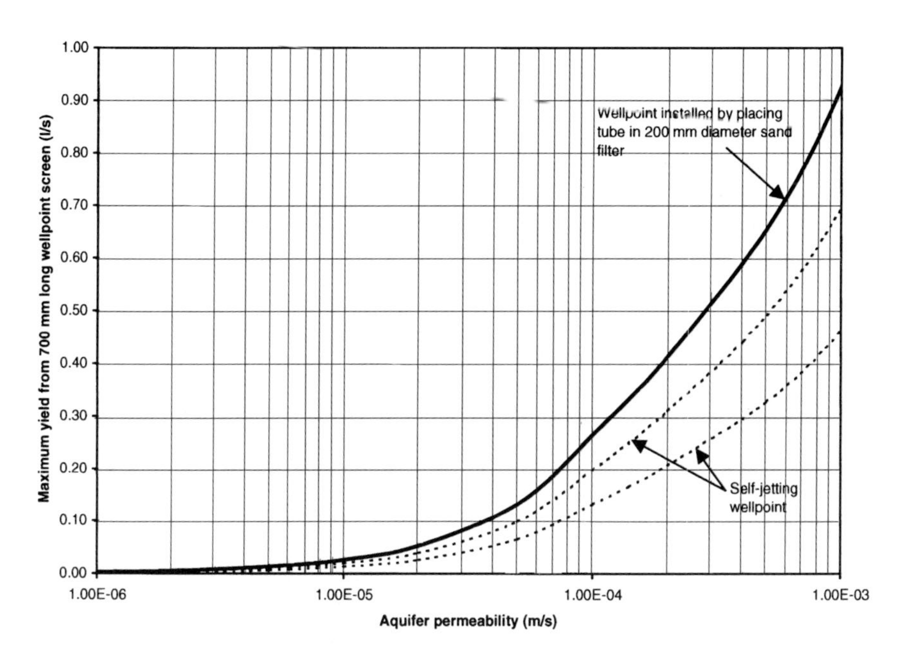

Figure 7.16 Maximum yield of wellpoints.

contingency. In practice, this process is rarely carried out. Wellpoint equipment is almost always used in one of a limited number of standard spacings. Methods of selecting wellpoint spacings are described below.

The estimated well yield can be used to select the capacity of the pumping equipment. For deep wells and ejectors the pumping equipment is located in each well, and is sized directly from the well yield. For wellpoint systems one pump acts on many wellpoints in concert. The designer has the choice of a few larger pumps, or a greater number of smaller units (see Section 9.6). When estimating the required pump capacity from calculated steady-state flow rate, an additional allowance must be made for the greater flow rate from storage release during the initial period of pumping (see Section 7.4).

7.5.2 Number of wells and well spacing

As described above, for deep well and ejector systems the number of wells required can be determined from the total discharge flow rate, divided by the predicted well yield, with some additional wells added as a contingency. This will then allow the average well spacing (the distance between adjacent wells) to be determined. Because well spacings of most groundwater lowering systems fall within relatively narrow ranges, comparison of the 'design' spacing with typical values can be a useful way of verifying a design.

For all systems the well spacing chosen will influence the time required to lower groundwater levels to the target drawdown. In general the closer the well spacing the quicker drawdown will be achieved. Because time is often as important a factor as cost in construction, groundwater lowering systems are often installed with wells at rather closer spacing than is theoretically necessary, to ensure drawdown is achieved within a few days or weeks.

Typical spacing of deep wells are in the range 5–100 m, although the great majority of systems use spacings in the range 10–60 m between wells.

In aquifers of moderate to high permeability (where maximum well yields are relatively high), the designer has the flexibility of choosing a larger number of closely-spaced low yield wells, or a smaller number of widely-spaced high yield wells. When potential well yields are large, high capacity pumps may not be available, and the output from each well may be controlled by the pump performance; this should be considered when estimating well numbers and spacings. If the aquifer extends to great depth it may be possible to use relatively few, very high capacity wells of very great depth at spacings of several hundred metres. This approach would require extensive pump test data, backed up by numerical modelling.

In low permeability soils the well yield will be relatively low, and the option of a small number of widely-spaced high yield wells is not available to the designer. To achieve the total discharge flow rate the wells will be at close spacings. If a well spacing of less than around 5–10 m is suggested by the design, or if the well yield is less than around 0.7 l/s, an ejector system could be considered as an alternative to deep wells.

Ejector systems can be used in two ways.

i They can be used in soils of moderate permeability as an alternative to low yield deep wells, or as 'deep wellpoints' to achieve drawdowns in excess of 5–6 m. When used in this way, well spacings are similar to those used for deep wells (5–10 m) or wellpoints (1.5–3 m).

ii In soils of low permeability they can be used as a vacuum-assisted pore water pressure control method. Because of the limited area influenced by each well, ejectors tend to be installed at close spacings (1.5–5 m) and operate at only a fraction of their pumping capacity.

In contrast to deep well and ejector systems, wellpoint spacings are rarely selected on a yield per well basis. The number of wellpoints is determined by first selecting the wellpoint spacing from a fairly narrow range; the number of wellpoints is then determined from the length of the line or ring of wellpoints. Typical wellpoint spacings are in the range 0.5–3.0 m (see Table 9.1 for spacings categorized by soil type). Closer spacings tend to reduce the time to achieve drawdown compared to wider spacings. Wellpoint spacing is controlled by different factors in soils of high and low permeability.

(a) In high permeability soils (such as coarse gravels) the total discharge flow rate will be large. Each wellpoint may be operating at high yields. A rule of thumb is that a conventional wellpoint cannot yield more than about 1 l/s, and the spacing is sometimes based on reducing the average wellpoint yield below 1 l/s. Wellpoints tend to be installed at close spacings in high permeability soils. If a spacing in the range 0.5–1.0 m is suggested by design calculations, wellpoint dewatering may not be the most appropriate method. High capacity deep wells at close spacings might be considered as an alternative.

(b) In low permeability soils (such as silty sands) the total flow rate will be low. Consideration of the maximum wellpoint yield may suggest that fairly wide (5–10 m) spacings may be possible. In reality, because of the limited area influenced by each wellpoint, such a system is likely to perform poorly and a very long time may be required to achieve drawdown. The solution is to install wellpoints at closer spacings of 1.5–3.0 m.

7.6 Other considerations

Although the prime concern of most groundwater lowering designs is to estimate the steady-state flow rate and the corresponding number and yield of wells that will be required, other issues sometimes need to be addressed during design.

7.6.1 Estimation of time-dependent drawdown distribution around well

As pumping continues, the zone of influence around a dewatering system will increase with time. This means that the area affected by drawdown will expand, initially rapidly, and then progressively more slowly as time passes. This is clearly a complex problem, and complete solution probably requires a suitably calibrated numerical model. However, there are some simpler methods that can be used to estimate the drawdown pattern around an excavation at a given time. These methods are outlined in this section.

In dewatering design, the time-dependent drawdown distribution may be required for the following reasons:

(a) To determine how far the zone of significant settlements will extend during the period of pumping. This is useful if the impact of potential side effects (see Chapter 13) is being assessed. This normally involves producing a plot of the drawdown versus distance at time t after pumping began. Successive plots at greater values of t can show the development of the zone of influence. Each plot of drawdown versus distance at a given time is known as an isochrone.

(b) To determine the time required to achieve the target drawdown in a particular part of the excavation. This is normally only an issue in low permeability soils where it may influence the construction programme. In moderate to high permeability soils experience has shown that most appropriately designed systems should achieve the target drawdown within 1–10 days pumping.

In many cases these calculations are unnecessary because the time to achieve drawdown and the risk of side effects is not a major concern. In other cases, the methods presented here can be used to *approximately* determine the time-dependent drawdown distribution. All these analyses assume that the aquifer is homogenous and that no recharge or barrier boundaries are present within the zone of influence (i.e. all pumped water is derived purely from storage release). Numerical modelling should be considered where aquifer boundaries are likely to be present.

If the groundwater lowering system consists of relatively widely spaced wells, then cumulative drawdown methods can be used. The theoretical approach (equation (7.16)) can allow the predicted drawdown at a selected point to be plotted against time. Alternatively, the distance–drawdown profile (or isochrone) can be determined at a given time after pumping commences. By repeating the calculation for different times a series of isochrones, showing the drawdown pattern at each time interval can be produced.

If the system consists of closely spaced wells in a regular pattern, the system can be analysed as an equivalent well or slot. Analyses which assume constant drawdown in the well or slot can then be used. The constant drawdown assumption is a reasonable one for most dewatering systems (apart from during the first few hours of pumping), and has given acceptably accurate results in practice.

For horizontal plane flow to an equivalent slot in a confined aquifer, the drawdown curve can be expressed as a parabola (Powrie and Preene 1994); shown in dimensionless form in Fig. 7.17. The drawdown s at distance x from the slot can be determined from this figure provided the drawdown s_w at the slot and the distance of influence L_0 are known. The drawdown at the slot is normally taken as the same as the drawdown inside the excavation (not the drawdown in individual wells), and L_0 is estimated from:

$$L_0 = \sqrt{\frac{12kDt}{S}} \quad (7.26)$$

where D is the aquifer thickness, k is the permeability, S is the storage coefficient and t is the time since pumping began.

For horizontal radial flow to an equivalent well in a confined aquifer, a time-dependent solution was developed by Rao (1973); this has been plotted in Fig. 7.18. The drawdown s at radius r from the centre of the well can be determined from this figure provided the drawdown s_w at the slot and the time factor T_r are known.

$$T_r = \frac{kDt}{Sr_e^2} \quad (7.27)$$

where r_e is the radius of the equivalent well, and all other terms are as defined previously.

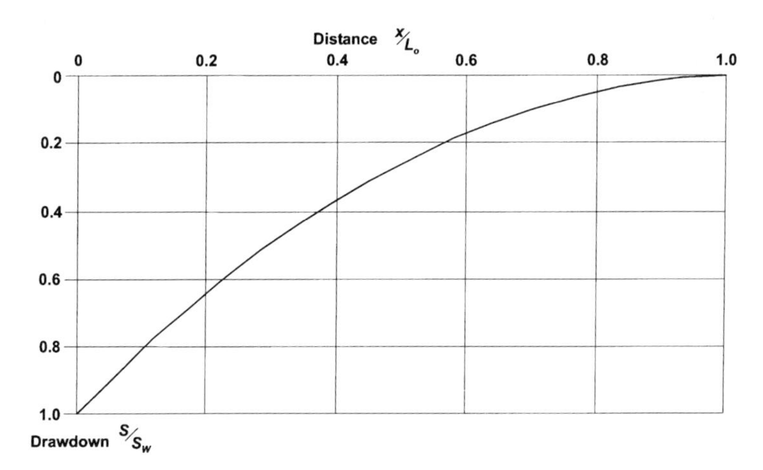

Figure 7.17 Dimensionless drawdown curve for horizontal plane flow to an equivalent slot.

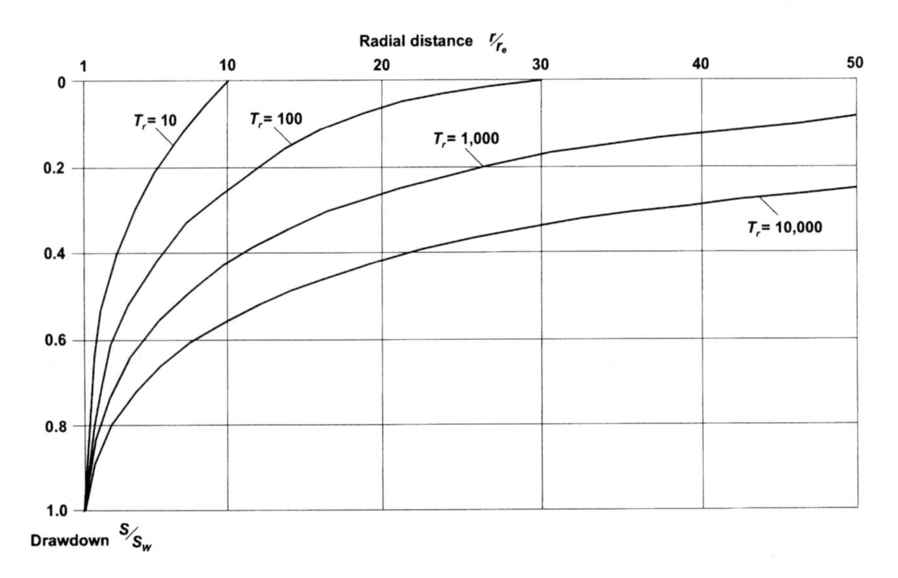

Figure 7.18 Dimensionless drawdown curve for horizontal plane flow to an equivalent well (after Powrie and Preene 1994).

These methods are only theoretically valid in confined aquifers, but can be used without significant error in unconfined aquifers where the drawdown is not a large proportion of the original saturated thickness.

Equations (7.26) and (7.27) use the storage coefficient S, which is appropriate for soils of moderate to high permeability. These methods can also be used where pore water pressure control systems are employed in low permeability soils, but the drainage characteristics of the soils may be expressed in terms of c_v, the coefficient of consolidation.

$$c_v = \frac{kE'_0}{\gamma_w} = \frac{kD}{S} \tag{7.28}$$

where E'_0 is the stiffness of the soil in one-dimensional compression and γ_w is the unit weight of water. For applications in low permeability soil equation (7.28) can be substituted into equations (7.26) and (7.27)

7.6.2 Estimation of groundwater lowering induced settlements

Ground settlements are an unavoidable consequence of the effective stress increases that result from groundwater lowering. In most cases the settlements are so small that there is little risk of damage or distortion to nearby buildings. However, if compressible soils (such as peat or normally consolidated alluvial clays and silt) are present, it is possible that damaging settlements may occur.

If the conceptual model has identified the presence of significant thicknesses of compressible strata within the zone of influence, it will be necessary to consider the magnitude of ground settlements which may result. Methods for estimating the settlements that will result from a given drawdown are outlined in Section 13.1. The drawdown which may be expected beneath individual structures can be estimated from the drawdown distributions given earlier in this section.

It is important to remember the aim of settlement assessments – to assess the risk of damage to structures and services, rather than to try and predict settlements to the nearest millimetre. This latter aim would be difficult to achieve since the stiffness and consolidation parameters of compressible soils are rarely known with sufficient accuracy. Predicted degrees of damage, corresponding to various levels of settlement, are given in Section 13.1. That section also discusses methods used to mitigate or avoid settlement damage resulting from groundwater lowering.

7.7 Numerical modelling

Numerical modelling of groundwater flow problems has been carried out for several decades, but it is only since the 1980s and 1990s that advances in personal computer (PC) technology has made this approach viable for a wide range of groundwater lowering problems.

PCs can be used in design in two ways:

i By use of spreadsheet programs to evaluate design equations that might previously have been carried out by hand. It is fairly easy to write routines for a spreadsheet to allow repeated sets of calculations to be performed as part of sensitivity or parametric analyses. Use of the PC dramatically speeds up this process compared with hand calculation.
ii By use of a groundwater numerical modelling package to solve complex groundwater flow problems that would not normally be amenable to solution by other means.

This section will concentrate on the use of groundwater numerical models applied to dewatering problems. It will describe the general approach that should be adopted, and will not go into detail of the modelling process, which is described by Anderson and Woessner (1992). While numerical modelling packages are becoming easier to apply, it is vital that anyone contemplating their use understands the theoretical basis and limitations of the program in question. In complex or unusual situations help should be obtained from an experienced groundwater modeller.

Essentially, a numerical model breaks down the overall problem, its geometry and boundary conditions, into a number of discrete smaller problems that can be solved individually. An iterative process is often carried out, whereby the solution to each smaller problem is adjusted until there is acceptable agreement at the boundaries between the smaller problems. It is sometimes stated that numerical models produce 'approximate' solutions. This is true in the mathematical sense, because there will be a small difference between an analytical solution and the numerical results, but in an engineering sense the numerical output is a pretty accurate reflection of the input data and the groundwater model that has been formulated.

The key point is that the numerical package is only following instructions given to it by the user. If there are errors in the input data or, more importantly, if the conceptual model (on which the groundwater model is based) is unrealistic, gross errors may result. It is a cliché but the phrase 'garbage in, garbage out' – meaning that the results can only be as good as the input and instructions – is very true for groundwater modelling. The conceptual model (see Section 7.2) is the critical starting point for any modelling exercise. If the conceptual model is not a good match for actual conditions then any output may be of questionable value.

It is important to select a numerical modelling package appropriate to the problem in hand. Some packages were originally developed for use in water resources modelling of large areas of aquifers for regional-scale studies. These packages can be useful for large dewatering works in highly permeable aquifers with large distances of influence, but may be less applicable for smaller-scale seepage problems. Another group of packages were developed for geotechnical problems such as seepage beneath cofferdams or through earth embankments; these may be appropriate for small-scale problems.

The modelling package used must be capable of solving the type of problem. Most packages can solve steady-state problems, but not all are designed for transient time-dependent seepage; this must be considered when selecting a modelling package. Most groundwater flow problems are three-dimensional to some degree, but for some cases the conceptual model may be able to simplify conditions to two-dimensional flow without significant error; some packages can model three-dimensional flow, but many are limited to two dimensions only.

The numerical modelling process can be divided into a number of stages:

(a) Development of the conceptual model. Even though it does not involve touching a computer, this is probably the most important stage of the numerical modelling exercise. The conceptual model must quantify the geometry, aquifer parameters and boundary conditions which will be used to define the numerical model. If the conceptual model does not reflect actual conditions, the modelling results are unlikely to be realistic or useful.

(b) Selection of software and setting up of numerical model. Once the conceptual model is defined, software capable of modelling those conditions can be selected. The software is then used to create the groundwater model (the set of instructions defining the relevant geometry, properties and boundary conditions) for the problem. Any errors or omissions in the input data and instructions will have an effect on the results produced by the model.
(c) Verification and calibration. These activities are essential to allow any errors in the input data or model formulation to be identified, and for the user to develop some degree of confidence in the validity of the output. The aim of verification is to answer the question 'has the model done what we intended it to do?' To answer this question the input data must be scrutinized for errors, and the output of the model must be compared with known analytical solutions. It is unlikely that an analytical solution will be available for the whole model, but it may be possible to simplify all or part of the model and compare it with results calculated by, say, flow net or equivalent well methods. If errors are detected, these are corrected and verification repeated until acceptable agreement is obtained. Following verification, the model should be calibrated against field data such as observation well readings in various parts of the modelled area. Calibration is a trial and error procedure whereby the model parameters and boundary conditions are varied (within realistic ranges chosen from the conceptual model) until there is acceptable agreement between the field data and the model output.
(d) Prediction and refinement. A verified and calibrated model can be used to predict the results of interest (flow rates, drawdowns, settlements, etc). Parametric and sensitivity analyses can be carried out to assess the effect on results of different well arrays or aquifer conditions. For larger or longer-term projects it may be possible to refine predictions by further calibration against results from monitoring of the dewatering system – this can be used as part of the observational method.

There are a number of ways that numerical models can be used in the design of groundwater lowering systems. Perhaps the most obvious is to use the model, following a comprehensive site investigation, as a direct design tool to finalize the dewatering system. The conceptual model is developed from the site investigation results, and the model is run repeatedly, adding, removing or relocating wells and pumping capacity until the target drawdown is obtained at specified points, or other design requirements are satisfied.

There is another way that numerical modelling can be used, as an aid both to design and site investigation. If a groundwater model is created at an early stage of the project (perhaps using a conceptual model based on the site investigation desk study) it can be used to crudely model the effect of possible ground conditions and construction options. Output from the model may highlight particular issues to be addressed by the ground investigation. Similarly, the effect of changing the size of excavation, depth of cut-off walls and so on can be investigated; this may be useful information for designers. Any potential side effects of groundwater lowering can also be quantified. The model is then developed, recalibrated and refined as additional data is gathered, and continues to provide information to designers on the effect of various options. This approach has been successfully adopted on a number of larger projects, including those where the observational method was used.

7.8 Design examples

Some design examples are presented in Appendix 7A to illustrate the application of the methods presented in this chapter. For ease of reference the relevant equation and figure numbers are noted. The examples do not merely cover the numerical aspects of design, but also discuss some of the issues over which 'engineering judgement' must be exercised.

Design example 1: Ring of relatively closely-spaced deep wells in a confined aquifer. This case is modelled as an equivalent well using radial flow equations.

Design example 1a: This analyses the case of example 1 using the alternative method of using shape factors for flow to equivalent wells in confined aquifers.

Design example 2: A line of partially-penetrating wellpoints alongside a trench excavation in an unconfined aquifer. This case is analysed using as an equivalent slot under plane flow conditions. The effects of assuming different aquifer depths and of including the contribution from radial flow to the end of the slot are assessed.

Design example 3: Ring of widely-spaced deep wells around a large excavation in a confined aquifer. The cumulative drawdown method is used to design the system.

Appendix 7A: Design examples

Four design examples, based on the methods of Chapter 7 are presented here. For ease of reference the relevant equation and figure numbers from the main text are also given.

Design example 1

This example is an application of the equivalent well method, to analyse a system of fully penetrating deep wells used to lower the piezometric level in a confined aquifer beneath a rectangular excavation. A sensitivity analysis is carried out to assess the impact on the calculated flow rate of various possible values of permeability.

Conceptual model

A rectangular excavation is to be made to a depth of 9 m. The details of the conceptual model can be summarized as shown in Fig. 7A-1:

- Excavation dimensions are: 35 by 15 m in plan, 9 m depth to deepest part of excavation. The excavation is to have vertical sides supported by sheet-piles. Dewatering is required for six months.
- A confined aquifer, consisting of a medium sand, extends from 10 m depth to 19 m depth. The confining layer above the aquifer is a stiff clay. Maximum piezometric level in the aquifer is 1 m below ground level.
- No pumping test was carried out during site investigation, but particle size distributions (PSD) and falling head test data can be used to estimate permeability.
- No recharge boundaries are believed to exist. Flow is likely to be radial to the array of wells, and can be idealized as a distant circular source.
- No compressible strata are believed to exist. Groundwater lowering related settlement is not anticipated to be a problem.

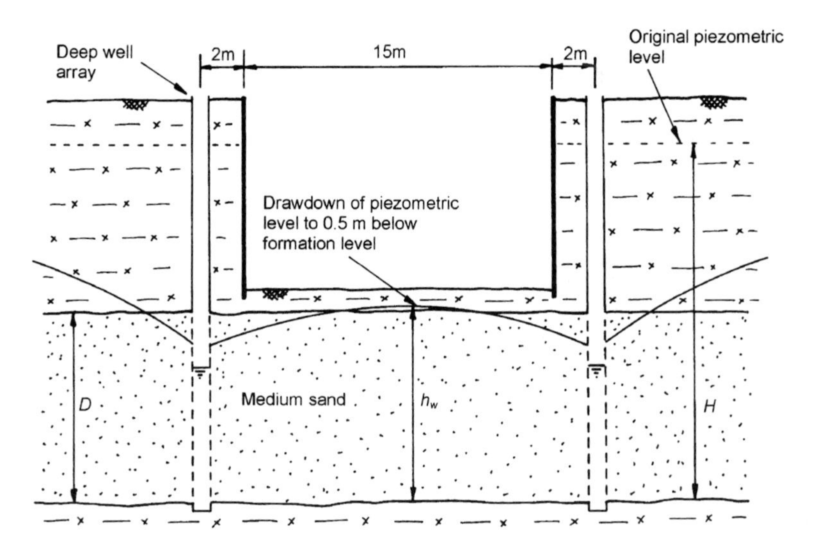

Figure 7A-1 Conceptual model for design example 1.

Selection of method

The presence of a confined aquifer at shallow depth beneath the excavation would result in a risk of base heave if the piezometric level is not lowered. The conservative case is to lower the piezometric level to 9.5 m below ground level (i.e. 0.5 m below formation level). This requires a drawdown of 8.5 m below original groundwater level.

Without assessing the available permeability data in detail, typical values of permeability given in Table 3.1 suggest the permeability k of a medium sand would be in the range 1×10^{-4}–5×10^{-4} m/s.

Inspection of Fig. 5.6 suggests that either deep wells or two-stage wellpoints would be suitable for this combination of drawdown and permeability. In this case the deep well method will be used, because the contractor wishes to excavate to full depth in one operation, and two-stage wellpointing requires a pause in excavation while the second stage is installed.

A system of relief wells (see Chapter 11) might have been considered. However, the anticipated permeability suggests the flow rate from the relief wells would have been too great to be handled by sump pumping without interfering with construction operations.

Estimation of steady-state discharge flow rate

The wells are to be installed in a regular pattern around the excavation. This geometry is amenable to solution as an equivalent well. If it is assumed that the wells will be fully penetrating, the total flow rate Q can be estimated from equation (7.3).

$$Q = \frac{2\pi k D(H - h_w)}{\ln\left[R_0/r_e\right]} \tag{7.3}$$

where

- k is the aquifer permeability. As described earlier, Table 3.1 suggests a likely permeability in the range 1×10^{-4}–5×10^{-4} m/s. The site investigation data include PSD data showing a 10 per cent particle size (D_{10}) of 0.1–0.3 mm. The sand is relatively uniform and Hazen's rule (equation (6.1)) can be used to estimate permeability as 1×10^{-4} to 9×10^{-4} m/s. These permeability estimates are broadly consistent with Table 3.1. The larger end of the range is rather greater than might be expected in a medium sand; the sample may have been affected by 'loss of fines' or may represent a coarser layer within the aquifer. Falling head tests in boreholes gave permeabilities in the range 1×10^{-8} to

1×10^{-6} m/s. Comparison with the soil description suggests these results are unrepresentatively low, and are likely to have been affected by silting up during the tests. The falling head test results are not used in subsequent assessments of permeability. It is probably reasonable to assume design values of permeability are between 1×10^{-4} and 9×10^{-4} m/s, and to carry out sensitivity analyses when estimating flow rate.

- D is the aquifer thickness: $D=19-10=9$ m.
- $(H-h_w)$ is the drawdown: $(H-h_w)=9.5-1=8.5$ m.
- r_e is the equivalent radius of the array of wells. Assuming that the wells are located 2 m outside the edge of the sheet-piles, the overall dimensions of the system will be 39 by 19 m. r_e can be estimated from either equation (7.1)

$$r_e=\frac{(a+b)}{\pi}=\frac{(39+19)}{\pi}=18.5\,\text{m} \tag{7.1}$$

or equation (7.2)

$$r_e=\sqrt{\frac{ab}{\pi}}=\sqrt{\frac{39\times19}{\pi}}=15.4\,\text{m} \tag{7.2}$$

Because r_e appears in a log term in equation (7.3), these two values will produce very similar estimates of flow rate, so a value of $r_e=18.5$ m will be used in subsequent calculations.

- R_0 is the radius of influence. In the absence of pumping test data, R_o can be estimated for equivalent wells from equation (7.14) (shown here for R_0, r_e and $(H-h_w)$ in metres and k in m/s)

$$R_0=r_e+3000(H-h_w)\sqrt{k} \tag{7.14}$$

The total flow rate Q can then be estimated from equation (7.3) by a sensitivity analysis, within the selected range of permeability. Using the units quoted above Q will be calculated in m^3/s; when commenting on results Q is normally quoted in l/s ($1\,m^3/s=1000$ l/s) to make the numbers easier to read and interpret.

k (m/s)	r_e (m)	R_0 (m)	Q (m^3/s)	Q (l/s)
1×10^{-4}	18.5	274	0.018	18
2×10^{-4}	18.5	379	0.032	32
5×10^{-4}	18.5	589	0.069	69
9×10^{-4}	18.5	784	0.12	120

Estimation of number of wells

For soils of permeability greater than 1×10^{-4}, q, the maximum yield of a well, can be estimated from equations (7.24) and (7.25), combined into the following equation (or alternatively taken from Figure 7.15):

$$q = \frac{2\pi r l_w \sqrt{k}}{15}$$

where l_w is the wetted screen length of wells and r is the radius of the well borehole. For each case of the sensitivity analysis the total wetted length (all wells in combination) can be estimated.

The number of wells and the corresponding well yield can then be estimated, once certain assumptions have been made about the dimensions of the well. In this case the diameter of the well borehole (not the diameter of the well screen) is taken to be 0.305 m. The wetted depth per well must also be assumed. The wells fully penetrate the aquifer and have a total screen length of 9 m. However, drawdown within the excavation is to 9.5 m below ground level, compared to the top of the aquifer at 10 m depth. The drawdown in the wells will be greater than the drawdown in the general excavation area, so the wetted length per well will be less than the aquifer thickness of 9 m. In this case the wetted length per well will be assumed to be 6 m (two-thirds of the aquifer thickness).

The number of wells is the total wetted screen length divided by the wetted screen length per well, with the answer rounded up to the next whole number. The nominal well spacing (assuming that the wells are evenly spaced) is then determined from the plan dimensions of the well system (39 by 19 m). The calculations are shown below.

k (m/s)	Q (m^3/s)	*Total l_w for all wells* (m)	*Number of wells and yield*	*Nominal well spacing* (m)
1×10^{-4}	0.018	28	5 no. at 3.6 l/s	23
2×10^{-4}	0.032	36	6 no. at 5.3 l/s	19
5×10^{-4}	0.069	48	8 no. at 8.6 l/s	15
9×10^{-4}	0.12	63	11 no. at 10.9 l/s	11

Empirical checks

It is vital that the basic design is checked against experience or the ‘normal’ range of dewatering systems.

In Section 7.5, it is stated that most deep well systems have a well spacing of between 10 and 60 m. The design fits within these limits, although if the permeability is at the higher end of the analysed range it can be seen that the well spacing is at the edge of the normal range – this implies that, if permeability is actually rather greater than the analysed range, then deep wells may not be the most appropriate method.

Another rule of thumb is that, where the aquifer depth allows, dewatering wells should generally penetrate to at least one and a half to two times the depth of the excavation. In this case the wells are 19 m deep, compared to an excavation depth of 9 m, so this condition is satisfied.

In Table 10.1, the minimum diameters of well bore for a given yield are listed. This table indicates that wells drilled at 305 mm diameter can accommodate pumps of capacity up to 15–20 l/s, within the range of the anticipated yield.

These empirical checks confirm, in the gross sense, the validity of the concept of the design.

Final design

Design calculations have produced a range of flow rates and well yields and spacings. Ultimately, a decision has to be made based on engineering judgement. Any records of local experience may help in forming an opinion. In this case the nominal system of six or eight wells would seem to be the most appropriate option. If the construction programme is tight, and cannot cope with delays, the larger system would be prudent. However, in this case it is assumed that there is time in the programme to install a few additional wells (if crude testing of the first wells shows that flow rates are higher than the design value). Accordingly, the nominal six well system is appropriate.

If the aquifer was unconfined, the calculated steady-state flow rates would need increasing to allow for the additional water from storage release in the early stages of pumping. However, the aquifer is initially confined, and after drawdown will only become unconfined local to the wells. Confined aquifers have very small storage coefficients, so the water released from storage will not be significant, and is not normally allowed for in design.

It is normally prudent to increase the number of wells by around 20 per cent, to provide some allowance for individual wells being temporarily out of service. In this case this increases the number of wells from six to eight. At the design yield of 5.3 l/s this give a total design capacity of around 42 l/s. In practice the system capacity will be slightly larger because submersible pumps are manufactured with discrete capacities. The pump chosen will normally have a slightly greater capacity than the design yield. These factors should all help provide a robust design, capable of coping with modest changes in the predicted flow rate.

Because the excavation is fairly narrow, it is likely that a well system of adequate capacity will achieve drawdown everywhere within the excavation area. In many cases no formal calculation is made of the drawdown distribution within the excavation. If necessary, the methods given in

Section 7.6 can be used to obtain *approximate* estimates of drawdown within the excavation.

Design example 1a

This example shows an alternative method for the estimation of steady-state flow rate from a system of deep wells in a confined aquifer. The well array is still modelled as an equivalent well, but the flow rate is determined using published 'shape factors'. The case analysed is exactly the same as in design example 1.

Conceptual model

As design example 1.

Selection of method

As design example 1.

Estimation of steady-state discharge flow rate

Flow to rectangular arrays of wells in confined aquifers can be determined by the shape factor method of equation (7.12).

$$Q = kD(H - h_w)G \tag{7.12}$$

where:

- k is the aquifer permeability. See design example 1 for discussion of selected values of permeability between 1×10^{-4} and 9×10^{-4} m/s.
- D is the aquifer thickness: $D = 19 - 10 = 9$ m.
- $(H - h_w)$ is the drawdown: $(H - h_w) = 9.5 - 1 = 8.5$ m.
- G is a geometry shape factor, obtained from Fig. 7.10. To determine the appropriate value of G for a given case, it is necessary to evaluate the following parameters:
  Array aspect ratio a/b: if the array of wells is of plan dimensions a by b, the aspect ratio a/b will have an effect on the flow rate. In this case $a/b = 39/19 = 2.1$.
  L_0/a: The ratio of distance of influence L_0 to the long dimension of the well array a. In the absence of pumping test data, L_0 can be estimated from either equation (7.13) or (7.15). In this case, the excavation is not very long and narrow and radial flow is likely to be the dominant flow regime. Therefore equation (7.13) is most appropriate.

$$L_0 = 3000(H - h_w)\sqrt{k} \tag{7.13}$$

Note that equation (7.14) (which includes the radius r_e of the equivalent well) should not be used with this method, because Figure 7.10 is based on distances of influence from the edge of the well array, not the centre. If the well array had been long and narrow (i.e. a had been much greater than b), plane flow would predominate and L_0 should have been estimated from equation (7.15).

The flow rate is then calculated from equation (7.12), using values of G from Fig. 7.10. The permeability sensitivity analysis of design example 1 has been repeated here.

k (m/s)	L_0 (m)	L_0/a	a/b	G	Q (m^3/s)	Q (l/s)
1×10^{-4}	255	6.5	2.1	2.4	0.018	18
2×10^{-4}	361	9.2	2.1	2.1	0.032	32
5×10^{-4}	570	14.6	2.1	1.9	0.073	73
9×10^{-4}	765	19.6	2.1	1.7	0.12	120

It is unrealistic to expect the flow rates calculated by two different methods to be precisely the same, but in this case there is good agreement between the methods used here and in design example 1a. When rounded to two significant figures (as above) the greatest difference between methods is only 3 l/s.

The remainder of the design process is carried out in the same way as for design example 1.

Design example 2

This example describes the design of a partially penetrating wellpoint system for trench works in an unconfined aquifer. The line of wellpoints is analysed as an equivalent slot under plane flow conditions, and the contribution from radial flow to the ends of the slot is assessed. A sensitivity analysis is carried out to assess the effect of varying the depth of the aquifer on the calculated flow rate.

Conceptual model

A narrow trench excavation is required to allow the laying of a shallow pipeline which extends for several hundred metres. The formation level for the trench is 3 m below ground level. The details of the conceptual model

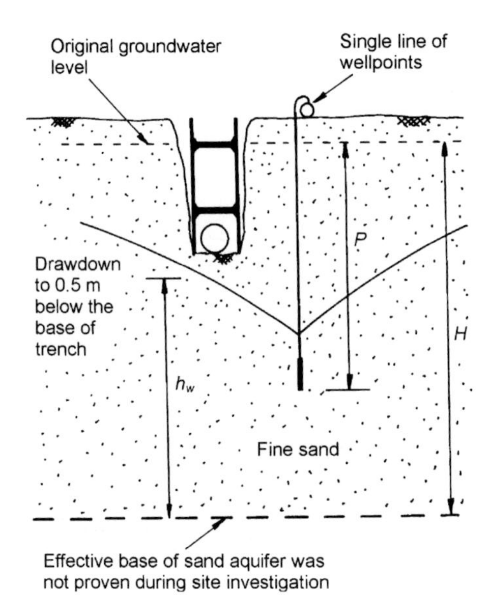

Figure 7A-2 Conceptual model for design example 2.

can be summarized as shown in Fig. 7A-2:

- The excavation is 1 m wide and 3 m in depth. The excavation has near vertical sides, with the pipelaying operatives working in the trench, protected by a 'drag box' – a temporary trench support system. The trench is several hundred metres long, but it is anticipated that no more than 20–30 m will be open at any one time.
- An unconfined aquifer is present beneath the site. The aquifer comprises a uniform fine sand, and groundwater level is generally 0.5 m below ground level.
- Back-analysis of previous groundwater lowering nearby indicates the sand is of approximate permeability 1×10^{-4} m/s. Boreholes at the current site indicate ground conditions are very similar to the nearby site, and analysis of PSD results from the current site (using Hazen's rule, equation (6.1)) confirm this permeability.
- The base of the aquifer was not determined during the investigation. The deepest borehole penetrated to 12 m depth and did not encounter any underlying impermeable stratum. Published geological maps of the area indicate that the stratum of fine sand may extend to 20 m depth or more.
- No recharge boundaries are believed to exist. Flow is likely to be plane to the line of wellpoints, and can be idealized as a distant source.
- No compressible strata are believed to exist. Groundwater lowering related settlement is not anticipated to be a problem.

Selection of method

The required drawdown (to 0.5 m below formation level) is 3 m (from 0.5 to 3.5 m depth), and the anticipated permeability is 1×10^{-4} m/s. Figure 5.6 suggests single stage wellpointing, which is the method commonly used for dewatering of pipeline trenches.

Although the trench is very long, only 20–30 m will be open or being worked on at any time; it will take around a week to excavate, lay and backfill this length of trench. In such circumstances it is not common practice to dewater the whole length of pipeline. Typically, pumping is maintained on a line of wellpoints of length equal to three or four times the weekly rate of advance (see Section 9.7). The wellpoints and pumping equipment are progressed forward to keep pace with the pipelaying. In this design it is assumed that wellpoints are pumped alongside 100 m of trench at any one time.

Estimation of steady-state discharge flow rate

For trench excavations, wellpoints are installed in closely-spaced lines parallel to the trench. This geometry is amenable to solution as an equivalent slot – although, as will be described later, it may be necessary to consider radial flow to the end of a line of wellpoints. For a trench depth of 3 m, wellpoints would normally be installed to the standard depth of 6 m. Even though the depth of the aquifer is not known, the wellpoints (and hence the equivalent slot) will be partially penetrating. The total flow rate Q for plane flow to a slot in an unconfined aquifer can be estimated from equation (7.11).

$$Q = \left[0.73 + 0.23(P/H)\right] \frac{kx(H^2 - h_w^2)}{L_0} \tag{7.11}$$

where:

- k is the aquifer permeability: k is taken as 1×10^{-4} m/s.
- x is the length of the slot: x is taken as 100 m, the length of the line of wellpoints.
- H is the depth from the original water table to the base of the aquifer and h_w is the depth from the lowered groundwater level (in the equivalent slot) to the base of the aquifer. Note that the drawdown of concern is in (or beneath) the trench itself, rather than at the wellpoints. In reality, the wellpoints will be so close to the trench there will be little difference in drawdown between the trench and the wellpoints. In calculations h_w is taken as the head beneath the trench (see Fig. 7A-2).
- P is the penetration of the slot below the original water table: $P = (6 - 0.5) = 5.5$ m.

- L_0 is the distance of influence. In the absence of pumping test data, L_0 can be estimated for equivalent slots from equation (7.15) (shown here for L_0, H and h_w in metres and k in m/s).

  $$L_0 = 1750(H - h_w)\sqrt{k} \quad (7.15)$$

  for $k = 1 \times 10^{-4}$ m/s and $(H - h_w) = 3$ m (drawdown from 0.5 to 3.5 m), L_0 is estimated to be 53 m.

If the depth of the aquifer was known, the total flow rate Q could then be estimated from equation (7.11). The problem of the unknown aquifer depth can be overcome by carrying out a sensitivity analysis – and of course, it is known that the base of the aquifer is more than 12 m below ground level. Using the units quoted above Q will be calculated in m³/s; when commenting on results Q is normally quoted in l/s (1 m³/s = 1000 l/s) to make the numbers easier to read and interpret.

Depth to base of aquifer (m)	H (m)	h_w (m)	Q (m³/s)	Q (l/s)
12	11.5	8.5	9.6×10^{-3}	9.6
14	13.5	10.5	1.1×10^{-2}	11
16	15.5	12.5	1.3×10^{-2}	13
18	17.5	14.5	1.4×10^{-2}	14
20	19.5	16.5	1.6×10^{-2}	16

This shows that, in this case, assuming a deeper base to the aquifer increases the flow by almost 70 per cent but, because the initial flow rate was modest, the actual increase in predicted flow is only 6.4 l/s. This small increase in flow will not significantly affect the design of the system. The effect of a deeper base to the aquifer would have been more problematic if the aquifer permeability was higher, because the initial flow rate would have been larger, and a 70 per cent increase would result in a much greater increase in flow.

The above calculations assume plane flow to the sides of the slot, but because the line of wellpoints is of finite length (100 m), there will be some contribution from radial flow to the ends (see Fig. 7.3(b)). The total flow rate to the ends of the slot is the same as the flow rate to a well of radius r_e equal to half the width of the slot. For a partially penetrating well in an unconfined aquifer, the flow rate from such a well can be estimated from equations (7.5) and (7.7), combined as:

$$Q = \frac{B\pi k(H^2 - h_w^2)}{\ln[L_0/r_e]}$$

where all the terms are as defined previously, apart from B, which is a partial penetration factor for radial flow to wells, determined from Fig. 7.9(b). This equation has been used to calculate the contribution from radial flow to the ends of the slots – the flow rate is theoretically split 50–50 at either end of the slot, but it is the total flow that is relevant now. In calculations, it has been assumed that the slot is a single line of wellpoints of width 0.2 m (a typical width of the jetted hole formed by a placing tube), so r_e was taken as 0.1 m. In this case the contribution from radial flow to the ends of the slot is small (between 8 and 15 per cent of the flow to the sides). Obviously, if the slot had not been so long, the percentage contribution from the ends would have been greater.

If it was anticipated that the wellpoint system would have been installed as a double-sided system, consisting of two parallel lines of wellpoints, the flow to the ends would have been greater. If the two lines of wellpoints were, say, 5 m apart, the radial flow calculation would have used $r_e = 2.5$ m, and the total flow to both ends of the slot would have been between 25 and 45 per cent of the flow to the sides. This highlights that it is more important to consider the flow to the ends of a double-sided wellpoint system than it is for a single-sided system.

Depth to base of aquifer (m)	*Q Plane flow to both sides only* (l/s)	*Q Radial flow to ends* (l/s)	*Q Total flow: plane plus radial* (l/s)
12	9.6	1.4	11
14	11	1.4	12
16	13	1.4	14
18	14	1.4	15
20	16	1.4	17

Determination of wellpoint spacing and pump capacity

It is likely that a single-sided wellpoint system (consisting of a line of wellpoints alongside one side of the trench only) will be appropriate for this excavation. This is because this case satisfies the conditions favourable to single-sided wellpointing set out in Section 9.7.

(a) A narrow trench
(b) Effectively homogeneous, isotropic permeable soil conditions that persist to an adequate depth below formation level (see Figure 9.1(a))
(c) Trench formation level is not more than about 5 m below standing groundwater level.

Table 9.1 gives typical wellpoint spacing in sands as 1 to 2 m. The length of the proposed wellpoint system is 100 m. Assuming an initial spacing of 2 m, fifty wellpoints will be pumped at any one time. If wellpoints of standard screen length 0.7 m are installed by placing tube, Fig. 7.16 suggests that for a permeability of 1×10^{-4} m/s each wellpoint would have a capacity of 0.26 l/s. This gives a maximum yield of 13 l/s for a fifty wellpoint system. This is less than the predicted flow rate if the aquifer is 20 m deep, so it would be prudent to install the wellpoints at 1.5 m centres, giving a system of sixty-seven wellpoints, with a maximum yield of 17 l/s, which is acceptable for the maximum predicted flow rate. On routine projects wellpoint spacings tend to be used increments of 0.5 m. If a spacing of 1.5 m was not satisfactory, the next case to be tried would be 1.0 m spacings. It is rare to consider spacings such as 1.4 m, 1.3 m and so on.

If the yield from each wellpoint is acceptable for the steady-state case, it is almost certainly adequate to deal with the additional flow rate from storage release during the initial drawdown period. This is because, until the steady-state drawdown develops, the wetted screen length of the wellpoints will be much greater, allowing them to yield more water.

The wellpoint pump(s) must also be selected. The choice of pump may be influenced by the equipment available locally, and sometimes much larger pumps than strictly necessary are provided, purely because they are close to hand. Whatever pumps are used, the pump capacity must be adequate for the predicted flow rate, not just the steady-state discharge, but also the additional water released from storage during initial pumping.

A simplistic calculation using equation (7.23) (and assuming the sand has a storage coefficient S of 0.2) estimates that the design steady-state distance of influence of 53 m will take around half a day to one day to develop. This gives a very crude estimate of the time during which water released from storage will be significant, and also roughly correlates with the time to achieve drawdown close to the line of wellpoints. A drawdown period of a day or so may seem fairly quick, but pipelaying is a progressive operation, moving forward the whole time. Each wellpoint may only be pumped for a week or so until it is turned off after the trench has passed. New wellpoints are continually being installed and commissioned ahead of pipelaying. A rapid drawdown period is essential to avoid the pipelaying operation moving too fast and advancing ahead of the dewatered area into 'wet' ground, where the target drawdown has not yet been achieved. This event should be avoided as it can waste a lot of time and money. This problem does occur from time to time, and, when it does, the pipelaying operatives are often forthright (to say the least!) in their criticism of the groundwater lowering operation. For 'static' or non-progressive excavations, there is often a little less pressure on achieving very rapid drawdowns.

In this case it is assumed that double acting piston pumps are available. In Section 12.1, a 100 mm unit is quoted as having a capacity of up to 18 l/s. This could handle the predicted steady-state flow rate, but may not be able to cope with the higher flows to establish drawdown. A larger 125 mm unit has a capacity of 26 l/s, around 50 per cent greater than the maximum predicted steady-state discharge, giving useful spare capacity to deal with water released from storage. If 125 mm pumpsets were not available, two 100 mm units could be provided as an alternative. Both units would be pumped during the initial drawdown period but later, when the flow rate has reduced, it may be possible to maintain drawdown using one pump only. The other pump would remain connected into the system as a standby. Even if large capacity pumps are available, the need for standby pumps must be considered; conditions when standby pumps are required are outlined in Section 9.6.

Design example 3

This example is an application of cumulative drawdown analysis (theoretical method), to analyse a system of fully penetrating deep wells used to lower the piezometric level in a confined aquifer beneath a rectangular excavation. Aquifer parameters determined from a pumping test are used to allow estimation of drawdown at specified locations around the excavation area.

Conceptual model

A rectangular excavation is to be made to a depth of 12 m. The details of the conceptual model can be summarized as shown in Fig. 7A-3:

- Excavation dimensions are 50 by 50 m in plan at formation level, with the deepest part of the excavation at 12 m depth. The sides of the excavation are battered back at 1 in 1.5, giving overall dimensions at ground level of 86 by 86 m. To try and keep any dewatering wells slightly closer to the deepest part of the excavation they will be installed on a bench in the excavation batters at a level of 2 m below ground level. The plan dimensions of the well array will be 80 by 80 m. The excavation was to be dewatered for a period of five months, and the construction programme required that drawdown be achieved within two weeks of commencing pumping.
- A confined aquifer, consisting of a sandy gravel, extends from 10 m depth to 26 m depth. The confining layer above the aquifer is a stiff clay. Maximum piezometric level in the aquifer is 5 m below ground level.
- A well pumping test was carried out, pumping at a rate of 15 l/s for seven days. The discharge during the test was limited by the capacity of the pump; if a larger pump had been available a greater flow rate would have been possible. Analysis of the pumping test data gave an aquifer permeability k of 6×10^{-4} m/s and a storage coefficient S of 0.001.
- No compressible strata are believed to exist. Groundwater lowering related settlement is not anticipated to be a problem.

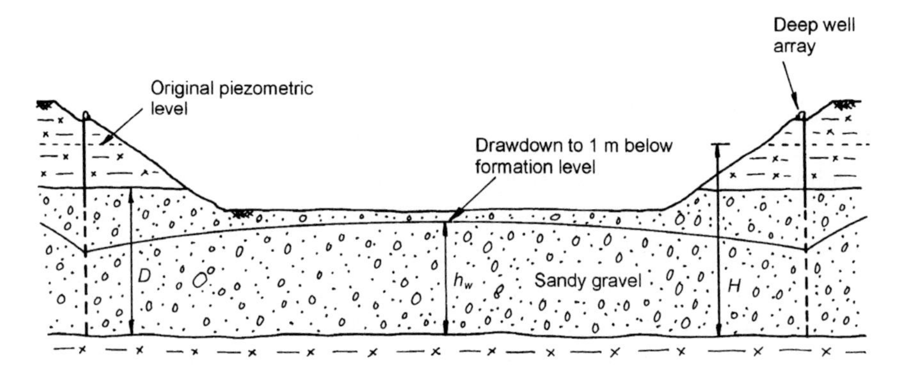

Figure 7A-3 Conceptual model for design example 3.

Selection of method

The excavation extends through the stiff clay aquiclude and into the upper few metres of the confined aquifer. The piezometric level in the confined aquifer will need to be lowered prior to excavation to prevent base heave during excavation through the clay, and then to provide a workable excavation when the excavation penetrates into the top of the aquifer.

The target drawdown is to lower the groundwater level to 1 m below formation level. This is 13 m depth, or a drawdown of 8 m below the original piezometric level.

For a drawdown of 8 m and the design permeability of 6×10^{-4} m/s, inspection of Fig. 5.6 suggests that either deep wells or two stage wellpoints would be suitable for this combination of drawdown and permeability. In this case the deep well method will be used, because the contractor wishes to excavate rapidly to full depth.

Estimation of steady-state discharge flow rate and estimation of number of wells

The cumulative drawdown method (using the Cooper–Jacob simplification) can be used in confined aquifers. It has also been successfully applied in unconfined aquifers where the final drawdown is less than around 20 per cent of the initial saturated aquifer thickness. In this case, because drawdown is required to 13 m depth compared with the top of the aquifer at 10 m depth, the initially confined aquifer will become unconfined. The aquifer thickness will be reduced by 3 m out of 16 m, or 19 per cent. Therefore, this problem will be analysed assuming confined behaviour throughout. The cumulative drawdown is calculated using equation (7.18).

$$(H-h)=\sum_{i=1}^{n} \frac{q_i}{4\pi kD}\left\{-0.5772-\ln\left[\frac{r_i^2 S}{4kDt}\right]\right\} \tag{7.18}$$

where:

- $(H-h)$ is the cumulative drawdown (at the point under consideration) resulting from n wells each pumped at constant flow rate q_i.
- k is the aquifer permeability: k is taken as 6×10^{-4} m/s.
- S is the aquifer storage coefficient: S is taken as 0.001.
- D is the original aquifer saturated thickness: $D = 26 - 10 = 16$ m.
- t is the time since pumping began. In this case the target drawdown is required within fourteen days. It is always prudent to design to obtain the drawdown a little quicker than planned – this allows for minor problems during commissioning. In design, we will aim to achieve the target drawdown within ten days. $t = 86{,}400$ seconds will be used in calculations.
- r_i is the distance from each pumped well to the point where drawdown is being estimated.

Equation (7.18) is valid provided $u = (r^2S)/(4kDt)$ is less than 0.05. In this case, taking r to be 113 m (the distance from the corners of the well array to the centre of the excavation) u is less than 0.05 after around two hours – therefore this method can be used for all t greater than two hours.

The method requires that the plan layout of the well array be sketched, and the x–y co-ordinates of each well be determined. The co-ordinates then allow the radial distances r_i (from each well to the point where drawdown is being checked) to be calculated. An initial guess is made of the number of wells and well spacing and the resulting x–y co-ordinates determined. In this case the initial guess was sixteen wells evenly spaced at 20 m centres (Fig. 7A-4).

A spreadsheet program is then used to evaluate equation (7.18) for the cumulative drawdown at selected locations within the excavation. For circular or rectangular excavations with evenly-spaced wells it is normally sufficient to determine the drawdown in the centre of the excavation, because drawdown everywhere else will be greater. This is the method used here. If the well array is irregular in shape (or if the depth of excavation is not constant) it will be necessary to determine the drawdown in a number of locations, to ensure the target drawdown is achieved at all critical locations.

The results from a spreadsheet calculating the drawdown in the centre of the excavation for a sixteen well system are shown below. The radial distance r_i, from each well (at location x_i, y_i) to the location (x_c, y_c) where the drawdown is being determined, is calculated from:

$$r_i = \sqrt{([x_i - x_c]^2 + [y_i - y_c]^2)}.$$

For simplicity, the flow rate q_i from each well has been assumed to be the same, but if it was intended to use pumps of different sizes in certain wells

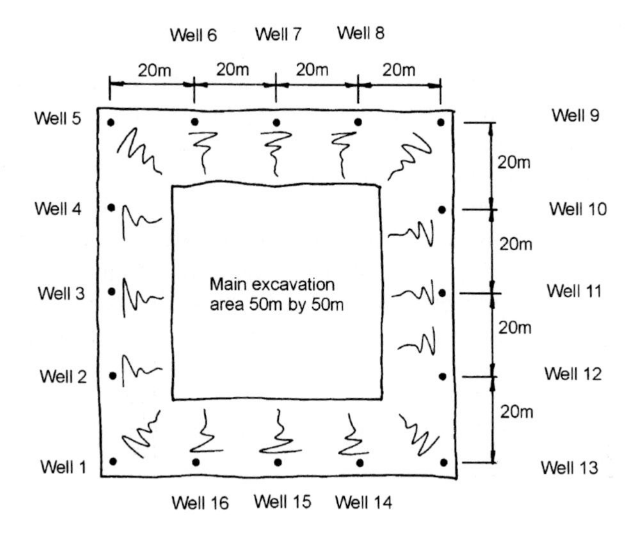

Figure 7A-4 Schematic plan of sixteen well system.

this can easily be incorporated in the calculation. In the spreadsheet different values of q_i were tried until the target drawdown of 8 m is just achieved in the centre of the excavation. The total flow rate is simply the sum of all the well flow rates.

Well	x *co-ordinate* (m)	y *co-ordinate* (m)	*Well flow rate* q_i (m^3/s)	*Radial distance* r_i (m)	*Drawdown* $(H-h)_i$ (m)
1	0	0	7.0×10^{-3}	56.5	0.50
2	0	20	7.0×10^{-3}	44.7	0.53
3	0	40	7.0×10^{-3}	40.0	0.54
4	0	60	7.0×10^{-3}	44.7	0.53
5	0	80	7.0×10^{-3}	56.6	0.50
6	20	80	7.0×10^{-3}	44.7	0.53
7	40	80	7.0×10^{-3}	40.0	0.54
8	60	80	7.0×10^{-3}	44.7	0.53
9	80	80	7.0×10^{-3}	56.6	0.50
10	80	60	7.0×10^{-3}	44.7	0.53
11	80	40	7.0×10^{-3}	40.0	0.54
12	80	20	7.0×10^{-3}	44.7	0.53
13	80	0	7.0×10^{-3}	56.6	0.50
14	60	0	7.0×10^{-3}	44.7	0.53
15	40	0	7.0×10^{-3}	40.4	0.54
16	20	0	7.0×10^{-3}	44.7	0.53
Total			0.112 m^3/s (112 l/s)		8.43

This calculation indicates that a system of sixteen wells, each discharging 7 l/s (total flow rate 112 l/s) will achieve the target drawdown in the centre of the excavation after ten days. During the pumping test the test well produced 15 l/s and could have yielded more if a larger pump has been used. It therefore makes sense to repeat the above calculations assuming fewer wells of greater discharge rate. The results of these calculations are summarized below:

No. of wells	Well spacing (m)	Well flow rate (l/s)	Drawdown in centre of excavation (m)	Total flow rate (l/s)
16	20	7.0	8.43	112
13	25	8.5	8.33	110.5
10	32	11.0	8.28	110
8	40	13.5	8.07	108

It is apparent that the target drawdown can be achieved by various combinations of well numbers and yields, but that the total flow rate remains approximately constant.

Final design and empirical checks

The number and yield of wells chosen for the final design will depend on a number of factors, including:

(i) The need for redundancy in a well system. Any system relying on relatively few wells is vulnerable to one or two wells suffering from damage or pump failure, leading to loss of drawdown and flooding or instability of the excavation. A system consisting of a greater number of wells will lose proportionately less drawdown if one or two wells are lost.

(ii) Each well must be able to yield the discharge flow rate q_i assumed in design. Because the pumping test well produced 15 l/s (and could have yielded more), and since all the current designs use q_i of less than 15 l/s, it is likely that all the current designs are feasible. In any case, the theoretical maximum well yield can be estimated from equations (7.24) and (7.25) (or from Fig. 7.15). Assuming a well borehole diameter of 0.305 mm and a wetted screen length of 10 to 12 m (i.e. a drawdown at the wells of 1 to 3 m below the excavation target drawdown level), the maximum well yield is estimated as 16 to 19 l/s. However, in practice, problems can occur if the dewatering wells are not designed, installed and developed in exactly the same way as the test well – this may cause the production wells to have lower yields than the test wells. Also, in fissured rock aquifers, well yields may vary significantly. Some wells can have high yields and yet others, poorly connected into fissures, may be almost 'dry'. In fissured aquifers the cumulative drawdown method needs to be applied with care.

In this case it is assumed that, due to the availability of pumps of suitable capacity, the nominal system of ten wells, each discharging 11 l/s each will be adopted. It is normal practice to apply an empirical superposition factor J; the system capacity is increased by a factor of $1/J$ (see equation (7.19)). This empirical factor allows for interference between wells, and also provides some allowance for additional drawdown around the wells and water released from storage when the aquifer becomes unconfined. Where aquifers become unconfined, and drawdowns are small (less than 20 per cent of the initial saturated aquifer thickness), the empirical superposition factor J is normally taken as 0.8 to 0.95. In this case, because the drawdown will reduce the thickness of the aquifer by almost 20 per cent, the maximum superposition factor of 0.8 will be applied, so the system capacity (and hence the number of wells) will need to be increased by $1/0.8 = 1.25$.

The final system design is, therefore for thirteen wells, of 11 l/s capacity each. Total system capacity is 143 l/s. Table 10.1 indicates that, to accommodate a pump of suitable capacity, a minimum well bore diameter of 300 mm is required. The corresponding well screen and liner diameter is 165 mm.

The design is then verified with some simple empirical checks. The well spacing for a thirteen well system is approximately 25 m; this is within the 'normal' 10 to 60 m range quoted in Section 7.5. The wells are intended to fully penetrate the aquifer, and so are 26 m deep, just over twice the depth of the excavation. Again this is consistent with guidelines given in Section 7.5. These empirical checks confirm, in a gross sense, the validity of the concept of the design.

References

Anderson, M. P. and Woessner, W. W. (1992). *Applied Groundwater Modelling*. Academic Press, New York.

Cedergren, H. R. (1989). *Seepage, Drainage and Flow Nets*, 3rd edition. Wiley, New York.

Chapman, T. G. (1959). Groundwater flow to trenches and wellpoints. *Journal of the Institution of Engineers, Australia*, October-November, 275–280.

Cooper, H. H. and Jacob, C. E. (1946). A generalised graphical method for evaluating formation constants and summarising well field history. *Transactions of the American Geophysical Union*, **27**, 526–534.

Dupuit, J. (1863). *Etudes Théoretiques et Practiques sur les Mouvement des Eaux dans les Canaux Decouverts et a Travers les Terrains Permeable*. Dunod, Paris.

Forcheimer, P. (1886). Uber die ergibigkeit von brunnen-anlagen und sickerschitzen. *Der Architekten-und Ingenieur-Verein*, **32**(7).

Hantush, M. S. (1964). Hydraulics of wells. *Advanced Hydroscience* **I**, 281–431.

Knight, D. J., Smith, G. L. and Sutton, J. S. (1996). Sizewell B foundation dewatering – system design, construction and performance monitoring. *Géotechnique*, **46**(3), 473–490.

Kruseman, G. P., and De Ridder, N. A. (1990). *Analysis and Evaluation of Pumping Test Data*. International Institute for Land Reclamation and Improvement, Publication 47, 2nd edition, Wageningen, The Netherlands.

Mansur, C. I. and Kaufman, R. I. (1962). Dewatering. *Foundation Engineering* (G. A. Leonards, ed.). McGraw-Hill, New York, pp 241–350.

Muskat, M. (1935). The seepage of water through dams with vertical faces. *Physics*, **6**, 402.

Nicholson, D. P., Tse, C.-M. and Penny, C. (1999). *The Observational Method in Ground Engineering*. Construction Industry Research and Information Association, CIRIA Report C185, London.

Powers, J. P. (1992). *Construction Dewatering: New Methods and Applications*, 2nd edition. Wiley, New York.

Powrie, W. and Preene, M. (1992). Equivalent well analysis of construction dewatering systems. *Géotechnique*, **42**(4), 635–639.

Powrie, W. and Preene, M. (1994). Time-drawdown behaviour of construction dewatering systems in fine soils. *Géotechnique*, **44**(1), 83–100.

Preene, M. and Powrie, W. (1993). Steady-state performance of construction dewatering systems in fine soils. *Géotechnique*, **43**(1), 191–205.

Preene, M. and Roberts, T. O. L. (1994). The application of pumping tests to the design of construction dewatering systems. *Groundwater Problems in Urban Areas* (Wilkinson, W. B. ed.). Thomas Telford, London, pp 121–133.

Preene, M. Roberts, T. O. L. Powrie, W. and Dyer, M. R. (2000). *Groundwater Control – Design and Practice*. Construction Industry Research and Information Association, CIRIA Report C515, London.

Rao, D. B. (1973). Construction dewatering by vacuum wells. *Indian Geotechnical Journal*, 3(3), 217–224.

Roberts, T. O. L. and Preene, M. (1994). The design of groundwater control systems using the observational method. *Géotechnique*, **44**(4), 727–734.

Rushton, K. R. and Redshaw, S. C. (1979). *Seepage and Groundwater Flow: Numerical Analysis by Analog and Digital Methods*. Wiley, Chichester.

Theis, C. V. (1935). The relation between the lowering of the piezometric surface and the rate and duration of discharge of a well using groundwater storage. *Transactions of the American Geophysical Union*, **16**, 519–524.

Weber, H. (1928). *Die Reichweite von Grundwasserabsenkungen Mittels Rohrbunnen*. Springer, Berlin.

Williams, B. P. and Waite, D. (1993). *The Design and Construction of Sheet-Piled Cofferdams*. Construction Industry Research and Information Association, CIRIA Special Publication 95, London.

Chapter 8

Sump pumping

8.0 Introduction

A sump might be defined as (Scott 1980):

> A pit in which water collects before being bailed or pumped out. The pump suction dips into a sump

By definition a sump is at a low level in relation to surrounding ground surfaces so that any water will flow to it due to gravity. For construction projects water is removed from sumps using suction or submersible pumps, not by bailing.

This chapter addresses the formation of pumping sumps and associated gravity drainage channels for the control of surface water and groundwater. Case histories are used to describe situations where sump pumping has been used appropriately, and where problems were created. The pumps suitable for sump pumping uses are dealt with in Chapter 12.

8.1 Applications of sump pumping

Sump pumping is the most basic of the dewatering methods. In essence it involves allowing groundwater to seep into the excavation, collecting it in sumps and then pumping it away for disposal. Sumps are provided for two separate purposes, though the form of a sump may be similar for either requirement:

(a) To collect surface water run-off channelled to it by means of collector ditches or channels for discharge to a disposal point or area (see Section 5.1).
(b) To collect and discharge pumped water due to lowering of the groundwater for a shallow excavation; also sump pumping may be required for gravity drainage to toe drains of battered slopes (see Section 4.4).

The preferred method of disposal of water collected by the sumps is by means of pumping. Typically, each sump is equipped with a robust, simple, pump – a 'sump pump' (see Section 12.3).

Sump pumping can be a very effective and economic method to achieve modest drawdowns in well-graded coarse soils (such as gravelly sands, sandy gravels and coarse gravels) or in hard fissured rock.

Unfortunately, under some conditions the use of sump pumping can lead to major problems. These problems arise primarily because the flow of water *into* the excavation can have a destabilizing effect on fine-grained soils. This can lead to fine soil particles being washed from the soil with the water – this is known as 'loss of fines'. Loss of fines can lead to ground movements and settlements because material is being removed from the soil, giving the potential for the formation and collapse of sub-surface voids (see Section 13.1). Disposal of the water can also create problems, because if loss of fines occurs the discharge water will have a high sediment load, which can cause environmental problems at the disposal point (see Section 13.4). These issues are discussed further in Section 8.6.

Powers (1985) lists soil types where the use of sump pumping has a significant risk of causing loss of fines. These include:

i Uniform fine sands,
ii soft non-cohesive silts and soft clays,
iii soft rocks where fissures can erode and enlarge due to high water velocities,
iv rocks where fissures are filled with silt, sand or soft clay, which may be eroded,
v sandstone with uncemented layers that may be washed out.

In these soil types, even the best engineered sump pumping systems may encounter problems. Potential problems can be avoided by employing groundwater lowering using an array of wells (wellpoints, deep wells or ejectors) with correctly designed and installed filters. Provided the wells are located outside the main excavation area, these methods have the advantage that they draw water *away* from the excavation, improving stability – avoiding the destabilizing flows into the excavation that are associated with sump pumping.

8.2 Surface water run-off

When an excavation exposes low permeability soils, all surface water, whether derived from rainfall or from any other source, should be controlled so that any occurrence of ponding is prevented. Movement of construction plant through surface water ponding will lead to deterioration of the surface, especially on clayey soils. This will inhibit efficient use of plant. It is good practice to form the surfaces of the construction areas so that they are not level; they should be gently sloped so that all surface water is shed to suitably sited and constructed collector channels or drains (see Section 5.1).

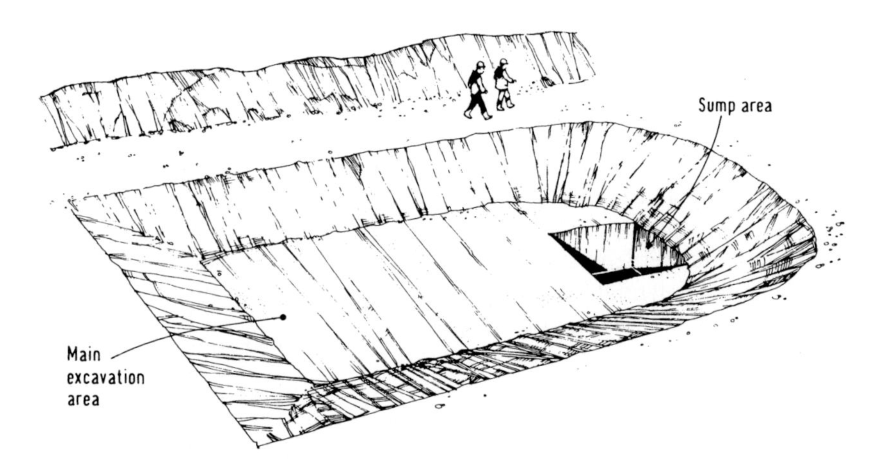

Figure 8.1 Typical sump within main excavation area (from Somerville 1986: reproduced by kind permission of CIRIA).

The fall on the bed of a collector drain should be sufficient to minimize silting up but not so steep as to cause erosion. Near the sump it may be prudent to increase the width of the drain to allow a flow velocity low enough to prevent erosion. The alternative is to provide check weirs at intervals along the line of the drain.

The surface water run-off within an excavation should be channelled to a conveniently located pumping sump as indicated schematically in Fig. 8.1. The forms of construction of the sump(s) are described below.

8.3 Pumping sumps

At all times the depth of a sump must be a generous amount deeper than the bed of collector drain(s) leading into it, and/or of the formation level of the excavation. A sump should be substantially larger than that needed to accommodate the pump(s). Surface water flowing to the sump is likely to transport fines. These are likely to be abrasive and capable of causing wear and damage to the pumping equipment. A sump of a generous size will allow some settlement of the larger (and probably more abrasive) fines. Adequate provision should be made for periodic servicing of the pumps and removal of accumulated sediment. The pump should be suspended so that the bottom of the unit is about 300 mm above the bottom of the sump to allow for some build up of sediment. This does not apply to smaller shallow sumps where the suction hose only is in the sump.

A common arrangement for a sump is to suspend the pump in a 200 litre drum or similar (Fig. 8.2(a)), with many holes punched through the sides of

(a)

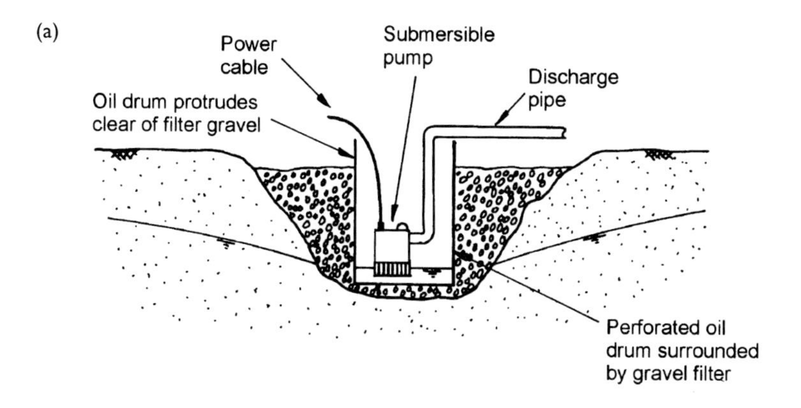

(b)

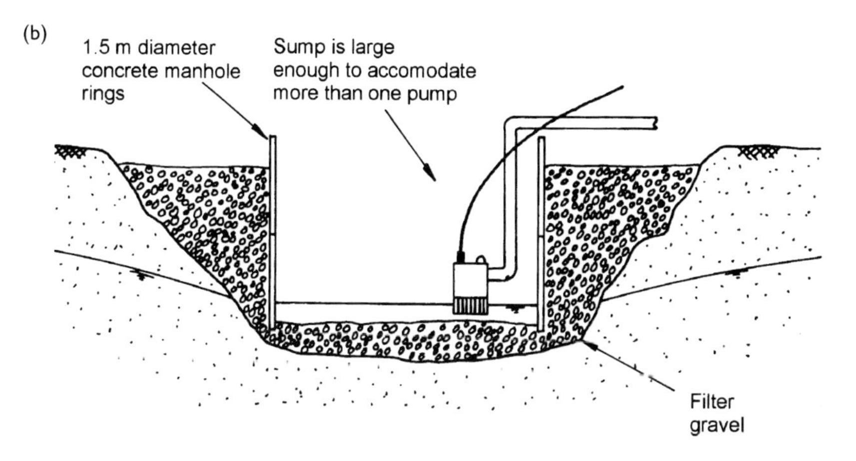

Figure 8.2 Typical forms of sump construction. (a) Small sump formed using a perforated oil drum, (b) large sump formed using concrete manhole rings.

the drum. Outside the drum is placed an annulus of fine gravel to act as a crude filter. Forms of sump construction are shown in Fig. 8.2.

The sumps should be dug to a greater depth than the main excavation and should be maintained in their original form throughout the construction period, though deepened if necessary as excavation proceeds. This will:

(a) allow placement of filter media that may be necessary to minimize loss of ground,
(b) keep groundwater below excavation level at all stages of the work,
(c) allow changes to be made in the construction scheme for the main excavation.

Most sumps are formed by excavation with the sides temporarily supported by sheeting for stability, before the body of the sump (i.e. the drum or similar) is placed. In certain circumstances jetted sumps may be preferred (see Fig. 8.3). A suitable size placing tube is jetted into the ground to the required depth. A disposable intake strainer and flexible suction hose is lowered into the placing tube in a manner similar to the positioning of a disposable wellpoint and riser pipe. Filter media is placed within the placing tube, around the strainer and riser pipe as the placing tube is withdrawn. The upper end of the riser pipe is connected to a suitable pump.

Generally the maximum effective depth to which a well-maintained vacuum assisted self-priming centrifugal surface pump will operate is about 6 m below the top of the sump. For excavations of greater depths it will be necessary to re-install the pumps at a lower level or to use a suspended submersible pump, which can be lowered down a lined shaft or perforated steel tube.

The need for sufficient pumping capacity is paramount since a greater pumping capacity is needed to initially dewater an excavation than is required to maintain the water level at a steady state in its finally lowered position. The pumping plant should be installed in multiple units so that

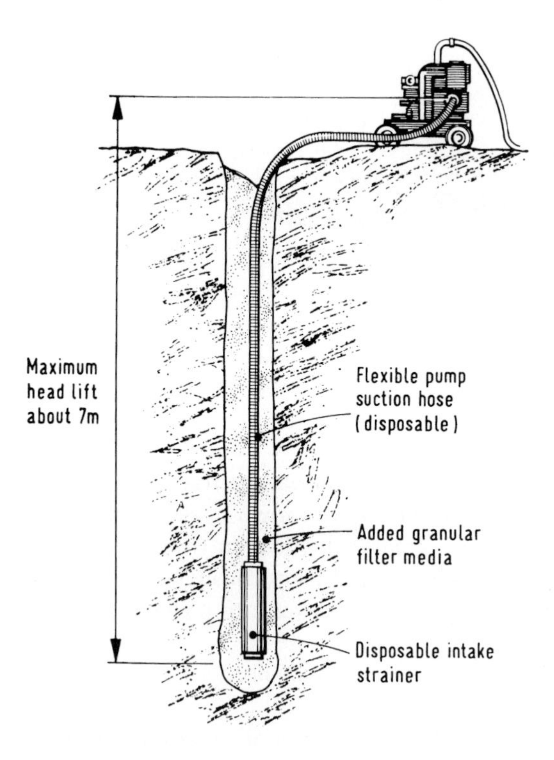

Figure 8.3 A jetted sump (from Somerville 1986: reproduced by kind permission of CIRIA).

the additional units required to give the increased capacity for the initial pumping load can be shut down as the required levels are reached. The spare pumpsets should be left in position to act as standby in case of breakdown or other emergency. Intermediate pumping may also be required in a subsequent backfilling state.

8.4 Drainage of side slopes of an excavation

Within an area of excavation toe drains and pumped sumps may be required to collect seepages of groundwater from the side slopes as well as rainfall run-off.

The collection of seepages from side slopes has been addressed in Section 4.4. The provision of a toe filter drain (Fig. 4.7) is important. The water pumped from the toe drainage system should be clear – that is, no fines. The presence of fines in the pumped seepage water would indicate that the filters are inadequate; that fines are being continually removed and eventually the slope will become unstable. The sumps of the seepage collection system should be as described above in Section 8.3. The expected seepage rates of flow should indicate the dimensions of the sumps that will be required.

8.5 Sump pumping of small excavations

Initially the discharge water from any pumping operation will be discoloured due to the presence of fines. Continuation of the discolouration of the water can be tolerated only if the pumped water is entirely derived from surface run-off. It cannot be tolerated if the discolouration continues when the flow is derived from groundwater – this is a warning of potential danger for it

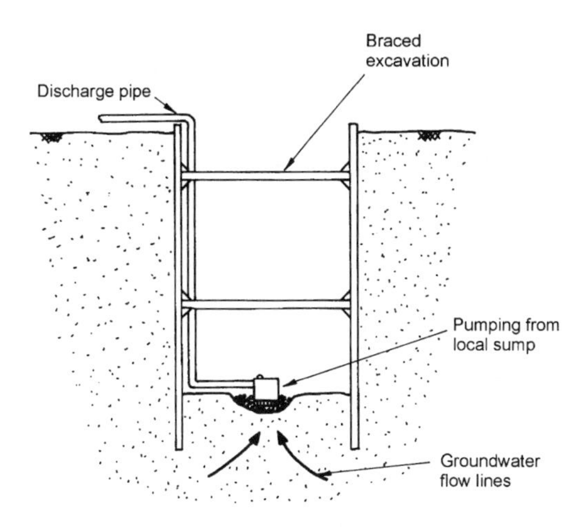

Figure 8.4 Sump pumping from within trench.

indicates continuing withdrawal of fines from the formation. This dangerous condition is often seen in small excavations and trenches of modest depths.

A considerable amount of excavations are carried out for trenching (see Irvine and Smith 1992). For these, sump pumping can be used safely in permeable soils such as gravels and clean sand/gravel mixtures. It is simple and economical and is particularly appropriate to use in trench-sheeted excavations – the sheeting tends to limit the inflow to be pumped (see Fig. 8.4) – provided the discharge water is clear. When dealing with finer grained soils (such as silty sand) a system of wellpoints adjacent to the side(s) of the trench should be used, instead of sump pumping.

8.6 Sump pumping problems

In practice sometimes a wellpoint or deep well installation fails to establish the total lowering required – perhaps because the wellpoints were not placed deep enough or were spaced too far apart. In such circumstances the safe and the correct procedure is to install a second stage of wellpoints (see Section 9.9).

There have been occasions where this correct procedure was not followed but instead efforts were made to achieve that final lowering – say another 0.5 m – by additional sump pumping. Fig. 8.5 shows a typical result of

Figure 8.5 Significant ground movement caused by inappropriate sump pumping. The wellpoint risers around the perimeter of the excavation had originally been installed vertically. Loss of fines due to poorly controlled sump pumping resulted in ground movements, distorting the wellpoints from the vertical.

the messy conditions created by this incorrect application. In this case the foundation slab for a valve chamber was required to be constructed at a modest depth below the water table. The proposed excavation into a very silty sand stratum was ringed by a wellpoint installation – the correct approach. Unfortunately the requisite amount of lowering was not achieved. The reason for this shortfall in lowering is not known. Perhaps the wellpoints were not installed deep enough; perhaps there were many air leaks in the pipework so a good vacuum was not established; perhaps the permeability of the soil was less than predicted so the potential achievable lowering was less. There would appear to have been no close observation of what was happening while sump pumping was continued. The sump pumping resulted in the continuous removal of fines from the sides of the excavation. It can be seen that the wellpoints and their risers (originally vertical) were moved towards the sump pumped excavation by the considerable movement of the surrounding ground.

Another example of an effect of sump pumping is shown in Fig. 8.6. A bund of predominantly granular material was placed to protect an excavation site beside a tidal estuary. A cut-off wall was formed to penetrate to an underlying stratum of low permeability and so exclude water in the estuary

Figure 8.6 Outwash fans due to sump pumping. The suction hoses for the sump pumps were mounted on the crude pontoon shown in the left of the photograph. This helped pumping continue as the water level within the excavation varied with the tide.

as the tide level rose. Prior to closure of the cut-off it was judged to be acceptable to carry out limited sump pumping of the excavation so that work inside the bund could be continuous even around the times of high tides. There came a day when accidentally pumping was continued longer than necessary. The photograph shows a series of outwash fans due to the transport of fine sand in prolonged seepage flows.

If the water flowing to the sumps is removing fines from the soil, the pumped water will have a significant sediment load of sand, silt and clay-size particles. Sump pumps are normally tolerant of some sediment in the water and are likely to continue to operate unless the sediment load is exceptionally high, when they can become choked with sand and silt. However, the discharge of the sediment-laden water at the disposal point is likely to cause environmental problems. If the water is discharged to a sewer the sediment may build up in the sewer, reducing capacity and causing the sewer to back up. If the water is discharged to a watercourse the sediment will have a harmful effect on aquatic plant, fish and insect life. It can be difficult to economically remove silt and clay-size particles from discharge water (see Section 13.4).

If the transport of fines cannot be controlled by the construction of sumps with adequate filters, and if the discharge water cannot be adequately treated, a change in dewatering method should be considered. The use of a well system (wellpoints, deep wells or ejectors) with adequate filters is normally a viable alternative to prevent loss of fines occurring.

8.7 Case history: sump pumping of large excavation

Quite large-scale projects have been dealt with adequately by sump pumping of relatively high permeability soils – such as alluvial gravels with some sand and having little or no 'fines'.

Morrison Construction Limited formed an open reservoir upstream of the city of Aberdeen beside the river Dee, to the instructions of Mott MacDonald, the Engineer appointed by the Client, Grampian Regional Council Water Services. The water surface area of the reservoir is of the order of 8 ha with a storage capacity of 240,000 m^3. The reservoir was created by forming a horse-shoe shaped embankment to marry with the existing flood embankment of the river Dee and all to similar height – the length of the reservoir embankment constructed by Morrison Construction Limited was 1,150 m. The flood plain soils beneath the reservoir are alluvial sandy gravels of very high permeability – one pumping test indicated a permeability of the order of 2.5×10^{-2} m/s. The levels of the groundwater were much influenced by the river level, not surprising in view of the very high permeability of the flood plain deposits!

The Engineer required the formation level of the floor of the reservoir to be some 1.5 m below normal river level. The floor and the sides of the reservoir

Figure 8.7 Dee Reservoir, central pumping sump (courtesy of Grampian Regional Council). The sump pumps comprise one electrically-powered unit, supplied from a portable generator, and two diesel-powered units.

were required to be lined with an impermeable membrane of low density polyethylene sheeting laid on a 50 mm thick sand bedding. Thus, in order to lay the impermeable membrane and satisfactorily joint adjacent sheets, it was necessary to lower the groundwater level to some 2 m below normal river level.

A large central pumping sump (Fig. 8.7) was excavated to a depth of about 4 m below the level of the underside of the membrane. Three drainage trenches (approximately 80–120 m long) were excavated to about two metres depth below the level of the underside of the impermeable membrane, sited to radiate out from the central sump to the riverside perimeter of the reservoir (Fig. 8.8).

Into these drainage trenches were placed 150 mm perforated uPVC drainage pipes and the trenches were backfilled with 75 mm cobble size material. This facilitated the general drawdown of the groundwater beneath the area of the reservoir by pumping from the one central sump; though there was some additional pumping towards the end of membrane laying from an auxiliary sump sited close to the inside toe of the embankment.

The discharge from two 200 mm vacuum assisted self-priming centrifugal pumpsets plus three 150 mm pumpsets – approximately 1,000 m^3/h – was

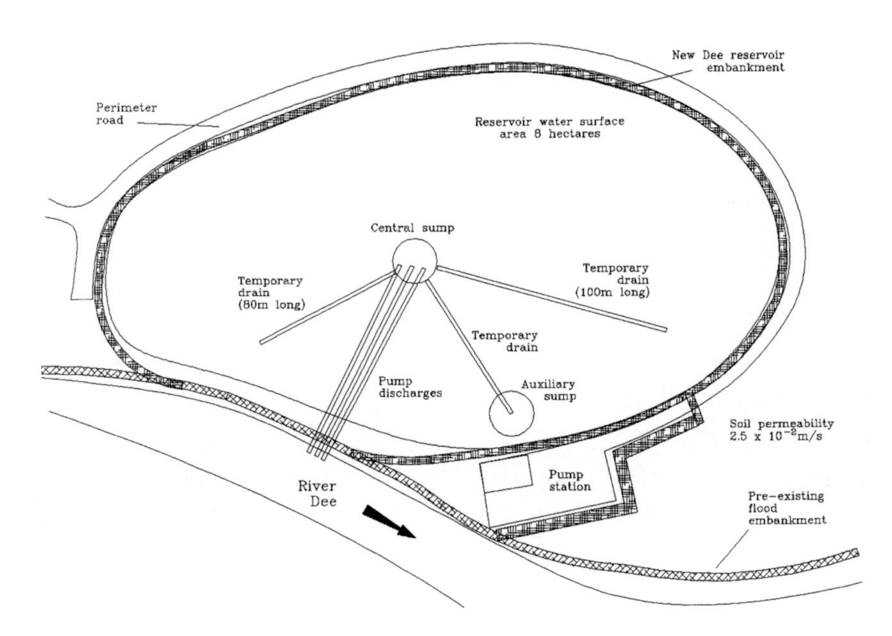

Figure 8.8 Dee Reservoir, plan showing positions of sump and temporary drainage trenches (courtesy of Grampian Regional Council).

discharged to the River Dee. This sump pumping installation enabled Morrison Construction Limited to lay the specified sand bedding and membrane satisfactorily.

References

Irvine, D. J. and Smith, R. J. H. (1992). *Trenching Practice*. Construction Industry Research and Information Association, CIRIA Report 97, London.

Powers, J. P. (1985). *Dewatering – Avoiding its Unwanted Side Effects*. American Society of Civil Engineers, New York.

Scott, J. S. (1980). *The Penguin Dictionary of Civil Engineering*, 3rd edition. Penguin, Harmondsworth.

Somerville, S. H. (1986). *Control of Groundwater for Temporary Works*. Construction Industry Research and Information Association, CIRIA Report 113, London.

Chapter 9

Wellpoint systems

9.0 Introduction

For small and medium sized construction projects of limited depth wellpointing is the most frequently used pumping method of control of groundwater. It is much used for shallow pipe trenching and the like. The deep well system is usually more appropriate for deep excavations and is addressed in Chapter 10.

This chapter describes the wellpoint pumping method. Current good practices, installation procedures and practical uses and limitations are discussed. Variations to the wellpoint system to cope with differing soil conditions are considered.

A variation on the conventional wellpoint system applicable to pipeline trenching in open country, often referred to as horizontal wellpointing is addressed in Chapter 11 together with other less commonly used groundwater lowering techniques including ejectors.

A wellpointing case history is described at the end of the chapter. It illustrates the success that can be achieved in dealing with difficult soil conditions by using adequate 'sanding-in' procedures to provide satisfactory vertical downward drainage and so achieve acceptable pore water pressure reductions.

9.1 Which system: wellpoints or deep wells?

There is a certain amount of confusion, probably due to some looseness of terminology, concerning the precise differences between wellpoint and deep well systems. Essentially, a wellpoint pumping system sucks groundwater up from a group of wellpoints to the intake of a pump unit and then pumps it to a disposal area. Its application is constrained by the physical limits of suction lift. In contrast, a deep well system consists of a group of wells, each having a submersible pump near the bottom of the well, which likewise pump to a disposal area; the method is not constrained by suction lift limitations. This is the key practical distinction between the two systems.

From consideration of both economic and technical criteria, it is likely that wellpointing will not be the most appropriate technique to use if the following conditions exist:

- Large excavations or where excavation depths are greater than 12–15 m.
- Where there is a pressure head in a confined aquifer below an excavation which should be reduced to preserve stability at the formation level.

For such conditions, a deep well system, sometimes supplemented by a system of relief wells, might be considered.

9.2 What is a wellpoint system?

A wellpoint system consists essentially of a series of closely spaced small diameter water abstraction points connected, via a manifold, to the suction side of a suitable pump.

The wellpoint technique is the pumping system most often used for modest depth excavations, especially for trenching excavations and the like (see Fig. 9.1). In appropriate ground conditions a wellpoint system can be installed speedily and made operational rapidly. The level of expertise needed to install and operate a wellpoint system is not greatly sophisticated and can be readily acquired. However, as with any ground engineering process, having experienced personnel to plan and supervise the works can be crucial in identifying and dealing with any change in expected ground conditions.

A wellpoint is a small diameter water abstraction point (the well screen), sometimes referred to as a 'strainer' (so called because of the wire mesh or other strainer of the self-jetting wellpoint) through which the groundwater passes to enter the wellpoint. They are installed into the ground at close centres to form a line alongside (Fig. 9.1), or a ring around (Fig. 9.2), an excavation. The perforated wellpoint is typically about 0.7–1.0 m in length and 40–50 mm nominal diameter. Each is secured to the bottom end of an unperforated pipe (the riser pipe) of slightly smaller diameter; 38 mm diameter pipe is commonly used. However, where the 'wetted' depth is limited due to the proximity of an impermeable surface (see Fig. 9.13(b)), it is preferable that the length of the wellpoint should be shorter (usually 0.3–0.5 m) to restrict the risk of air intake at maximum drawdown. Often it will be necessary to install wellpoints at closer centres to compensate for the lesser screened length of each short wellpoint (see Section 9.7).

Each wellpoint is connected to a header main (typically of 150 mm diameter) that is placed under vacuum by a wellpoint pump (Fig. 9.2). The header main is normally made of high impact plastic, although steel pipe is sometimes used, especially when there is a risk of damage from construction activities. The pipe is typically supplied in 6 m lengths and is joined on site by simple couplings that allow a certain degree of skew to allow the header

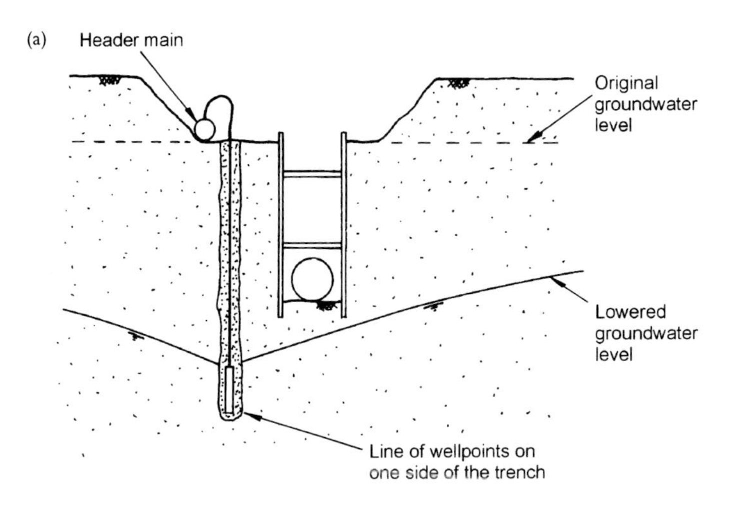

(b)

Figure 9.1 Single-sided wellpoint system. (a) Section through trench with line of wellpoints close to one side of trench, (b) Single-sided wellpoint system for a shallow pipeline trench (courtesy of Sykes Pumps).

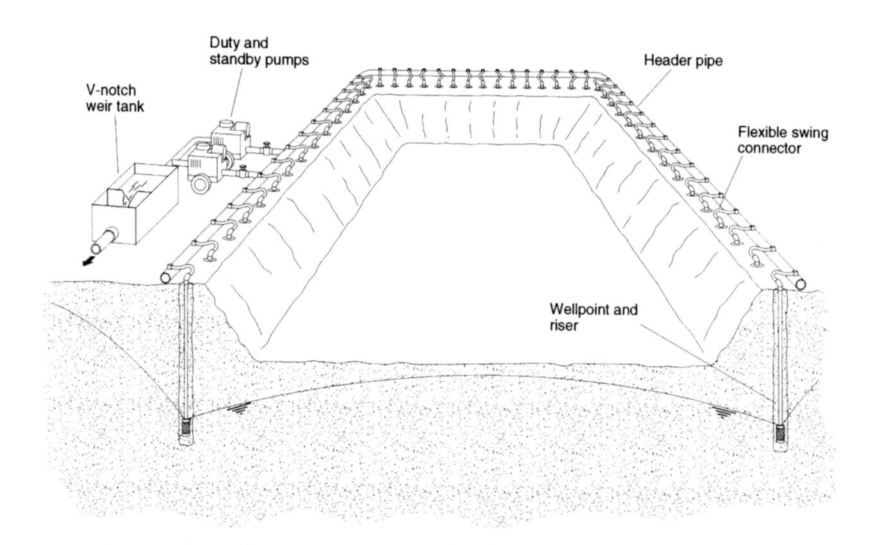

Figure 9.2 Wellpoint system components (from Preene *et al.* 2000; reproduced by kind permission of CIRIA).

main to be laid around gentle curves if necessary. For sharper curves, 90° and 45° bends are available, as well as tee pieces, blanking ends, etc.

The header main has connections for wellpoints at regular intervals, perhaps every metre.

The individual wellpoint riser pipes are connected to the header main via 'swing connectors' sometimes known simply as 'swings'. The swing connectors provide some flexibility in connecting the wellpoints (which are unlikely to be installed in a precise straight line) to the relatively rigid header main. Before the advent of readily available plastic hose the connectors were made from a series of metal pipework bends which, when swung around, gave the necessary articulation – hence 'swing connectors'. Nowadays swings are typically formed from flexible plastic hose (of 32–50 mm diameter), with each swing incorporating a trim valve (Fig. 9.3) to allow the abstraction rate of each wellpoint to be regulated if necessary (see Section 9.6).

The applied vacuum from the pump sucks the groundwater from the surrounding ground through the wellpoint screens, into the riser pipes, through the swing connectors and thence up into the header main and so to the pump intake. The pump then forces the water through the discharge main for ultimate disposal. Thereby the system effects local lowering of the groundwater within the area that it encompasses.

Ideally, the header main should be just above the static groundwater level to minimize the amount of suction lift. This may entail shallow trenching (Fig. 9.4). If it is done, there is the added advantage that the excess jetting

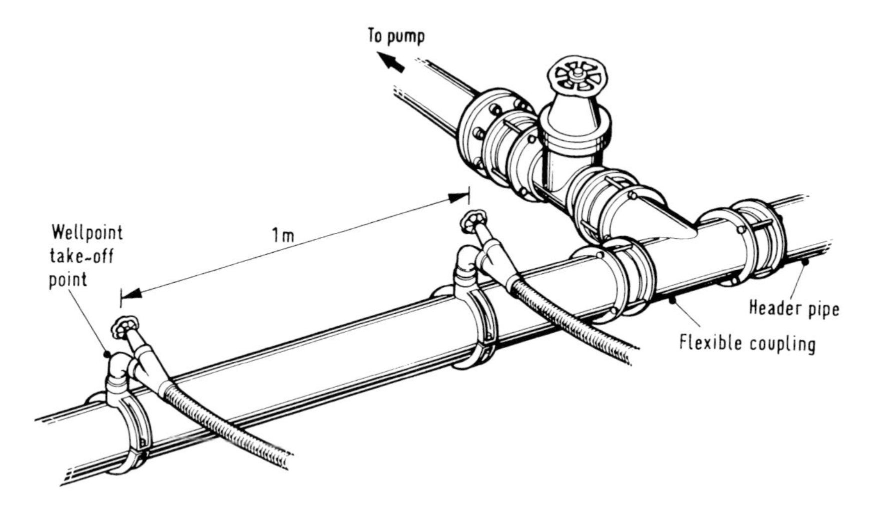

Figure 9.3 Flexible connection from wellpoint riser to suction manifold via trim value (from Somerville 1986; reproduced by kind permission of CIRIA).

water from installation of the wellpoints tends to be contained within the header trench rather than cause general flooding of the site.

Ideally, the suction intake of a wellpoint pump should be at the same level as the header main. Often this will require that a small pit be dug to lower the pump body to the requisite level.

The amount of lowering that can be achieved by a wellpoint (or shallow well) system is limited by the physical constraints of suction lift. It is generally in the range of 4.5–6 m though 7 m drawdown is not unknown. It depends very much on the efficiency of the water/air separation device on the pump and the vacuum efficiency of the total installation (i.e. the air-tightness of the system pipework). In addition it depends on the structure and permeability of the soil mass. The width of the excavation is also pertinent to achieving the required amount of lowering at its centre. Where the land area available for construction is not constrained, the wellpoint system can be used for deeper excavations by means of a multistage wellpoint installation (Fig. 9.5).

9.2.1 Types of wellpoint

The wellpoints are vital components of every installation. There are two types of wellpoints:

(i) *The self-jetting wellpoint* (Fig. 9.6(a))
These are known as self-jetting since they can be installed without the use of a placing tube. The wellpoint and riser are metal and therefore are rigid,

Figure 9.4 Wellpoint header main installed in shallow trench.

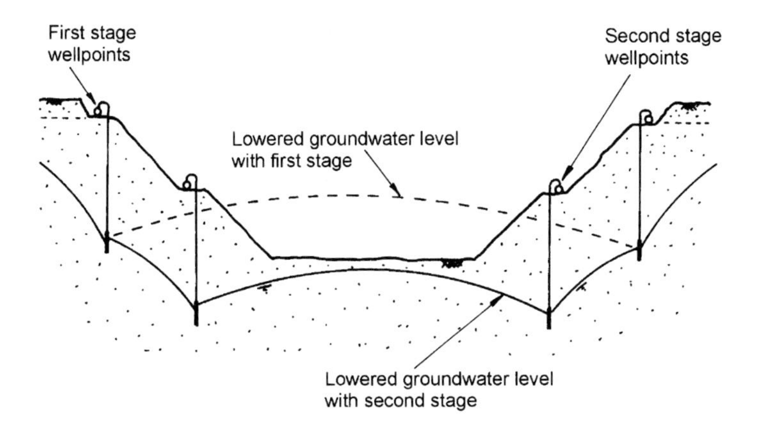

Figure 9.5 Two stage lowering using wellpoints.

with the wellpoints connected to the bottom of the riser pipe. This type may be recoverable for subsequent reuse.

The wellpoint and riser are installed using high pressure water supplied to the top of the riser pipe from a clean water jetting pump. There is a hollow jetting shoe below the wellpoint screen. Near the lower end of the shoe a horizontal pin is located; above the pin is a lightweight loose fitting ball. When the water pressure is applied to install the wellpoint, the ball is displaced downwards to allow the passage of the high pressure water flow but the pin retains the ball within the jetting shoe. When pumping starts the applied vacuum sucks the ball up onto a shaped spherical seating and thereby seals the lower end of the wellpoint, so that water from the surrounding ground can only enter through the screen section. The unperforated riser should extend to near the lower end of the wellpoint screen to minimize the potential for air intake at maximum drawdown.

Whilst the self-jetting wellpoint can be subsequently extracted for reuse, it is not uncommon for the riser pipes to be damaged during extraction and so need straightening or even replacement before reuse. In addition, the wellpoint screens may need 'desanding' before being suitable for reuse.

(ii) *The disposable wellpoint* (Fig. 9.6(b))
The wellpoint and riser is usually of plastic materials and therefore inert to corrosion. They are installed by means of a placing tube or holepuncher using similar high pressure water jetting techniques for installation as for the self-jetting wellpoint.

While many purpose manufactured plastic wellpoints have been marketed for some time, current practice is to adapt low cost thin walled convoluted uPVC perforated land drainage pipe to form disposable wellpoints with woven mesh stocking or 'coco' wrapping. The latter forms a good filter

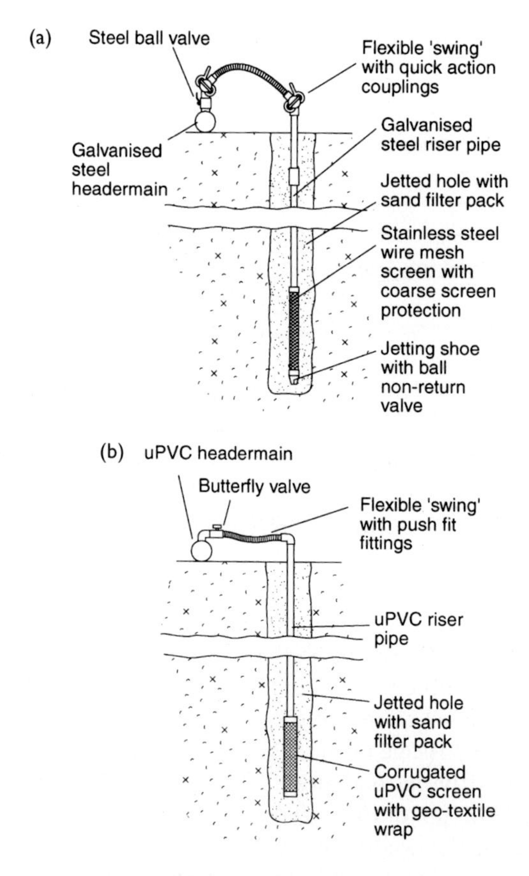

Figure 9.6 Disposable and reusable self-jetting wellpoints (from Preene *et al.* 2000: reproduced by kind permission of CIRIA). (a) Self-jetting wellpoint, (b) disposable wellpoint.

and in conjunction with the normal sanding-in is very effective, even in difficult silty soils. As with the self-jetting wellpoint, the riser (typically made of uPVC ducting) should extend to near the bottom of the screened length. The bottom end of the disposable wellpoint is sealed. Generally the riser pipe is a nominal 6 m long but it can be longer.

The disposable wellpoints are not recoverable but some of the plastic riser pipes can sometimes be recovered for future reuse. The disposable wellpoint is very appropriate to use for long duration pumping duty.

9.3 Wellpoint installation techniques

Self-jetting wellpoints and placing tubes used with disposable wellpoints are installed using high pressure water supplied from a high pressure jetting pump.

The jetting pump (see Section 12.2) most commonly used supplies water via the jetting hoses at a rate of about 20 l/s and at a pressure of 6–8 bar. The flexible jetting hoses are usually standard 63 mm fire hoses with instantaneous male and female connections. There are other more powerful jetting pumps and hoses that are often used for the holepuncher, pile jetting, and other similar heavy duty uses.

The procurement of an adequate supply of jetting water for installation must be resolved for each site. A continuous supply of clean water for jetting is essential for efficient placing of all types of wellpoints, whichever installation technique is used. If the source of jetting water is restricted, consideration must be given to obtaining water by pumping from the first few wellpoints as each is installed.

Generally the volume of jetting water required for installation per wellpoint with 6 m long riser will be of the order of 1–1.5 m^3 but will depend greatly on soil conditions. However, it can be in the range 0.5–35 m^3, the later figure being applicable to jetting in a very permeable river gravel.

Jetting in compact sands and gravels, and especially in an open gravel, may be difficult and slow, mainly because the displaced or slurrified soil particles tend not to be washed from the jetted hole to ground surface due to rapid dissipation of pressurized water into the open permeable formation. However, there is some compensation because in such permeable soils sanding-in is unlikely to be required.

The wellpoint system is very flexible. For instance subsequent to the initial installation, extra wellpoints can be placed speedily to deal with localized trouble spots. It is especially useful for shallow depth trench excavations where very often the pumping period is expected to be of short duration.

9.3.1 Installation of self-jetting wellpoints

The wellpoint and its riser pipe are assembled to the required length and a flexible jetting hose is attached to the top of the riser pipe via a jetting adapter. The other end of the flexible jetting hose is connected to the supply outlet from the high pressure jetting pump.

It is prudent to form a small starter hole into which to position the wellpoint; or, for a line of closely spaced wellpoints to form a trench and so restrict general flooding of the site area. Prior to turning on the supply of jetting water, at the required wellpoint position the jetting crew must upend the wellpoint and riser with jetting adapter and jetting hose all connected (Fig. 9.7). There is considerable weight in the riser and hoses – this is a strenuous and tricky task.

The jetting pump forces high pressure water down the metal riser pipe, past the ball at the lower end of the wellpoint. A strong jet of water emerges from the base of the wellpoint. The high pressure water slurrifies the granular soil immediately below the bottom of the wellpoint enabling downward

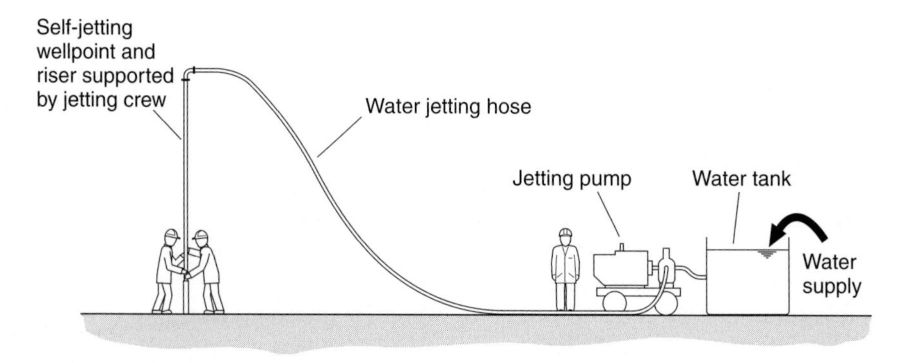

Figure 9.7 Installation of self-jetting wellpoint (from Preene *et al.* 2000: reproduced by kind permission of CIRIA).

penetration to be made until the required depth is achieved. The slurrified soil is washed to the surface with the jetting water and emerges in the annulus around the riser pipe. This turbulent flow of water and soil is colloquially known as the 'boil'. On reaching the required depth, the output from the jetting pump is throttled back during sanding-in. Then, and only then, should the water be turned off completely and the jetting adapter and hose be disconnected.

In suitable soils, such as clean sands, the installation of self-jetting wellpoints is speedy – in some cases it may be only a matter of a few minutes per wellpoint for the actual jetting to depth. In such circumstances it is, therefore, very cost effective.

As a rough guide where the Standard Penetration Test (SPT) N values are less than about $N=25$–30, both self-jetting wellpoints, and lightweight manhandleable placing tubes (for installing disposable wellpoints) can be used. Either technique requires minimal mechanical plant, but has a greater need for manual labour force.

In denser gravels one of the controlling factors is the sand content of the soil. If the soil is sufficiently sandy to allow the jetting water to return to the surface as a 'boil', jetting will probably be feasible. If gravels have little sand content and allow the jetting water to dissipate the 'boil' will be lost, and jetting will be difficult if not impossible.

9.3.2 Installation of disposable wellpoints by placing tube

For the installation of disposable wellpoints a similar process is mobilized using a placing tube. High pressure water from the jetting pump is applied to the placing tube via the appropriate fitting at the top of the tube (Fig. 9.8). The high pressure water emerges from the bottom of the placing tube and slurrifies the soil so that penetration to the required depth is achieved. The 'boil' emerges around the outside of the placing tube. The placing tube is the temporary casing into which the disposable wellpoint and riser pipe are then installed centrally and sanded-in.

Modest cranage will be needed to handle the placing tube (e.g. a backhoe loader such as a JCB 3C or a small 360° hydraulic excavator). However, if

(a)

Figure 9.8 Installation of wellpoints using placing tube (courtesy of Dewatering Services Limited). (a) Steel placing tube is suspended from excavator, (b) placing tube is jetted into the ground, (c) plastic disposable wellpoint is installed.

(b)

Figure 9.8 Continued.

access for plant is restricted, where jetting conditions are easy a lightweight 100 mm placing tube that can be manhandled may be used.

The placing tube itself is a robust open-ended jetting tube of nominal 100 or 150 mm casing having a wall thickness of 4–5 mm. The fittings at the top of the tube are arranged such that not only high pressure water can be applied from a jetting pump but there is in addition, the facility to apply compressed air to assist downward penetration through difficult ground conditions.

(c)

Figure 9.8 Continued.

When an excavator is used to handle the placing tube, the top of the tube should be fitted with a platform or anvil such that if difficult ground conditions are encountered (e.g. a thin layer of stiff clay or random cobbles) the bucket of the machine can be used to press down on the platform when installation progress is slow. The anvil also serves to protect the cap and various fittings at the top of the tube.

On reaching the required depth, the jetting water is turned off, the cap at the top of the tube is removed, the wellpoint and its riser inserted to depth centrally inside the placing tube and the placing tube is withdrawn as sanding-in progresses.

As a rough guide the placing tube technique for wellpoint installation is appropriate where the SPT N values are about $N=40$ or below. Where many random cobbles are expected and the SPT values are about $N=35$–40 and where SPT values are above $N=40$ the holepuncher technique should be considered.

In recent years there has been a tendency to extend the use of the placing tube technique to replace the more labour intensive use needed for the self-jetting wellpoint and the lightweight placing tube techniques.

9.3.3 Installation using the holepuncher and heavy duty placing tube

The holepuncher technique has been used successfully to progress through soils having SPT N values below $N=30$ and up to about $N=65$. While it is often used to install wellpoints in hard or difficult ground conditions, it is also used for the installation of deep wells. For wellpoint applications the holepuncher is typically suitable for installations up to 15 m deep.

The holepuncher (often called a 'sputnik') is a simple robust form of wash boring equipment having in addition a drop hammer driving facility. It consists of an outer heavy duty casing (usually 200–250 mm nominal bore but can be up to 450 mm bore or even 600 mm for deep well installations) and an inner wash pipe which has a weighted head and can be used also as a drop hammer. Usually the top of the wash pipe has two intake ports for supply of high pressure jetting water and also has a facility for a compressed air supply to be connected. The 'boil' of washings rises to ground level in the annulus between the inner wash pipe and the outer casing (Fig. 9.9).

Cranage will be needed for the use of a holepuncher. The crane must have free fall on two hoist lines. If compressed air is used to assist installation, the minimum compressor capacity needed is of the order of 55 l/s at about 10 bars.

The forerunner to the holepuncher was the heavy duty placing tube, which is still used by some organizations. This equipment similarly consists of a thick walled casing (usually 150–250 mm bore) with a box at the head containing lead weights. There are two jetting water connections and a compressed air connection – again similar to the holepuncher – provided just below the top. The inner jetting tube and lead weighted head cannot be lifted independently of the outer casing, unlike the holepuncher. Hence, the displaced soil spoil is backwashed to ground level outside the casing tube. Similarly a crane with free fall hoist lines is required to handle the heavy duty placing tube.

The wellpoint and its riser pipe are installed and sanded-in using the same procedures as for the placing tube method.

There are particular safety hazards associated with using a holepuncher or heavy duty placing tube. The tube itself is a heavy and unwieldy device not easily operated by crane drivers who have never used one before. The volume of jetting water and compressed air applied inputs a lot of energy into the ground; this can result in gravel and cobble fragments being ejected from the top of the tube with great velocity, to land some distance away.

Figure 9.9 Holepuncher for installation. (a) Schematic view of holepuncher (from Somerville 1986; reproduced by kind permission of CIRIA). The inner wash pipe is raised and lowered during jetting, to allow the weighted head to the used as a drop hammer. (b) Installation in progress (courtesy of WJ Groundwater Limited). The holepuncher is suspended from a crane. (c) Close up of top of holepuncher (courtesy of Sykes Pumps). The holepuncher has been jetted in to depth, the return water can be seen washing up between the inner pipe and outer casing. The jetting pump is visible in the background.

(c)

Figure 9.9 Continued.

Especially in urban areas or on confined sites, the risks to operatives and the public should be carefully assessed when considering using the holepuncher technique.

9.3.4 Installation by rotary jet drilling

The rotary jet drilling technique (Fig. 9.10) is a recently developed adaptation of jetting by placing tube. An excavator-mounted drill mast with hydraulic rotary head and swivel allows the placing tube to be rotated as it is jetted in. The rotary head can also be used to apply downward force to the placing tube to aid penetration. This system has been used in a range of conditions

(a)

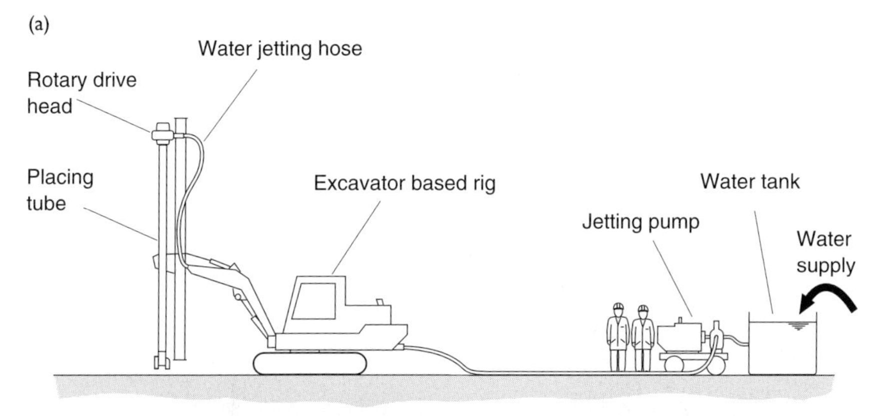

(b)

Figure 9.10 Rotary jet drilling rig. (a) Schematic view of rig (from Preene *et al.* 2000: reproduced by kind permission of CIRIA). (b) Installation in progress (courtesy of WJ Groundwater Limited).

including clays, sands, sandy gravels and weak rock such as weathered sandstones. It is sometimes used as an alternative to a holepuncher, to avoid some of the safety hazards associated with the latter method.

9.3.5 Installation through clay strata

In cohesive soils overly water-bearing granular soils, it may be expedient to form a pre-bore through the cohesive soil layers using a flight auger (Fig. 9.11) and then, having penetrated the clay soils, change to a water jetting technique.

Figure 9.11 Hydraulic auger attachment for pre-boring through cohesive strata (courtesy of WJ Groundwater Limited).

Installation by jetting is likely to be difficult if clay layers have to be penetrated since, due to the cohesiveness of clays, the size of the jetted annulus outside the wellpoint and riser or placing tube is likely to be very restricted (or almost non-existent) and so constrain the backwash of soil particles to the surface.

A technique used in North America to install self-jetting wellpoints in these conditions is 'chain jetting'. A chain is attached to the lower end of the wellpoint and wound around the riser to ground level prior to jetting in. The lifting and surging of the wellpoint and riser is effected using the chain not the riser pipe. The chain will increase the size of the hole made through the clay and thereby create an annulus through which granular soils from below the clay layer can be washed to the surface and into which subsequently the sanding-in filter media can be installed from the base of the wellpoint to ground level.

When jetting in mixed soils (such as glacial till) the clay impediment may be further exacerbated by the presence of cobbles and boulders. If it is necessary to install wellpoints in such soil conditions a holepuncher or rotary jet drilling rig is likely to be preferred.

Where it is difficult to install wellpoints by jetting, a site investigation borehole drilling rig may be appropriate. In such a case boreholes may be drilled to the required depth using temporary boring casing. The wellpoint and riser are then installed centrally in the borehole and the annulus backfilled with a sand filter as the casing is withdrawn. This onerous variation on installation techniques is more likely to be appropriate to a pore water pressure reduction project but may be needed for a groundwater lowering requirement in a highly stratified soil formation.

9.3.6 The merits of jetted hole installations

A water jetting technique using a self-jetting wellpoint, a placing tube, a holepuncher or rotary jet drilling establishes holes without side smear and so provides a more efficient drainage hole than a hole made using a site investigation boring technique or a continuous flight auger technique. Hence, from the superiority of the resulting water abstraction properties apart from cost considerations, the water jetting technique is to be preferred wherever practicable.

9.3.7 The need for filter media

During pumping, apart from the initial pumping period, fines should not be continuously withdrawn from the surrounding ground – so the discharge water should be clear. If this were not so the installation is faulty for assuredly, continuous withdrawal of fines will lead to instability.

Prevention of continuous movement of fines can be achieved by means of a column of filter media around each wellpoint and its riser pipe by sanding-in. This is very similar to the provision of a filter pack around the screen of a deep well (see Section 10.2) but the grading of the filter media for sanding-in of a wellpoint is not as critical as it is to a deep well installation.

The wellpoints must have sufficient flow capacity through the wellpoint screen to provide adequate water abstraction from the surrounding soil. This should not be a problem when using adequately designed wellpoints. Where there is the need to drawdown to a level close to an impermeable interface it is prudent to use short (say 0.3–0.5 m length) wellpoints to reduce the risk of drawing in air and the consequent need for repetitive adjustments to the trim valves. In a high permeability soil, the reduced length of wellpoint will restrict its potential water passing capacity and a compensating reduction in spacing between wellpoints will be needed. Thus the number of wellpoints required will be greater.

The conventional coco or woven mesh stocking wrapping to the disposable wellpoints provides a filter additional to that of the sanded-in media.

9.3.8 *Sanding-in*

Efficient sanding-in of wellpoints and risers helps prevent the removal of fines. The dewatering efficiency of individual wellpoints will be improved by providing preferential down drainage paths from any overlying perched water tables that may be due to variations in vertical permeability in heterogeneous strata. Generally a washed sharp sand, similar to a medium coarse concreting sand will be a satisfactory filter or formation stabilizer. Sanding-in around the wellpoint and its riser pipe is essential in silty soils and fine sands but may be omitted when installing wellpoints in non-silty coarse sands, gravelly sands and sandy gravels. In such soils an effective natural filter pack can be formed by the washing action of the jetting process (see Section 10.2).

The procedure for sanding-in a self-jetting wellpoint and its riser is as follows. Upon reaching the required depth, throttle back on the jetting pump so that the water emerging in the 'boil' from the wellpoint hole at ground level just gently 'bubbles'. This upward flow of water should maintain the oversized jetted hole temporarily while sanding-in is progressed. If this oversized hole is not maintained because the jetting pump has stopped, the soil may collapse around the wellpoint and the integrity of the filter column will be impaired. In general the technique for the placing of the filter media is very basic, shovelling sand into the annulus which is being kept open by the 'bubbling' rising water until the annulus is filled to ground level.

Sanding-in is vital to achieve an efficient installation in layered or other non-homogeneous soils. Figure 9.12 shows sanding-in columns of filter media exposed in the face of an excavation. The near vertical face of the excavation is sustained by capillary tension effects in fine-grained soils.

It has been known for only two or three shovels of sand to be placed in a wellpoint annulus – this is totally inadequate!

Disposable wellpoints installed by placing tube are sanded-in in a similar manner. After the water is turned off and the top cap removed from the placing tube on reaching the required depth, the wellpoint and riser are installed centrally within the placing tube. The filter sand is placed around the wellpoint and riser within the placing tube as it is withdrawn. The level of the top of the filter sand should always be above the level of the bottom of the placing tube as it is being withdrawn. Some installation crews sand-in by adding sand around the outside of the placing tube as it is withdrawn.

The soil structure is important – is the soil mass nearly homogeneous or is it anisotropic? If the soil structure is markedly anisotropic there will be potential for encountering a series of perched water tables or local trouble

Figure 9.12 Exposed sanding-in columns.

spots. Adequate sanding-in is vital to successful drainage of anisotropic soils. Also where a stratum of soft clay is penetrated by a self-jetting wellpoint the clay may squeeze in before the vertical column of filter sand is placed – the filter sand may bridge at the top of the clay layer.

9.4 Spacing of wellpoints and drawdown times

The theoretical number of wellpoints required for a particular project, and the associated spacings will be indicated by the calculations outlined in Chapter 7. However, the spacing of wellpoints for simple trenching excavations is often determined from past experience of working in similar soils.

In practice the spacing between wellpoints tends to be influenced by the spacing of the take-off points on the suction header main supplied by the manufacturer. These are mostly at 1 m centres; so usually actual spacings will be at 1, 1.5 and 2 m centres. A spacing of 0.5 m centres can be achieved with two parallel header mains, each with 1 m take-off spacings. It can be difficult to actually install wellpoints at 0.5 m spacings in a single line; jetting in one wellpoint may blow the adjacent wellpoints out of the ground. In such cases it can be appropriate to install the number of wellpoints equivalent to one line at 0.5 m spacings, but actually lay the wellpoints out in two parallel lines, each at 1.0 m spacing.

Table 9.1 Typical wellpoint spacings and drawdown times

Soil type	*Typical spacings (m)*	*Drawdown (days)*
Fine to coarse gravel	0.5–1.0	1–3
Clean fine to coarse sand and sandy gravel	1–2.0	2–7
Silty sands	1.5–3.0	7–21

Note: If there is a risk of residual overbleed seepage, wellpoint spacings should be at the closer end of the range.

While the spacing selected to effect drawdown will depend on permeability and soil structure; programme time requirements are often of primary importance. If rapid drawdown is needed wellpoints should be installed at closer centres. Typical spacings and approximate drawdown times for a single stage installation are given in Table 9.1.

9.5 Sealed vacuum wellpoint system

Gravity drainage to a wellpoint installation in fine-grained soils of permeability less than about 1×10^{-5}–5×10^{-5} m/s is usually very slow. The rate of drainage can be improved by sealing the annulus around the wellpoint riser at the topsoil zone, with puddled clay or a cement/bentonite mix having a putty like consistency. This allows the suction action of the pump to generate a vacuum in the entire filter column, increasing the hydraulic gradient between the soil and the wellpoint. This is often described as vacuum wellpointing, and is one of the methods of pore water pressure control used in fine-grained soils (see Section 5.4).

This technique can be quite effective in stratified soils, provided that proper sanding-in has been achieved from the bottom of the wellpoint to the underside of the clay plug. On occasions it has been used effectively to reduce the moisture content of low strength clay soils and so thereby increase slope stability. The vacuum tank unit pumpset (see Section 12.1) would be the appropriate type to use for this particular duty. Ejector systems (see Section 11.1) might be considered for use in place of a vacuum wellpoint system.

The wellpoints should be closely spaced because of the limited effect of individual wells in low permeability soils. If this technique is applied to low permeability soils in a loose condition, the side slopes of excavation should be protected from sudden disturbances – such as use of any vibration techniques – as these might cause liquefaction.

9.6 Wellpoint pumping equipment

Wellpoint equipment sets are available from many specialist manufacturers. The two important aspects to be considered before ordering equipment are:

(a) Long-term reliability.
(b) Overall cost, including maintenance costs.

Unfortunately some suppliers do not conform to these simple and sensible guidelines. Some supply cumbersome equipment that requires cranage for handling. Other suppliers go to the other extreme, supplying flimsy equipment that is readily susceptible to damage and so has only a short useful life.

A typical medium sized wellpoint installation will incorporate a 150 mm duty pump plus a standby pump connected to 150 mm sized header main (which acts as a suction manifold) connected to the trim valves of each individual wellpoint via its riser. The individual component parts have been described earlier.

9.6.1 *Duty or running pumps*

The total connected pump capacity must be sufficient to deal with the greater rate of abstraction required during initial drawdown. In unconfined aquifers of moderate to high permeability the calculated steady-state pumping rate is unlikely to be adequate to achieve acceptably rapid drawdown. In such circumstances the initial rate of pumping may be up to twice the calculated equilibrium rate of pumping required to maintain drawdown. It is common practice on start up to operate both the running and the standby pumpsets to achieve a fast drawdown.

When dealing with high flow rates or a large installation it is prudent to provide multiple pumpsets in the system rather than a single large output running pump since this affords greater flexibility. This often results in better fuel economy and is economical on the provision of standby pumps. Chapter 12 describes the types of pumps suitable for use with wellpoint systems.

For systems running with a single duty pump, the ideal position for the pump is at the middle of the suction header main so that water is being abstracted from an equal number of wellpoints to either side of the pump station. Where two duty pumps are used they may be positioned adjacent to each other – thus at the end of each set's header line – and a standby pump may be positioned and connected so as to pump from either set. This is acceptable only if the pumping load required of either running pump is significantly below its rated capacity.

Generally wellpoint installations are operated using diesel driven units. However, if the running pump is to be electrically powered (perhaps in order to reduce noise levels), it is often acceptable and economical that the standby pump be diesel powered. Since groundwater lowering pumpsets operate 24 h per day fuel is a significant cost factor. Fuel consumption should be highlighted in the build up of cost estimates. The costs of alternative sources of power (such as mains electric power) should be compared if their provision is practicable.

The fuel consumption of the double acting piston pump is more modest than that of the centrifugal pump; also the wear and tear is less because of the slower rate of movement. However, the maximum amount of lowering that can be achieved using a piston pump is generally limited to about 4.5 m. An efficient system operated with a vacuum assisted self-priming centrifugal pumpset should achieve an additional 1 m or more of drawdown in similar soil conditions.

These are general indications for guidance only. Actual achievements will depend upon the permeability of the soils, the mechanical efficiency of the individual pumpsets and of all the pipework connections of the installed system. Air leaks must be minimized because they can have a very significant effect on the amount of suction available at the wellpoints and so affect the lowering that is achievable.

9.6.2 *Standby pumps*

In general, groundwater lowering systems should operate continuously 24 h per day, seven days per week. Hence it is imperative that the installed system incorporates facilities to ensure that pumping is indeed continuous. Generally, this can be provided by connecting an additional (standby) wellpoint pump to the suction header and discharge mains with suitably placed valves for a swift change over of operation from the running pump to the standby unit as and when required. This might be necessary in the event of an individual running pump failure or maintenance stoppage to check oil levels, etc.

A judgement should be made before commencing site work, of the likely effects of a cessation of pumping. The decision should be based on the answers to the following two questions:

i Will a cessation of pumping cause instability of the works? If the answer to this question is 'yes' the provision of standby pumping and/or power facilities is essential.

ii Will a cessation of pumping create only a relatively minor mess that can be cleaned up afterwards at an acceptable cost both in terms of inconvenience and delay? If the answer to this question is 'yes' and there is no safety risk, a judgement has to be made whether or not to accept the cost savings by not providing standby pumps whilst recognizing that a risk is being taken.

9.6.3 Operation of a wellpoint system and 'trimming'

The amount of lowering of the water level that can be achieved by a wellpoint system is governed by:

1 First and foremost – the physical bounds of suction lift. Elevation above sea level and to a lesser extent ambient temperatures, have some input into this limitation. As ground elevation above sea level increases, available suction decreases. Also, ambient temperatures and ground elevation are of significant relevance to the rating of the engine required to drive the pump to achieve the necessary rate of pumping.
2 The hydraulic efficiency of the total wellpoint installation – this includes the pipework sizing above and below ground as well as the adequacy of the wellpoint sanding-in procedures and the air/water separation unit on the pumpset
3 Air leaks. These can drastically reduce the amount of vacuum that is available to withdraw water from the soil via the abstraction points.

As suction is applied to self-jetting wellpoints, their ball valves are seated and the groundwater is sucked through the wellpoint screen only. The bottom of the riser pipe terminates near the bottom of the wellpoint screen so as to minimize or delay the intake of air when maximum drawdown is being achieved.

The bottom end of a disposable wellpoint is sealed and so similarly the groundwater is sucked through the screen to the bottom of the riser pipe and thence to the suction header main.

As the water level at each wellpoint is drawndown to near the level of the top of the screen there will be a risk of entraining air with water and thereby reducing the amount of available vacuum. The trim valves (Fig. 9.3) at each individual wellpoint connection to the header main enable the experienced operator to adjust the amount of suction such that the intake is predominantly water and the amount of air intake is minimized. This is necessary to ensure that the wellpoint system operates at the maximum achievable drawdown.

The adjustment of the valves on each wellpoint is known as 'trimming' or 'tuning' and is an important part of operating any wellpoint system. A poorly trimmed system may achieve significantly less drawdown than system that has been trimmed correctly. Trimming is probably more of an art than a science, but can normally be mastered with experience.

As the system is trimmed (i.e. the wellpoint valves are closed or throttled) the vacuum shown on gauges on the pump and header main should increase. This is because the amount of air entering the system via the wellpoint screens is reducing, allowing the pump to generate more vacuum. When trimming a system it can be very satisfying to see the vacuum increase as a result of your efforts. However, it must be remembered that the aim is not to maximize vacuum but to maximize drawdown and flow rate. It is important not to become obsessed with obtaining greater and greater vacuums. An overzealously trimmed system (where all the wellpoint valves are almost closed) will have a very impressive vacuum but will pump little water and generate little drawdown – this is a classic mistake made by novices when first attempting trimming!

A wellpoint that is pumping problematical volumes of air will probably be producing 'slugs' or 'gulps' of air and water. The momentum of each slug of water will make the wellpoint swing connector jump up and down, perhaps quite violently – this is known as 'bumping'. A 'bumper' is trimmed by slowly closing the valve until the flow is smooth, and then re-opened slightly – the last action is vital to avoid over-trimming. A small steady flow of air along the swing connector is acceptable (provided the rest of the system is functioning well); it is probably counter productive to attempt to trim a system so that no air at all enters the wellpoints.

Wellpoint equipment sets supplied by British, North American and Australian suppliers usually include trim valves to regulate the rate of water/air flow to each wellpoint. However, suppliers based on the continent of Europe generally do not supply trim valves, unless specifically asked to do so. The reason for this is not clear. South East Asian suppliers also tend to omit trim valves and to follow continental European philosophy. The inclusion of trim valves is added cost but the authors consider that the potential for increased lowering by adjustment of trim valves, especially for a multi-stage wellpoint installation, is cost-effective.

9.7 Wellpoint installations for trench excavations

A considerable amount of shallow depth trenching for pipelaying is carried out worldwide. Irvine and Smith (1992) give overall guidance to good trenching practices and include many specific recommendations concerning support of the excavations and other safety guidelines, both with and without the need for groundwater control. This book does not address the many methods of trench support. However, it is important that for each and every trench pumping installation, the basic concepts described should be observed; though depending upon local conditions, some adjustments may be appropriate.

9.7.1 Single-sided wellpoint installations

The pipelaying contractor will much prefer having the header main and associated wellpoints only on one side of the trench, because this will allow uninterrupted access for his plant and equipment on the other side of the spread (Fig. 9.1). The basic question is, how to assess whether a single-sided installation will be adequate. Water levels must be reduced below formation level. The critical point is the bottom of the trench on the side remote from the line of wellpoints of Fig. 9.1. If the drawndown phreatic surface is below that point, a single-sided installation is adequate.

A single-sided system will be suitable only if a sufficient depth of permeable soil exists beneath formation level and if there are no significant layers or lenses of impermeable material in the water-bearing soils, especially above the level of the base of the wellpoints.

The slope and shape of the drawdown surface is very relevant, and primarily depends upon the soil permeability. The lower the permeability, the steeper the slope of the drawdown cones that individual wellpoints will establish. The spacing between wellpoints must be close enough to ensure that their individual cones of depression interact and so produce a general and fairly uniform lowering.

The dewatering contractor has no control over soil conditions on site but does have control over the spacing between wellpoints. If the proposed excavation depth is modest and the depth of the unconfined water-bearing stratum is considerable, the wellpoints can be installed at wide spacings on long riser pipes. The use of long riser pipes has the merit of minimizing the possibility of a wellpoint sucking air and so reducing the available vacuum. This may be offset by increased difficulty in the handling of long risers during installation. Generally this variation is more suitable for an installation that is to be pumped for a significant period of time than to a progressive type installation, when the pumping period will be of short duration.

Experienced engineering judgement is needed when deciding on wellpoint spacings and whether a single-sided installation will be adequate. The conditions favourable to the use of a single-sided wellpoint installation for trench excavation are:

(a) A narrow trench.
(b) Effectively homogeneous, isotropic permeable soil conditions that persist to an adequate depth below formation level (see Fig. 9.1a).
(c) Trench formation level is not more than about 5 m below standing groundwater level. The actual depth achievable depends upon the permeability of the water bearing soil beneath formation and the overall pumping efficiency of the installation.

For trenching in excess of about 5 m deep, it is possible to use a lower stage of wellpointing as described in Section 9.8, provided that there is sufficient depth of permeable soil beneath final formation level and the width of the wayleave is sufficient. The sides of the trench excavation must be safely supported throughout.

Single-sided systems work well in isotropic soil conditions. In practice soils are often heterogeneous and anisotropic. In such conditions double-sided systems, where wellpoints are installed on both sides of an excavation, may be more effective.

9.7.2 Double-sided wellpoint installations

The conditions likely to require the use of a double-sided wellpointing (Fig. 9.13(a)) for trench excavation are:

(a) A relatively wide trench.
(b) Trench formation level 4.5–6 m or more below standing groundwater level.
(c) Impermeable stratum close to formation level, or low permeability layers or lenses present above formation level.

If there is an impermeable stratum (such as a clay layer) at or close to formation level (Fig. 9.13(b)), even if a double-sided wellpoint system is employed, there may be some residual seepage or overbleed near the interface between the overlying permeable stratum and the underlying impermeable soil. In this situation particular thought must be given to the wellpoint depth and spacing. The wellpoints must be 'toed into' the impermeable stratum – in effect they are installed to penetrate a few hundred millimetres into the clay, to form a sump of sorts around each wellpoint to maximize drawdown. This requires careful installation – overzealous 'toeing in' will result in the wellpoint screen being installed too deep into the clay, thereby becoming clogged and ineffective. The wellpoints should be installed at rather closer spacings than normal to try and intercept as much of the overbleed seepage as possible.

Even if these measures are adopted, some overbleed seepage is likely to pass between the wellpoints and enter the trench. It is essential to control the overbleed so that pore water pressures do not build up and that the water flow does not continuously transport fines and risk instability of the sides of the trench. Some form of sand-filled permeable bags, granular drainage blanket or geotextile mesh should be placed as indicated in Fig. 9.13(b). This will allow the water to flow into the excavation without the build up of pore water pressures and movement of fines. Thus stability of the trench will be preserved. The water flowing to the trench must be continuously removed by conventional sump pumping to prevent standing water building up in the working area.

There is a hybrid of single and double-sided systems that can be used when trenching in extensive permeable strata where an impermeable layer exists above formation level (see Fig. 9.14). These conditions could be dealt with by means of a double-sided wellpoint installation but this could be an

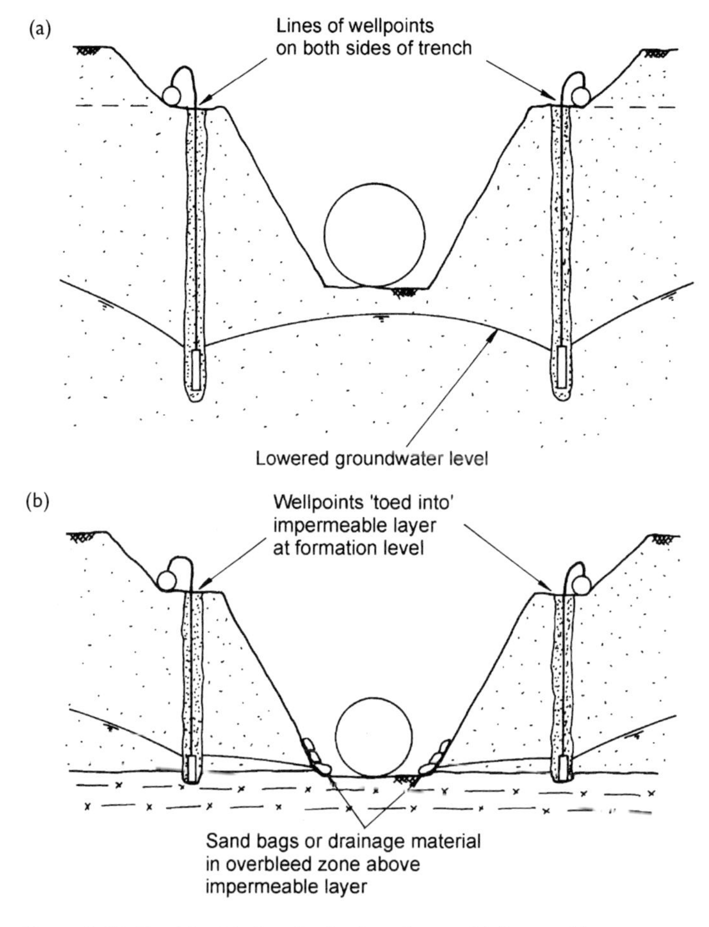

Figure 9.13 Double-sided wellpoint installation. (a) Permeable stratum extends below formation level, (b) impermeable stratum present near formation level.

encumbrance to the contractor's excavation and pipelaying activities. The alternative is to have a single-sided wellpoint installation on one side of the trench excavation with vertical sand drains on the other side. The sand drains consist of a series of holes jetted at diameters and spacings similar to the wellpoints and penetrating to below the low permeability layer. These are backfilled with sand of the same grading as that used for the sanding-in of the wellpoints. These will provide downward drainage of the perched water on the other side of the trench. This should reduce troublesome overbleed on the side of the trench remote from the wellpoints.

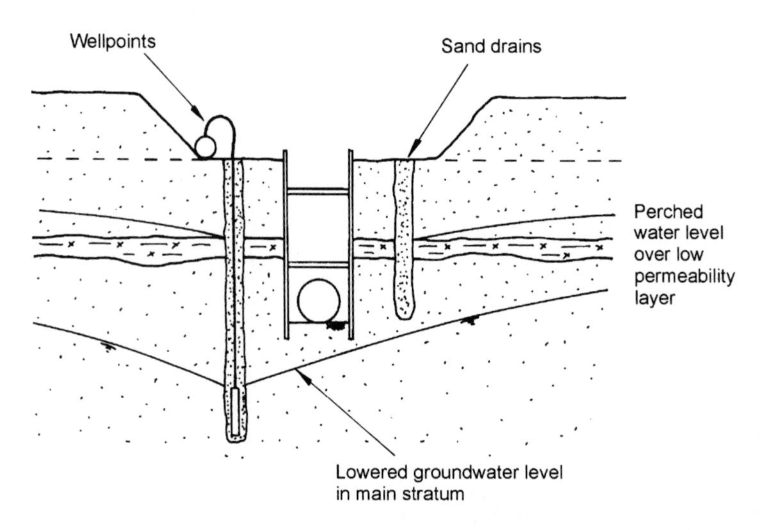

Figure 9.14 Single-sided wellpoint installation with sand drains to aid control of perched water table.

9.7.3 *Progressive installation for trench works*

The same principles applicable to wellpoint systems for static excavations, are pertinent to rolling trench excavations. There are additional complications because of the need to progress the groundwater lowering system to keep pace with the trench works. The initial lowering syndrome to establish drawdown, is always at the head of the progressively advancing installation.

Generally the equipment length for a wellpoint system for a rolling trench excavation should be about three to four times as long as the planned weekly advancement of the trench (Fig. 9.15(a)). This allows for time for the extraction of wellpoints, risers and manifolds behind completed work, their progressive installation ahead of the work (Fig. 9.15(b)) and for – the least certain factor – the time to establish drawdown of the groundwater ahead of trenching excavation and subsequent pipelaying and backfilling.

Where the soil is of low permeability, the time for drawdown will be protracted, so in such soils, it is prudent to allow for a greater length of operating equipment to be installed and operating ahead of the length of open excavation.

9.8 Wellpointing for wide excavations

Where a wide excavation is required, a perimeter wellpoint installation alone may not be sufficient to establish adequate lowering at the middle of

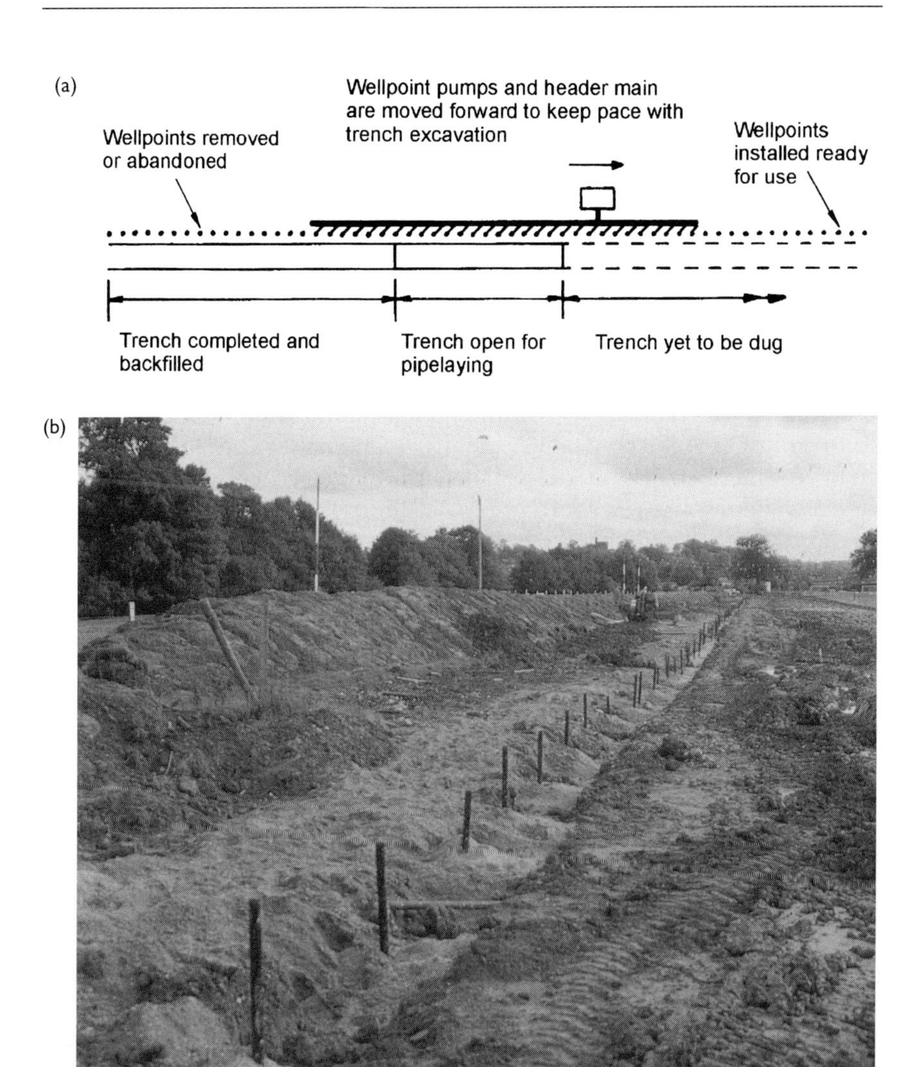

Figure 9.15 Progressive wellpoint system for trench works. (a) Method of wellpoint progression, (b) disposable wellpoints installed in advance of trench excavation.

the excavation. The lower the permeability, the more likely it is that the amount of lowering at the centre of the excavation will be insufficient. It might be prudent, even necessary, to install additional wellpoints within the excavation area.

9.9 Wellpointing for deeper excavations

9.9.1 *Long risers and lowered header mains*

If the excavation width is modest, but the required drawdown is greater than achievable by a single stage, and the thickness of permeable soils beneath formation level is significant, drawdown can be increased by installing wellpoints on longer than normal 6 m risers to an adequate depth below formation level.

The additional drawdown is achieved by initially pumping the wellpoints from pumps and header mains laid at about standing groundwater level. Pumping of this initial installation will establish some lowering of the water level. Excavation is then made down to the lowered water level and a second suction header main is installed at the lower level. This lower main is pumped while the upper suction main is still active. Progressively each individual wellpoint (having shortened its length of riser) is disconnected from the upper header main and connected to the lower header main. Pumping should be continued from the original upper main until all wellpoint risers have been shortened and connected into the lower main. Using the standby pump from the upper header main to begin pumping on the lower header main can reduce the number of pumps required for this method. When all the wellpoints have been connected to the lower main the upper pump can be moved down to act as a standby unit for the lower main. The upper header main is then dismantled and removed.

This process could be described as a two-stage installation having only one stage of installation of the wellpoints and risers (see Section 9.10 for description of a case history where this procedure was used).

9.9.2 *Multi-stage wellpointing*

Multi-stage wellpoint installations (Fig. 9.5) can be used for deep excavations as an alternative to deepwells (see Chapter 10) or ejectors (see Section 11.1). If a multi-stage wellpoint system is used it is necessary to make sufficient allowance for side slopes or batters to excavate safely to formation depth and berms to support header mains of lower stages.

The recommended site procedure for the installation of multi-stage sets is as follows:

(a) Excavate to about the standing water level; it may be possible to excavate to around 0.5 m below the water level.
(b) Install and connect first stage wellpoint system around the perimeter, making due allowance for subsequent slope batters and berms.
(c) Pump continuously on the first stage system.
(d) Excavate to about 0.5 m above the lowered water level.

(e) Install and connect the second stage wellpoint system around the perimeter that likewise allows for excavation batters and berms for subsequent stages.
(f) Pump continuously on the second stage system as well as continuing to pump on the first stage.
(g) Again excavate to about 0.5 m above the level to which water has been lowered by pumping on both first and second stages.
(h) Continue the sequence of excavation, wellpoint installation and pumping until formation level is reached.

This sequence of operations will entail short halts in excavation as each further intermediate stage level for wellpoints and header main, etc. is installed for the further lowering of the groundwater.

At each stage, in addition to installation of the wellpoint system, a number of observation wells should be installed to monitor the lowering being achieved and to indicate the depth to which the next stage excavation can be taken.

Often, when pumping on multi-stage systems of more than two stages, the first stage pumping output declines rapidly as the third stage is brought into operation. When this happens the top stage pumps can be stopped and transferred down to the fourth stage if there is one: or can be connected to the third stage as standby units. The wellpoints and risers of the top stage should not be extracted since it may be necessary to reactivate the upper stage(s) as the structure is being built up and backfill placed. The recovery of the groundwater level must be controlled so that at every stage of the building of the works, there is no risk of flotation of the partially completed structure due to an uncontrolled rise of the groundwater.

Three and four stage wellpoint installations have been operated satisfactorily for deep excavations. There have been, and will be in the future, many projects where a multi-stage wellpoint system is an economic solution to a particular groundwater lowering problem. Consideration should also be given to the fact that the degree of expertise required for installation and operation of a wellpoint system is less sophisticated than that appropriate for a deep well or ejector system. Hence, there may be occasions or geographical locations where the use of a wellpoint system will be the more appropriate choice for a deep construction project.

9.10 Case history: Derwent outlet channel, Northumberland

In the early 1960s, the Sunderland & South Shields Water Co. awarded to John Mowlem & Co. Ltd. a contract for the construction of an earth dam about 15 miles to the south west of Newcastle on Tyne, in the valley of the river Derwent (see Rowe 1968; Buchanan 1970; Cashman 1971). This required the formation of a spillway outlet channel about 240 m long and

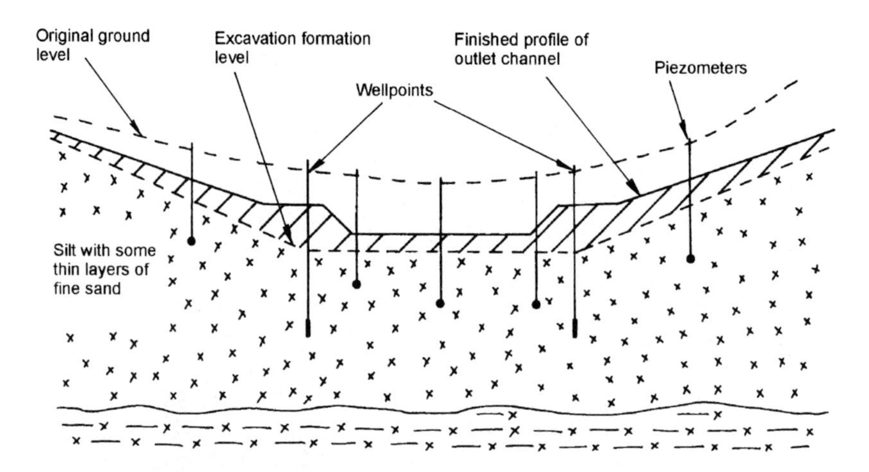

Figure 9.16 Typical cross section of Derwent outlet channel (courtesy of Northumbrian Water Limited).

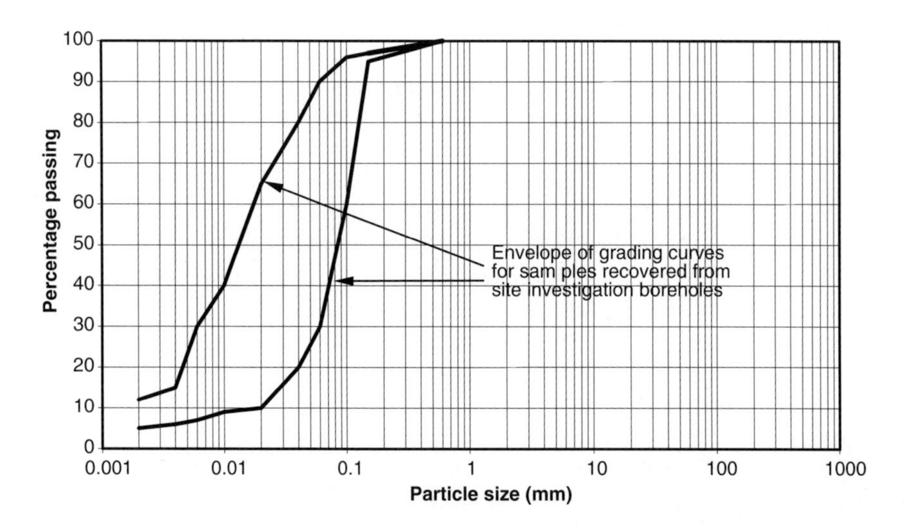

Figure 9.17 Envelope of gradings of glacial lake soils, the Derwent outlet channel (courtesy of Northumbrian Water Limited).

approximately 8 m deep on average. Laminated glacial lake silts of low permeability were revealed at the upstream end (Fig. 9.16).

Grading curves for soil samples obtained from site investigation boreholes confirmed the existence of difficult to stabilize soils (Fig. 9.17). In addition the varved soil structure was complicated.

Figure 9.18 Initial excavation (showing unstable conditions) for Derwent outlet channel (courtesy of Northumbrian Water Limited).

Initially an excavation was made to modest depth below the piezometric surface in the glacial lake deposits. This aided assessment of the excavation problems likely to be encountered and to determine appropriate methods for achieving a stable excavation for the construction of the outlet channel. The conditions exposed were not encouraging (Fig. 9.18). There were many outwash fans due to seepages that continuously transported fines. Also, the four piezometer tubes seen in Fig. 9.18 extended to some height above the excavated level because the pore water pressure at depth was artesian relative to the excavation floor. The excavation surface was so unstable that duckboards were needed in order to be able to take the measurements in these four piezometers.

Careful undisturbed sampling of these glacial lake soils revealed, within the layered soil structure of these deposits, some very thin layers of more permeable material that could act as preferential drainage layers to influence the reduction of pore water pressures of the total soil mass. It was agreed by the client, his specialist advisors and the contractor that a trial wellpoint installation, with careful attention to sanding-in should be undertaken.

Since the amount of lowering of the groundwater required to reach formation level was greater than that which could be achieved with a single-stage wellpoint installation, the wellpoints were installed at 1.8 m centres on

Figure 9.19 Workable conditions, following wellpoint pumping, under which the channel was actually formed (courtesy of Northumbrian Water Limited).

extra length riser pipes (see Section 9.9). Pumping from this initial installation effected about 3 m of lowering of the water level within a period of about a week. This created a surface of sufficient firmness to support an excavator and 2.75 m deep trenches were then opened beside the two lines of wellpoints. A duplicate suction header main was laid at this lower level and progressively each riser length was shortened and connected into the lower active pumping main. By this means lowering to formation level for the outlet channel was achieved. The vertical column of sanding-in material around the wellpoints provided the all important downward drainage for multi-layer perched water tables.

In Fig. 9.19 the excavator in the centre was sited at approximately the level of the duckboards to the four piezometers shown in Fig. 9.18. This was also the level at which the first, upper header main was placed.

The lower header main is visible in the foreground of the photo. As excavation to formation level moved forward, this main was progressively supported on scaffolding. The channel has been shaped to the required formation and side slopes (see Fig. 9.16) and as shaping proceeded a sand blanket was quickly laid to permit the drainage of pore water but at the same time preventing the removal of fines.

The principles observed were:

- Shape the excavation to just below formation level and form the required side slopes.
- As the side slopes are formed, immediately and progressively blanket them to allow relief of water seepages, thus preventing any build up of pore water pressures.
- Do not allow continuous transportation of fines, this can only lead to slope instability.
- Remove all seepage waters by controlled (i.e. filtered) sump pumping.

The message is – study and understand the soil structure and observe the basic installation guide lines. Thereby the almost impossible can become quite possible.

References

Buchanan, N. (1970). Derwent Dam – Construction. *Proceedings of the Institution of Civil Engineers*, **45**, 401–422.

Cashman, P. M. (1971). Discussion on Derwent Dam. *Proceedings of the Institution of Civil Engineers*, **48**, 487–488.

Irvine, D. J. and Smith, R. J. H. (1992). *Trenching Practice*. Construction Industry Research and Information Association, CIRIA Report 97, London.

Preene, M. Roberts, T. O. L. Powrie, W. and Dyer, M. R. (2000). *Groundwater Control – Design and Practice*. Construction Industry Research and Information Association, CIRIA Report C515, London.

Rowe, P. W. (1968). Failure of foundations and slopes in layered deposits in relation to site investigation practice. *Proceedings of the Institution of Civil Engineers*, Supplement, 73–131.

Somerville, S. H. (1986). *Control of Groundwater for Temporary Works*. Construction Industry Research and Information Association, CIRIA Report 113, London.

Chapter 10

Deep well systems

10.0 Introduction

A deep well system consists of an array of bored wells pumped by submersible pumps. The wells act in concert – the interaction between the cone of drawdown created by each well results in groundwater lowering over a wide area. Because the technique does not operate on a suction principle, greater drawdowns can be achieved than with a single stage wellpoint system. This chapter addresses the temporary works deep well groundwater lowering system and describes good practices to be used during installation and operation.

The principles of the method, and the stages in well design are discussed. The methods used for well drilling, installation, development and operation are outlined and some practical problems with the operation of deep well systems are presented. The vacuum deep well and bored shallow well systems, which are variants on the deep well method are described briefly. The chapter ends with a case history of a large-scale temporary works deep well system.

10.1 Deep well installations

A deep well system consists of a number of wells pumped by submersible pumps. Each well consists of a bored hole (typically formed by a drilling rig) into which a special well liner is inserted. The liner consists of plastic or steel pipe of which a section is slotted or perforated to form a well screen to allow water to enter; other sections consist of unperforated pipe (the well casing). Generally, deep well systems are installed in drift deposits and the annulus between the borehole and the well screen/casing is backfilled with filter media or formation stabilizer to form what is known as the filter pack.

The wells are generally sited just outside the area of proposed excavation (although for very large excavations wells may be required within the main excavation area as well as around the perimeter). A deep well system has individual pumps positioned near the bottom of each well; usually the pumps are borehole electro-submersibles (see Section 12.4). The well screen

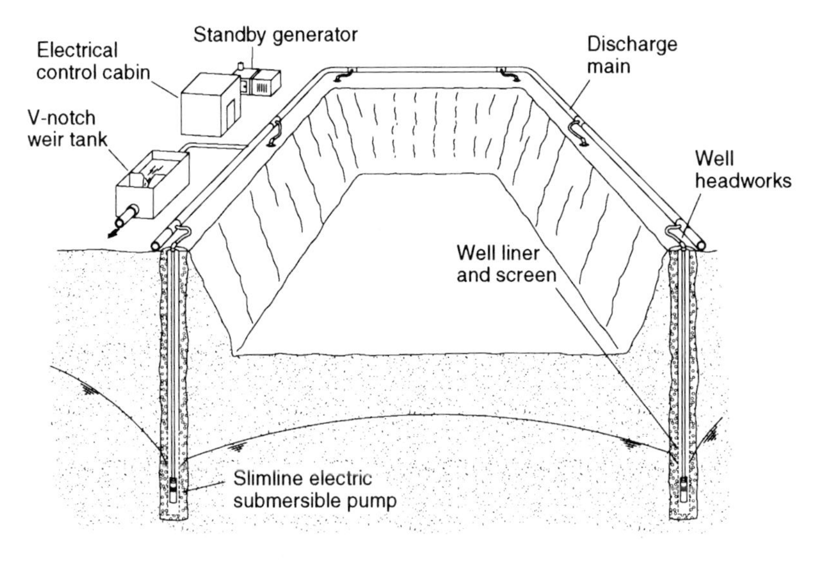

Figure 10.1 Deep well system components (from Preene *et al.* 2000; reproduced by kind permission of CIRIA).

and casing provides a vertical hole into which a submersible pump attached to its riser pipes can be installed (and also recovered, as and when required).

A typical deep well system (Fig. 10.1) consists of several wells acting in concert. Each well creates a cone of depression or drawdown around itself, which in a high permeability aquifer may extend for several hundred metres. The interaction between the cone of drawdown from each well produces the drawdown required for excavation over a wide area. Apart from pumping tests, deep wells are rarely used in isolation or individually; the method relies on the interaction of drawdowns between multiple wells.

The components making up an individual temporary works well are shown in Fig. 10.2. Generally, for most temporary works requirements:

1 The well screen and casing sizes will be in the range of 150–300 mm diameter. The well screen and casing are typically plastic, with steel being used only rarely.
2 The drilled borehole sizes will be in the range of 250–450 mm diameter.
3 The well depths will be in the range of 10–35 m. Occasionally, wells are drilled to greater depths, especially for shaft or tunnel construction projects.
4 The soils through which the wells are bored are usually drift deposit soils. Occasionally weak rock formations are encountered (such as a weak sandstone). The screened length in a rock formation may not need a well screen and filter pack so the required size of bore may be smaller.

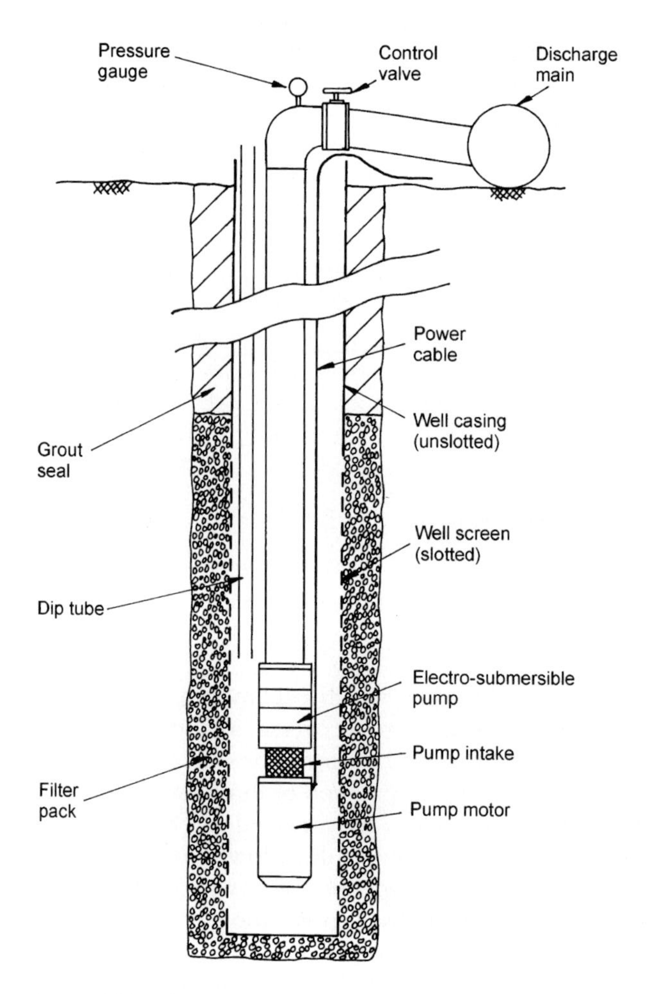

Figure 10.2 Schematic section through a deep well.

The vital feature of the deep well system compared to the single-stage wellpoint is that the theoretical drawdown that can be achieved is limited only by the depth of well and soil stratification. The wellpoint method (see Chapter 9) is limited by the physical bounds of suction lift. In contrast, the drawdown of a deep well installation is constrained only by the depth/level of the intake of the pump(s) – provided, of course, that the power of the pump is adequate to cope with the total head from all causes. Hence, the rated output of the installed pumps should match the anticipated well yield.

The energy costs of operating a deep well installation are likely to be competitive due to the greater efficiency of borehole pumps compared with the total system efficiency of a multi-staged wellpoint installation.

The well screens, pumps and other materials are similar to those used for water supply wells. However, since the working life of a temporary works well will almost always be significantly less than the life of a water supply well, temporary works wells can be constructed by rather cheaper and simpler methods. Also, the onerous health regulations to control the risk of water contamination during construction of water supply well installations are mostly inappropriate to temporary works wells.

The initial cost of installing a deep well system is significant. A high standard of expertise in the design and control of installation procedures is required to ensure that the appropriate good practices are implemented throughout and to promote optimum and economic performance.

10.2 Design of wells for groundwater lowering

There are three major factors to be considered when designing an individual temporary works well:

1 The depth of the well and screen length.
2 The diameter of the borehole and well screen.
3 The filter media to be used.

The design of a well will generally be done after the design of the overall groundwater lowering system has estimated the total flow which must be pumped to obtain the required drawdown for the excavation (see Section 7.4). The specification of well yield and spacing is outlined in Section 7.5. The following sections will describe some of the practical issues associated with the design of individual deep wells.

10.2.1 Depth of well

In essence the well must be deep enough to:

1 Be able to yield sufficient water so that all the wells in concert can achieve the required flow rate, and hence the required drawdown.
2 Be of sufficient depth to penetrate the geologic strata in which groundwater pressures are to be lowered (this is especially important if deep confined aquifers are present).

The design of a well to ensure adequate yield is described in Section 7.5. The depth requirements are generally met by a consideration of the soil stratification at the site. There are certain rules of thumb about well depth, built up over many years of practice, and well designs should be compared with these to check for gross errors. If the aquifer extends for some depth below the base of the excavation being dewatered, the wells should be between 1.5 and 2 times the depth of the excavation. Wells significantly shallower than this are unlikely to be effective unless they are at very close spacing (analogous to a wellpoint system).

If the geology does not consist of one aquifer that extends to great depth, this may affect well depth. If the aquifer is relatively thin screen lengths will be limited and a greater number of wells will be required. If there is a deeper permeable stratum that, if pumped could act to 'underdrain' the soils above (see Section 7.3), it may be worth deepening the wells (beyond the minimum required) in order to intercept the deep layer.

It is always prudent to allow for a few extra wells in the system over and above the theoretically calculated number. This allows for some margin for error in design but also means that the system will be able to achieve the desired drawdown if one or two wells are non-operational due to maintenance or pump failure.

10.2.2 Diameter of well

In general, the diameter of the well will be chosen to ensure that the borehole electro-submersible pump (see Section 12.4) to be used will fit inside the well screen and casing, and that any necessary filter media can be placed around the well screen. This will allow the necessary drilled diameter of borehole to be determined. Sometimes, when working in remote locations or in developing countries, the selection has to be made in reverse, beginning with the borehole diameter that can be drilled by locally available equipment, and then working backwards to the size of pump that can be accommodated.

The starting point for determining the size of the bore is the diameter of the pump to be installed in the well screen. Having determined the expected individual well yield (see Section 7.5) at the steady-state rate of pumping, a pump should be selected from a manufacturer's or hirer's catalogue that has a maximum rated performance between 110 and 150 per cent of the steady-state flow rate at the anticipated working head. This allows for some additional pumping capacity to help establish the drawdown.

The pump manufacturer's catalogue will list the minimum internal diameter of well screen necessary to accommodate the pump to be used, assuming the wells are perfectly straight and plumb. In practice, most wells deviate from the ideal alignment, and using a slightly larger screen diameter reduces the risk of a pump getting stuck down a well. Some general guidance on well screen diameters is given in Table 10.1. The recommended minimum well screen diameters are generally larger than those quoted by the pump manufacturers. Even so, if a well has a large amount of deviation, even a very small pump may become jammed at the tight points in the well.

Table 10.1 Recommended well screen and casing diameters

Maximum submersible pump discharge rate (l/s)	*Recommended minimum internal diameter of well screen and casing*[a] (mm)	*Recommended minimum diameter of boring*[b] (mm)
5	125–152	250–275
10	152–203	300–325
15	165–250	300–375
20	180–250	300–375
25	203–300	325–425
44	250–350	375–475

Notes

a Diameter will depend on external dimensions of pump used.

b Minimum diameter of boring is based on nominal filter pack thickness of 50 mm. Slightly smaller diameters may be feasible if a natural filter pack can be developed in the aquifer.

Knowing the required minimum internal diameter, select an appropriate well casing and screen from the manufacturer's catalogues. The standard sizes of well screens are unlikely to exactly match the internal diameter needed for the pump, so the next available size up of well screen and casing is used. When selecting a well screen and casing, care must be taken to ensure that the wall thickness of the base pipe forming the screen and casing is adequate to withstand collapse pressures from soil and groundwater loadings. There have been occasions when thin-walled plastic well screens have collapsed during development and pumping. In most of these cases, further problems were avoided by installing additional wells with thicker-walled screens – the screen manufacturer should be able to provide guidance. Collapse of steel well screens is rare, but these are not commonly used for temporary works wells on cost grounds.

Typically, the wells used for groundwater lowering will have a 'filter pack' of sand or gravel placed in the annulus between the borehole wall and the well screen. Preferably the annular thickness of the filter pack should be about 75 mm (but never less than about 50 mm or more than 100–150 mm). Thus, the minimum size of the borehole to form the well should be the external diameter of the well screen and casing (including at joints where the diameter may be greatest) plus twice the thickness of the filter pack. This diameter is unlikely to exactly match the standard sizes of drilling equipment, so the borehole should typically be drilled using the next size up of drill bits, taking care to check that this does not result in filter pack thicknesses in excess of 150 mm.

10.2.3 Design of filter media and slot size

A temporary works groundwater lowering well (Fig. 10.2) typically consists of a well screen and casing installed centrally inside a borehole formed by a drilling rig. The annulus between the screen and the borehole wall is filled with granular filter media to form a 'filter pack'.

The filter media must be selected (based on the particle size distribution of the aquifer material) to meet the following two conditions:

i To be sufficiently coarse so that the filter pack is significantly more permeable than the aquifer, to allow water to enter the well freely.
ii To be sufficiently fine so that the finer particles are not continually withdrawn from the aquifer.

The selection of any filter media has to be a compromise between these two conflicting conditions. Concentrating on condition (i) will give a high yield well, but with an increased risk of continuously pulling sand or fines into the well. Concentrating on condition (ii) will prevent movement of fine particles but may restrict well yield.

An additional requirement is that the material chosen must be suitable for placement in wells (see Section 10.5) with minimum segregation – filter media of uniform grading are preferred for this reason.

There have been numerous theoretical and practical studies of design methods for granular filters in relation to water supply wells and for dams. CIRIA Report C515 (Preene *et al.* 2000) summarized suitable criteria for design of granular filters for dewatering purposes as follows:

1. $D_{15\text{filter}} > 4 \times D_{15\text{aquifer}}$. This satisfies condition (i) above. For widely graded materials this should be applied to the finer side of the filter grading envelope and the coarser side of the aquifer grading envelope. Additionally, the filter material should contain no more than 5 per cent of particles finer than 63 μm.
2. $D_{15\text{filter}} \leqslant 5 \times D_{85\text{aquifer}}$. This satisfies condition (ii) above, and is known as Terzaghi's filter criterion. For widely graded materials this should be applied to the coarser side of the filter grading envelope and the finer side of the aquifer grading envelope.
3. $U_{\text{filter}} < 3$. This allows the filter to be placed without risk of segregation. U is the uniformity coefficient ($U = D_{60}/D_{10}$) – very uniform materials (consisting of only a small range of grain sizes) have a low U. If U is greater than three there is a risk of segregation during placement, and the filter material should be placed carefully by tremie pipe (see Section 10.5).

Application of these criteria to an aquifer grading would produce a filter grading envelope as shown in Fig. 10.3. A filter material that falls within the envelope is then selected from those available.

Application of these criteria to real aquifers results in a relatively narrow range of materials that are suitable for use as granular filter media. At one end of the range a relatively low permeability silty sand aquifer might need

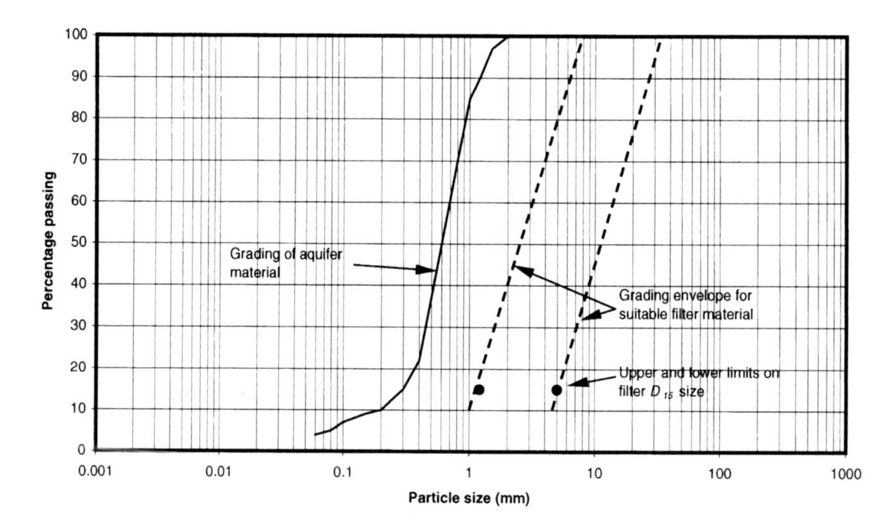

Figure 10.3 Aquifer and filter grading envelopes.

a 0.5–1.0 mm filter sand, while at the other end a very high permeability coarse gravel might need a 10–20 mm pea gravel. Materials outside these ranges are rarely used. Rounded uncrushed aggregates are generally considered to have higher permeabilities than the same grading of angular crushed material. For this reason rounded materials are preferred for use as filter media.

The slot size (the minimum width of the slot) of the well screen should be chosen to match the filter, and avoid large percentages of the filter material being able to pass into the well. In general, the slot size should be approximately equal to $D_{10\text{filter}}$. There is another condition that must be considered when specifying the slot size on the well screen – there must be enough total area of slots (known as the 'open area') to allow the desired flow to pass through the screen. Open area is defined as the total area of slots or apertures expressed as a percentage of the total area of well screen. In general terms the open area is much greater for well screens with larger slot sizes than smaller ones. The slotted well screens commonly used for temporary works wells typically have open areas of between 5 and 20 per cent for fine and coarse slots respectively. The manufacturer's catalogues should indicate the actual percentage open area for the slot size and screen type in use.

Driscoll (1986) recommends that sufficient open area be provided to ensure that the average screen entrance velocity (well discharge divided by total area of screen apertures) is less than 0.03 m/s. Parsons (1994) has argued that this 'entrance velocity' approach has little theoretical basis.

Nevertheless, it is an established rule of thumb which produces wells which perform at least adequately in many situations.

When the filter is placed it may be relatively loose, and is likely to compact a little during development. Accordingly, the filter pack when placed must extend to some height (normally at least 0.5–1 m) above the top of the slotted well screen to ensure that aquifer material does not come into direct contact with the well screen if the filter pack compacts. Sometimes permanent tremie tubes are left in place to allow the filter pack to be topped up by the addition of more material.

There are some types of granular soils when it is not necessary to introduce artificial filter media, but where it is possible to directly develop the aquifer, remove the finer particles and form a natural filter pack in the aquifer immediately outside the well screen. This method can be employed in coarse well graded soils such as sandy gravels. It can give cost savings by allowing a well borehole to be drilled at a smaller diameter for a given well screen size – the space for a filter pack is not needed, and the aquifer material is allowed to collapse directly onto the well screen. According to Clark (1988) soils may be appropriate for natural filters if $D_{40aquifer} > 0.5$ mm and $U_{aquifer} > 3$. Natural filters are not appropriate for uniform fine-grained soils, where there are not sufficient coarse particles in the soil to form an effective filter structure. Slot size must be chosen carefully when proposing to use natural filters. CIRIA Report C515 (Preene *et al.* 2000) suggests a slot size of $D_{40aquifer}$ to $D_{50aquifer}$ is acceptable in most cases, but if the maximum yield is required from very widely graded soils a slot size in the range $D_{60aquifer}$ to $D_{70aquifer}$ might be considered.

If wells are installed in weak rock (for example weathered chalk) where flow is predominantly from fissures, a filter pack is not needed to prevent movement of fine particles. However, it is good practice to fill the annulus between the well screen and borehole wall with a coarse permeable filter gravel which acts as a formation stabilizer. The formation stabilizer prevents weaker blocks of aquifer rock from collapsing against and distorting the well screen. Formation stabilizers should be highly permeable to allow free passage of water; 10–20 mm pea gravel is often used.

In addition to conventional granular filter media placed during well installation, pre-formed filters (fitted at manufacture) are also available. The two main types are:

1 Resin-bonded screens. Specific granular filter media are bonded directly to the slotted well screen. The grading of the filter media appropriate to the aquifer is chosen in the normal way, and then applied to the screen in the factory.
2 Geotextile screens. Geotextile mesh filter is wrapped around slotted base pipe. The most hydraulically efficient designs provide a high open area by placing a very coarse spacer mesh under the filter mesh itself to

allow water to flow around the outside of the base pipe beneath the filter mesh. Guidance for selection of the size of openings in the filter mesh can be obtained from the manufacturers. Once selected, the mesh is normally applied to the screen in the factory. Granular filters or formation stabilizers may be used in conjunction with geotextile screens.

As discussed earlier, well filters are designed against two conflicting criteria – the need to be highly permeable and the need to prevent continuous movement of fines. Add to this the problems that aquifers are variable and heterogeneous (a filter that works on one well may not work on one at the other end of the site) and that suitable filter materials may not be available locally, and it is obvious that experienced judgement is needed for filter design. The problem of aquifer variability can be addressed by carrying out a thorough site investigation. The problem of limited availability of suitable filter material (especially on remote sites) may require less than ideal filters to be used. In such cases, a series of trial wells using local materials may be appropriate. If the locally available material does not give acceptable well performance the use of pre-formed (e.g. geotextile or resin-bonded) filters might be considered, since shipping costs for these materials may be less than for bulky filter gravels.

10.3 Constructing deep wells

The methods used to construct wells for temporary works groundwater lowering purposes have much in common with those used to form water supply wells, but there are some differences in techniques and equipment. A general appreciation of well drilling methods can be gained from some of the publications in the water supply field (Stow 1962 and 1963; Cruse 1986; Rowles 1995).

There are four main stages in forming a groundwater lowering well:

1 Drilling of borehole.
2 Installation of materials (screen, casing, gravel, etc.).
3 Development of well.
4 Installation and operation of pump.

These are described in the following sections.

10.4 Drilling of well boreholes

Many methods are available for the formation of well boreholes. The technique selected will depend upon the type of equipment available in the territory and the expertise of the well boring organization selected and the soil and groundwater conditions anticipated.

There are three main methods used commonly to form well boreholes:

1 Cable tool percussion drilling and variations on this technique, generally requiring the use of temporary boring casing to support the sides of the borehole.
2 Wash boring or water jet drilling, sometimes with the addition of compressed air. This method is very similar to the installation of wellpoints by jetting using a holepuncher.
3 Rotary drilling, either by direct or reverse circulation. The drilling fluid may be water or air with specialist additives which help support the sides of the borehole – this is typically employed when drilling drift deposits.

For temporary works wells cable tool percussion drilling is limited to well depths of about 20–35 m, but has been used down to 50 m depth on occasion – for water supply wells the method has been used to considerably greater depths. Wash boring and holepuncher equipment have been used to install wells to about 35 m depths. Rotary drilling rigs can cope with significantly greater depths if necessary.

The wash boring and rotary drilling techniques, when used appropriately, generally provide more productive wells and require less development. The effectiveness (i.e. potential yield) of a rotary drilled hole depends greatly on the properties of the drilling fluid used and the adequacy of the removal of the drilling residues on completion of the hole – this often gives rise to much debate between groundwater lowering specialists, drilling contractors and drilling additive suppliers about the relative merits of the various marketed drilling slurries. The merits of various types of drilling fluids and additives are discussed by Driscoll (1986). Adequate development of the well on completion is vital to ensure its best performance (see Section 10.6).

Throughout the well drilling operation, the arisings from the drill hole should be observed and logged to determine if soil conditions are as expected. If not, it may be prudent to determine whether the design of subsequent deep wells needs to be varied. For instance the arisings might reveal an unexpected layer of impermeable soil within the wetted depth. However, do not ignore the fact that information gleaned from well boreholes is rarely of as high quality as that obtained from dedicated site investigation holes. Any judgements made on the basis of observing the well borehole arisings can only be somewhat speculative.

Whatever drilling technique is used, it is essential that wells be relatively straight and plumb otherwise there will be severe operational problems with the submersible pumps. Water supply wells are typically required to be drilled to a verticality tolerance of 1 in 300. This tolerance is probably unnecessary for temporary works wells. Verticality requirements are often not explicitly specified for temporary works wells, but if they are then 1 in 100 appears to be a more reasonable requirement.

10.4.1 Cable tool percussion drilling

The original cable percussion drilling rig made use of a reciprocating mechanism known as a 'spudding arm' or 'walking beam' to repeatedly lift and lower a bit (or chisel) suspended on the end of a wire rope inside telescoping sizes of temporary boring casings. The action of the bit breaks up the soil or rock at the base of the borehole to form a slurry with the groundwater entering the well. As the slurry builds up it deadens the percussive action of the bit; the bore has to be bailed periodically to remove the slurry. The largest rigs were powerful and progress could be made, albeit slowly, through even boulder and rock formations. Inevitably, the arisings are not much better than a mashed up slurry of mainly indistinguishable consistency. This type of rig is still in use by some drillers, and is especially prevalent in developing countries.

Some time later there came a variation on this technique, the early tripod bored piling rig which also made use of temporary boring casings of telescoping sizes. The cable passes over a pulley at the top of the tripod and is raised and lowered by a winch linked by a clutch to a small donkey engine. To make progress through granular formations a sand pump or shell is used and to penetrate a clay stratum, a clay cutter is used; chisels are used if cobbles or boulders are encountered. Tripod rigs became more mobile and towable. Self erecting tripod rigs are commonly used for site investigation in the United Kingdom, where there are colloquially known as 'shell and auger' rigs, although they are more correctly known as light cable percussion rigs.

While light cable percussion rigs are most commonly used for 150–200 mm diameter site investigation boreholes, the larger models (of 2 and 3 tonne winch capacity) are widely used to drill temporary works wells. Figure 10.4 shows a 2 tonne rig being used to drill a 300 mm diameter well bore to a depth of approximately 25 m.

Whatever type of cable tool percussion rig is used, drilling will generally be slower than by the rotary method. This is mainly due to the time required to install and remove the temporary casings. The cable tool percussion method is most suitable for drift deposits such as sands, gravels and clays. Progress can be very slow indeed if used to drill rock formations, although the method has been successfully used in some soft rocks such as weathered Chalk or highly weathered Triassic Sandstone.

When using temporary boring casing to penetrate a multi-layered soil structure, there is risk of smearing of the side of the hole as the casing is being installed and withdrawn. This could mask or block the permeable strata, especially if the individual soil layers are thin. Also, when using percussion boring methods there is some risk that the boring cuttings and slurry will clog the water-bearing formation, so subsequent well development should be carefully monitored and supervised (See section 10.6).

Figure 10.4 Light cable percussion boring rig.

10.4.2 Wash boring and/or jet drilling

The technique of jetting was first developed for the installation of well-points – relatively small diameter holes (see Chapter 9) – and has been extended to the making of larger holes for the installation of wells and piles.

It essentially entails the use of a jet of high pressure water being applied at the bottom of the wash pipe or drill pipe such that the soil at the level of the water jet is slurrified (i.e. put into an almost liquid state). The soil particles tend to be flushed back up to ground level, leaving a bored hole that has been washed relatively clean with little side smear as might result from cable percussion drilling.

One of the most common applications of this method is the use of a holepuncher or heavy duty placing tube used to jet wells into place. Figure 10.5 shows a 300 mm diameter holepuncher in use. Holepunchers have been used up to diameters of 600 mm and depths of 36 m, although depths of more than 20 m are uncommon.

Figure 10.5 Holepuncher used for installation of deep wells (courtesy of T. O. L. Roberts).

10.4.3 Rotary drilling: direct circulation

This method uses a drill bit which is attached to the bottom end of a rotating string of hollow drill pipe (sometimes known as drill 'rods'). Bits are typically a tricone rock-roller bit consisting of moveable cutters, or fixed drag bits; the action of the bit breaks up the soil in the base of the bore into small cuttings. A continuous supply of drilling fluid is pumped down the drill pipe to cool the rotating bit. The fluid rises back to the surface in the annulus between the drill pie and the borehole walls and flushes the cuttings to the surface. Hence, the drill fluid must have sufficient 'body' and rising velocity to retain the cuttings in suspension throughout their upward travel to the surface and also to provide a degree of support to the borehole to prevent collapse.

The bit and the drilling fluid are each vital components to the process and an understanding of their interdependent functions is necessary to successful usage of the process (see Driscoll 1986). The drilling fluid may be water-based with long-chain polymer compounds added to the water to form a fluid with better support and cutting transportation properties. Water-based drilling fluids are sometimes known as 'muds', a term that dates from when bentonite slurries were used to drill wells – bentonite is now rarely used in drilling fluids for water wells. If air (supplied by a powerful compressor) is the drilling fluid small amounts of additive may be used to form a 'foam' or 'mist' to carry the cuttings out of the bore. Many modern drilling additives are naturally biodegradable, so that any residues left in the ground after development will decay and are less likely to act as an impediment to flow of water into the well.

On reaching the required depth of bore, the drill string and bit are withdrawn, with the borehole left filled with drilling fluid prior to installation of the well screen and filter pack. If possible, it is good practice prior to installation to try and displace the fluid in the borehole and replace with a clean, thin solution of fluid.

Rotary drilling rigs are available in a variety of sizes, from small site investigation units that can be towed behind a four-wheel drive vehicle to rigs suitable for shallow oil wells that need to be delivered on several articulated trucks and have to be assembled by crane. For groundwater lowering deep wells, self-contained rotary rigs mounted on trucks or on crawler or four-wheel drive tractors (Fig. 10.6) are most commonly used. Such rigs have drilled wells to depths of up to 100 m.

10.4.4 Rotary drilling: reverse circulation

This is a variation on rotary drilling often used to drill large diameter water supply wells, but used only rarely to construct temporary works deep wells. The principle of operation is the same as for direct circulation rotary drilling, but the direction of flow of drilling fluid is reversed.

The mixture of drilling fluid and cuttings in suspension flows up the hollow drill pipe with a fast rising velocity because the internal diameter of the drill pipe is limited. The borehole is kept topped up with drilling fluid which allows the return flow of fluid to descend slowly in the annulus – of

Figure 10.6 Truck mounted rotary drilling rig (courtesy of British Drilling and Freezing Company Limited).

greater cross sectional area than the drill pipe – between the drill pipe and the borehole wall. The level of the drilling fluid must always be a few metres above the level of the groundwater to help maintain a stable bore. As with the direct circulation method, on reaching the required depth of bore, the drill string and bit are withdrawn, and the borehole is kept topped up with fluid.

10.5 Installation of well materials

Once the bore is complete the well casing and well screen are placed in the hole. The casing and screen consist of threaded pipe, typically of between 150 and 300 mm diameter, supplied in lengths of between 2.5 and 6 m. The diameter of casing and screen required is determined as described in Section 10.2. The casing sections are plain (unperforated) and the screen sections are perforated (generally by slotting) and for temporary works wells are typically made from uPVC or thermoplastic – steel screen and casing are used only rarely.

Prior to installation, the drilling crew should be instructed on the number and sequence of casing and screen lengths to be installed to ensure that the screens are located within the water-bearing horizons. The screen and casing is installed in sections by lowering into the borehole using the drilling rig's winch. A bottom cap is fitted to the first length of screen or casing which is lowered into the bore, further sections are then added until the string of screen and casing is installed to the required depth. It is often a good idea to place a few hundred millimetres of filter gravel in the base of the bore before commencing installation of screen and casing. This prevents the bottom length sinking into any soft sediment or drilling residue remaining in the base of the hole.

When the screen and casing are in position the filter pack (see Section 10.2) must be placed in the annulus between the well screen and the borehole wall. The filter pack is formed from filter media generally consisting of uniformly sized gravels, although coarse sands are sometimes used. The filter material may be supplied in bags or in bulk loads. Ideally, to ensure correct placement, the gravel should be installed using one or more tremie pipes. The tremie pipe would typically be of uPVC or steel sectional pipe of 50 mm internal bore (although sizes down to 32 mm bore have been used to place uniform sands). The filter media are poured slowly into a hopper at the top of the tremie, perhaps washed down by a gentle flow of water. Patience is vital in this operation – adding the filter media too quickly will cause the tremie pipe to block or 'bridge'. If this occurs the tube will have to be removed and flushed clear, causing delays and inconvenience. The tremie tube is raised slowly, keeping pace with the rising level of filter media.

It is sometimes acceptable to place the filter media without a tremie pipe, by pouring into the annulus from the surface. This is only acceptable provided the filter media is very uniform and will not segregate as it falls down the bore and if the wall of the borehole is very stable (e.g. if supported by temporary casing). Again, care must be taken to ensure the filter media are added slowly. If too much is added at once a blockage or 'bridge' may occur; these can be difficult to clear and may result in the screen and casing having to be removed and the complete installation recommenced.

If temporary drill casing is used to support the borehole, this must be removed in sections as the filter media is added. The level of the filter media should not be allowed to rise significantly above the base of the temporary casing, otherwise a 'sand lock' may occur between the temporary drill casing and the well screen. This can cause the well screen to be pulled out when the temporary casing is withdrawn. The best practice is to alternately add some filter media, pull the temporary drill casing out a little, add more filter media, pull the drill casing further, and so on, monitoring the level of the filter media continuously. Boreholes drilled by rotary methods without temporary casing do not have this constraint, and filter media can be placed in one continuous, steady operation.

The filter pack is sometimes brought up to ground level, but increasingly, it is good practice to place a very low permeability grout seal above the filter pack (see Fig. 10.2). This reduces the risk of aquifer contamination by surface water (or water from other aquifers) passing down the filter pack into the aquifer (see Section 13.3). The grout may be neat cement or cement-bentonite (see Driscoll 1986 for guidance on grout mixes and placement). If the grout were placed directly on top of the filter pack, some grout may be lost into the filter media, compromising its permeability. To avoid this, a 1–2 m thick layer of bentonite pellets should be placed on top of the filter pack, and allowed to swell before placing the grout.

10.6 Well development

Development is a process, carried out between completion of the well and installation of the pumps, with the aim of removing any drilling residue or debris from the well, and maximizing the yield of clean sand-free water. If development is not carried out, not only might the yield be low, but also the borehole electro-submersible pump will be damaged as a result of pumping sand-laden water.

The most commonly used forms of development are intended to induce two-way groundwater flow between the well and the aquifer to remove any loose particles from the filter pack and aquifer immediately around the well. This will increase the aquifer permeability locally and will remove any potentially mobile soil particles that might damage the operational pumps. It is most important that the development generates two-way flow (alternately into and out of the well) to help dislodge mobile soil particles that may be loosely wedged in soil pores – this is much more effective than continuous pumping. Development can only be effective in wells that have appropriately designed and installed filter packs. No amount of development can correct a well with a filter pack that is too coarse or is discontinuous due to installation problems.

There are a wide range of development techniques (see Driscoll 1986) used on water supply wells, but in practice most temporary works wells are developed by one of three techniques:

(1) Airlift. Air from a compressor is used to lift water from the well up an eductor tube formed from plastic or steel pipe of 75–150 mm diameter

(Fig. 10.7(a)). This is a robust method of pumping and, with suitable equipment, can transport significant volumes of sand with the water. Reversal of flow is achieved by lowering the air line past the bottom of the eductor tube and delivering a short blast of air into the well. Alternatively, airlift pumping can be used for a few seconds to raise a 'slug' of water to just below ground level, at which point the air is turned off causing the water to fall back down the well, inducing flow out of the well and into the aquifer. The airlift method is most effective when water level in the well is near the surface, and becomes less efficient for deeper water levels.

(2) Surge block. A tight fitting block is lowered into the well screen (Fig. 10.7(b)) and pulled sharply upwards using a tripod and winch (such as a light cable tool percussion rig). As the block moves upward, water will be forced out of the well above the block, and drawn in below – thereby achieving reversal of flow. There are a number of ways that the block can be surged using either many short strokes or fewer long ones

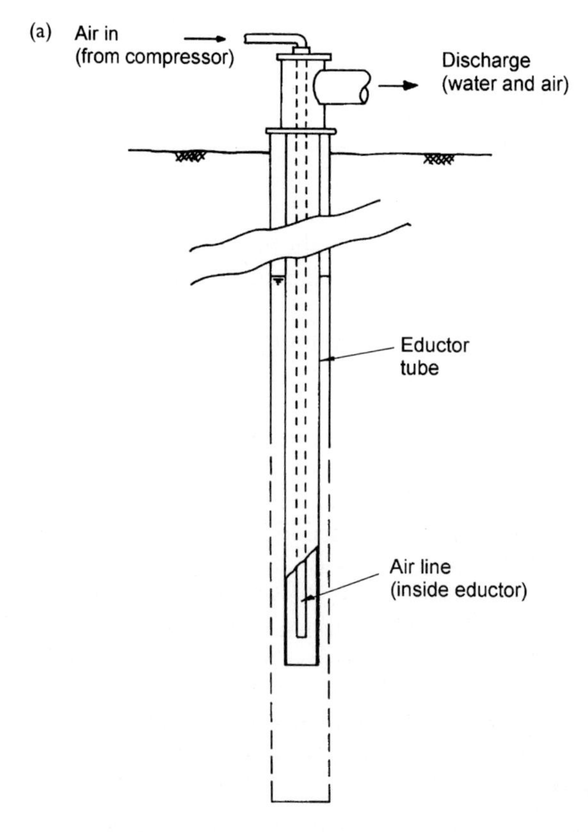

Figure 10.7 Well development methods. (a) airlift with eductor tube, (b) surge block.

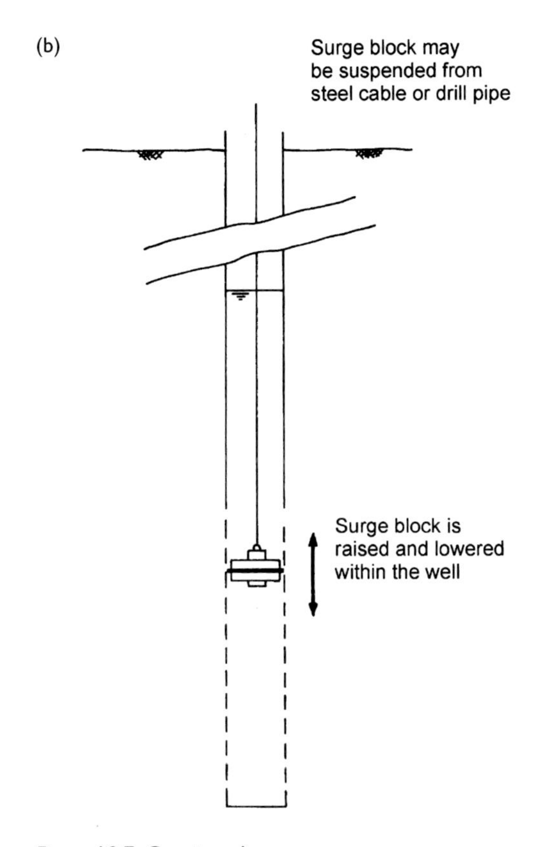

Figure 10.7 Continued.

(see Driscoll 1986); some methods of surging are known as 'swabbing the well'. If a special surge block is not available, a weighted shell or bailer of slightly smaller diameter than the well screen is sometimes used. The sediment and debris that builds up in the well will need to be removed by airlift pumping or bailing.

(3) Jetting. Less commonly used, this method involves lowering a jetting head (mounted on drill pipe) down inside the well screen. The drill pipe and jetting head are slowly rotated and high pressure water is pumped down the drill pipe and jets horizontally at the screen via small nozzles in the head. The jetting generally forces flow into the aquifer, and so may need to be alternated with airlift pumping to get flow reversal and also to remove any sediment or debris generated by jetting.

Development is usually discontinued when the well no longer yields sand or fine particles when pumped by the airlift. This is normally monitored by observing the discharge water, which initially will appear very dirty or discoloured but will become clear as development proceeds. However, a note of caution should be sounded. Sometimes, the discharge water may appear clear but still contain small but significant amounts of fine sand – enough to cause problems for the pump, and to create voids around the well. The best way to check for this is to take a sample of water in a clean white plastic tub (of the sort used to hold soil samples). Any sand will be clearly visible in the bottom of the tub. A specialist sediment sampling container called an 'Imhoff cone' can also be used to check for sediment in the water, but this device is intended for much higher sediment loads, and in many ways inspection of a sample in a white tub is a more sensitive method.

The question often asked is – how long should it take to develop a well? There is no simple answer to this, but experience suggests that many wells in granular aquifers will take between 6 and 12 h to be effectively developed. Certainly, if a well still yields copious amounts of sand after more than 2 to 3 days (say 20 – 30 h development) the well is unlikely to improve and a replacement well (possibly with a different filter media) should be considered.

For wells drilled in carbonate rocks, such as chalk, wells may be developed by acidisation. This is a technique commonly used in water supply wells (see Banks *et al.* 1993) where acid is introduced into the well to dissolve any drilling slurry and to improve the well-aquifer connection by dissolving aquifer material in fissures locally around the well. In practice, several tonnes of concentrated acid are introduced into the well. The reaction between the acid and the carbonate rock generates large volumes of carbon dioxide gas, which must be carefully controlled by fitting a gas tight head plate (equipped with specialist valves and pressure relief devices) to the top of the well. Acidisation is not a straightforward procedure and should be planned and undertaken by experienced personnel, with particular emphasis on ensuring that necessary health and safety measures are in place. Acidisation is not commonly used to develop temporary works wells, but the method has been used on some deep groundwater lowering wells into the chalk beneath London, including for the Jubilee Line Extension tunnel project in the 1990s.

If wells are operated for long periods of time (several months or years) and are affected by encrustation or clogging well performance may deteriorate. In such cases, well performance may be improved by periodic redevelopment using one of the methods described above (including acidisation, which has been used on heavily encrusted wells).

10.7 Installation and operation of deep well pumps

Following development, the pumps can be installed. The following sections deal exclusively with the procedures used with borehole electro-submersible pumps (see Section 12.4), by far the most commonly used type in deep well systems.

10.7.1 Installation of borehole electro-submersible pumps

The pump is connected to the riser pipe, which is a steel or plastic pipe which carries the discharge from the pump up to ground level (see Fig. 10.2). The riser pipe is typically between 50 and 150 mm diameter for the smallest and largest pumps used for temporary works. The riser is supplied in sections with threaded or flanged joints – extra sections are added until the pump has been lowered to the design level (see Fig. 12.6). The electrical power cable connected to the pump is also paid out as the pump is lowered into position. The cable should be kept reasonably taut and be taped or tied to the riser pipe every few metres. If this is not done then any slack in the cable may form 'loops' down the well, interfering with the subsequent installation of dip tubes.

If steel riser pipe is used, it may be strong enough to support the weight of the pump, riser pipe and water column. Otherwise (and in any case if plastic riser pipe is used) the weight must be supported by straining cables or ropes from the pump, tied off at the wellhead. It is good practice to install a dip tube of 19–50 mm diameter to allow access for a dipmeter to be used to monitor the water level in the well.

For groundwater lowering applications the pump is normally installed near the base of the well so that potential drawdown is maximized. The base of the pump should be at least 1 m above the base of the well to avoid the pump becoming stuck in any sediment that may build up at the bottom of the well. When the pump is installed, a headworks arrangement – typically including a control valve and pressure gauge – is attached to the top of the riser pipe and connected to the discharge pipework.

The best sort of valve for use at the wellhead is the gate valve. It is preferable because its slide plate is at right angles to the water flow and its position can be finely adjusted by its screw mechanism to regulate the flow. Its sensitivity of opening adjustment is good and when fully open it offers less resistance to flow than any other type of valve. The other type of valve sometimes used for well output control is the butterfly valve. It is less costly than the gate valve (size for size) but the sensitivity of opening control is less than that of the gate valve.

Most borehole electro-submersible pumps are powered by a three-phase supply; if the phases are connected in the wrong order the pumps can run (or rotate) backwards (i.e. anticlockwise rather than clockwise). Pumps running backwards move very little water, so on installation it is important to 'check the direction' of each pump in turn. One method is to start the pump with the control valve fully closed and observe the pressure gauge.

If the needle indicates a high pressure (consult the pump manufacturer's data sheets for the expected value) the motor is running in the right direction. But if the indicated pressure is low the motor may be running backwards. Turn the pump off, isolate the pump switchgear; have an electrically competent person change over two of the phase wire connections and restart the pump against a closed valve. The pressure gauge should register full pressure indicating that the motor is now running in the right direction.

10.7.2 Adjustment of electro-submersible pumps

For groundwater lowering projects the pumps generally run continuously, with the operating level in individual wells only a little above the level of each pump intake. Generally, the flow from each pump is adjusted to achieve this condition by manipulation of the control valve on each wellhead. Partial closure of the valve will impose just sufficient back pressure (extra artificial head) to constrain the well output and maintain the optimum operating level.

If the well operating level is too close to the level of the pump intake, the pump will tend to 'go on air' from time to time – this is when the water level in the well reaches the pump intake, and air is drawn into the pump. In this condition the pump will race when 'on air' and the riser pipe will tend to vibrate. This must not be allowed to continue in the long-term because under such conditions the pump motor will be damaged severely. The symptoms of 'going on air' can be detected by observing the needle of the pressure gauge at the wellhead sited upstream of the control valve (see Fig. 10.2), and by observing vibration of the riser pipe. When the discharge flow contains some air (rather than water only) the gauge needle will repeatedly drop to zero. If these erratic variations in pressure gauge readings are observed, close the control valve gradually until the pressure gauge needle remains steady – a very slight fluctuation in the needle position is acceptable. Alternatively, if the gauge needle remains steady, the well operating level may be above the design level. If this condition is suspected, gently open the control valve until the needle starts to flicker and then close the control valve by a very small amount.

10.7.3 Use of deep wells on tidal sites

On sites where groundwater levels vary tidally, the pumped level in each well may vary too, although generally these variations will be small (less than a few metres). It is suggested that the control valve be adjusted for the low tide condition only and not altered unless the amount of lowering required at the high tide condition is critical. If this is the case, frequent adjustment of all operating levels will be required. This is tedious and liable to be overlooked on occasions, especially during night shifts. If the maximum lowering is required at high tides as well as at low tides, one possible solution is the installation of additional wells to be operated only for a few hours either side of high tide.

10.7.4 Use of oversized pumps

Occasionally it may not be possible to equip a well with the appropriate output pump and so a pump rated at an output greater than necessary is installed in the well. Often, in such circumstances, the lack of sensitivity on the control valve will make it difficult (sometimes even impossible) to adjust the well output so that continuous pumping is practicable. There are two alternative expedients that can be employed in such circumstances:

(a) At the wellhead fit an additional small bore pipe and valve to bypass the main control valve. Operation involves closing the small by-pass valve completely and adjusting the main valve to achieve best possible coarse adjustment; then gradually opening the by-pass valve to make the best fine adjustment. This mode of operation makes feasible continuous pumping.
(b) Modify the pump controls for intermittent running with automatic stop/start. This requires sensing units in the pump electrical controls linked to upper and lower level sensor electrodes in the well.

10.7.5 Level control switches

If the pumps are to be operated intermittently, sensor electrodes in the well are used to trigger the pump to start when the water level is high, and stop when the water level is too low. Typically, the lower level electrode is set to switch the pump motor off just before the water level is drawn down sufficiently to cause the pump to 'go on air'. The upper sensor is set to switch the pump on when the water level in the well has recovered to a predetermined level (which must obviously be below excavation formation level). The average level of lowering achievable by this mode of pumping will be approximately the mean level between the two sensors. The distance between the upper and lower sensor electrodes should be the minimum practicable to allow the best possible amount of lowering by intermittent pumping. The disadvantage of this is that it will cause more frequent starts of the pump motor, increasing the risk of motor failure.

10.7.6 Problems due to frequent start-up of motor

Most borehole electro-submersible pumps are operated using direct-on-line electrical starter controls. With this arrangement the starting current of an electric submersible pump motor is greater than the running current by around a factor of up to six. The heat generated on start up is significant.

Thus, if the motor is frequently switched on and off (for example, due to the use of level controls), with time the motor windings may tend to become overheated eventually resulting in failure of the motor. There is a risk of this happening if there is on/off switching more than about six times per hour. In fact, even six starts per hour is a lot, and the number of starts should be minimized if at all possible.

10.7.7 Encrustation and corrosion

Deep wells operating for long periods may sometimes be affected by encrustation due to chemical precipitation or bacterial growth in the well screen, filter pack, pump or pipework. Encrustation may reduce the efficiency of the well (reducing yield and increasing well losses) and may increase the stress on the pump, leading to more severe wear and tear and, ultimately, failure.

In certain types of groundwater corrosion of metal components in pumps and pipework may be severe. Recognition of such conditions is vital when specifying equipment – for example maximizing the use of plastic pipework can reduce the problems, leaving only the submersible pumps made from metal.

The nature of these problems, and appropriate avoidance and mitigation measures are discussed in Section 14.8.

10.8 Vacuum deep well installations

If deep wells are installed in a low permeability aquifer and well yields are low, well performance can be enhanced somewhat by sealing the top of the well casing and applying a vacuum. The vacuum is generated by an exhauster unit (a small vacuum pump) located at ground level. Typically the exhauster unit would be connected to a manifold allowing it to apply vacuum to several wells. Because air flow to the exhauster will be low (once vacuum is established) the pipework connecting the exhauster and the wells need only be of small diameter, perhaps 50 mm or less. For the method to be effective the top of the well filter pack must have a bentonite or grout seal, otherwise air may be drawn through the gravel pack into the well, preventing the establishment of a vacuum. A typical vacuum deep well is shown in Fig. 10.8.

The top of the well needs to be sealed airtight between the well casing and the pump riser pipe, power cable and vacuum supply pipe. Purpose-built well head seals are available, but plywood disks and copious amounts of sealing tape have also proved effective.

The application of vacuum to deep wells is unlikely to be a panacea for a poorly performing system, especially if poor yields are associated with

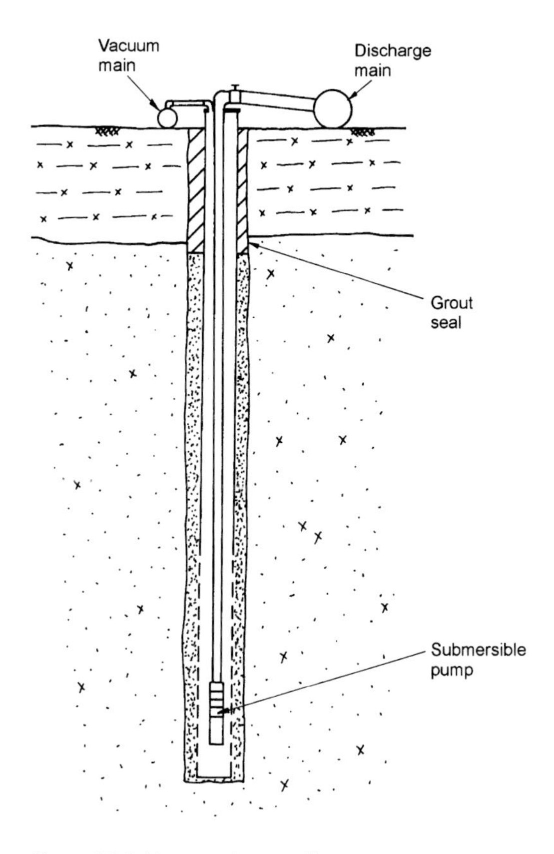

Figure 10.8 Vacuum deep well.

inappropriate filter gravel gradings or inadequate well development. The yield from the wells will not be transformed wholesale, but, at best, will increase by 10–15 per cent, occasionally more. Even if a vacuum equivalent to 8 m head of water can be maintained in the wells, do not expect the draw-down outside the wells to increase by a similar amount. It may be that the vacuum is mainly used to overcome well losses at the face of the well. The application of vacuum can cause operational problems including increased risk of collapse of well screens or sand being pulled through the filter pack.

10.9 Shallow well installations

The bored shallow well system is a synthesis of the deep well and wellpoint systems. Bored shallow wells are constructed in the same way as a deep well system but use the wellpoint suction pumping technique to abstract the water (Fig. 10.9). Hence, the amount of lowering that can be achieved is subject to

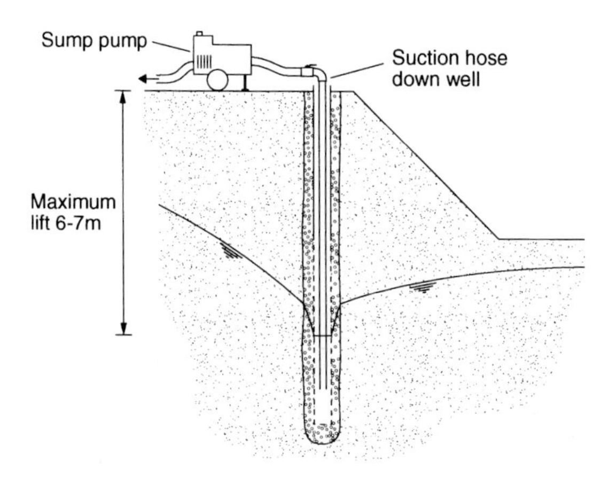

Figure 10.9 Shallow well system (from Preene *et al.* 2000; reproduced by kind permission of CIRIA).

the same limitations with a wellpoint system, namely that drawdowns in excess of 6 m below pump level are difficult to achieve. The method is most useful on congested urban sites, and where the soil permeability is high. Because the wells are of larger diameter than wellpoints, they generally have a greater yield and so can be at greater centres. This means that the wells create fewer constraints on the activities of the steel fixers, shutter erectors and other trades needing to carry out work within the dewatered excavation. The shallow well method is known by some as 'jumbo' wellpointing, since the wells can be thought of as widely spaced, grossly oversized wellpoints.

10.10 Case history: Tees Barrage, Stockton-on-Tees

In the early 1990s a barrage was constructed across the River Tees at a site between Stockton-on-Tees and Middlesborough (see Leiper and Capps 1993; Franklin and Capps 1995). The barrage was constructed by Tarmac Construction Limited on behalf of the Teeside Development Corporation. The barrage substructure (a reinforced concrete slab 70 m wide, 35 m long and 5 m thick) was formed in a large construction basin (Fig. 10.10), which provided dry working conditions while the river flow was diverted to one side during the works.

The construction basin was approximately 70 m by 150 m in plan dimensions, and was contained within bunds containing sheet-pile cut offs; formation level of the basin was at −8 mOD. The bunds were formed 'wet' across the river, and it was intended that the water that remained trapped in the basin when the bunds were closed would be removed by sump pumping. Ground investigations indicated the soil in the floor of the basin to be

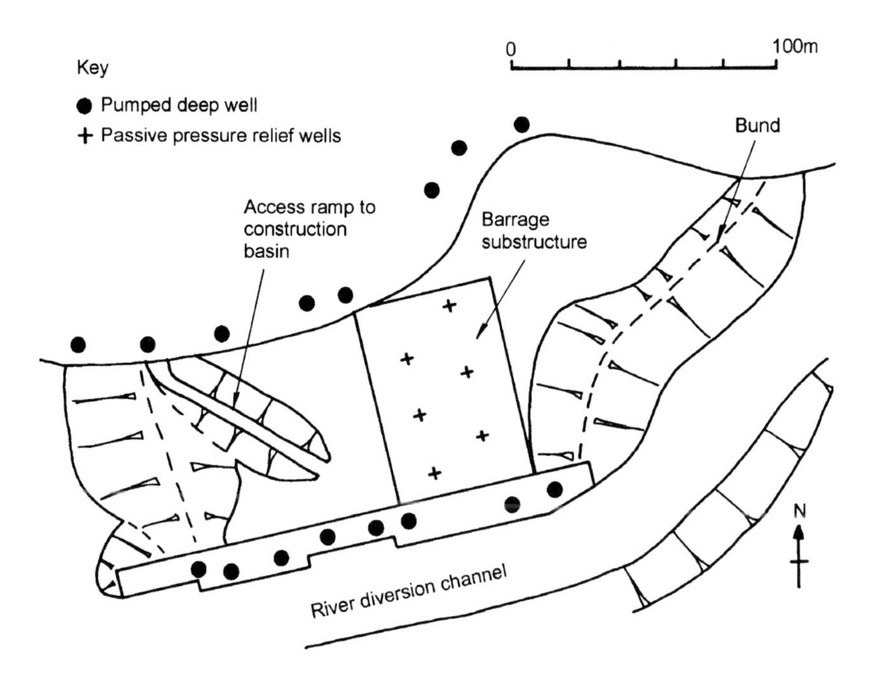

Figure 10.10 Schematic plan of construction basin for Tees Barrage (after Leiper and Capps 1993). Deep wells are located to the north and south of the basin. Pressure relief wells are located in the deepest part of the basin.

very stiff glacial clay, an ideal material on which to found the barrage. However, a few metres below the floor of the basin a confined aquifer of glacial sand and gravel was present (see Fig. 10.11) with a mean piezometric level of +1 to 2 mOD (up to 10 m above the floor of the excavation). The high piezometric level in this aquifer (and the possible presence of gravel lenses in the glacial clay) meant that, when the basin was pumped dry, if the piezometric pressure was not lowered significantly, there would be a risk of heave of the base of the excavation.

The solution adopted was to install a system of deep wells around the perimeter of the bund to lower the piezometric level down to formation level and ensure that factors of safety against heave were acceptably high. Because there was limited time available for additional ground investigation when the problem was identified, the system design was finalized using the observational method (see Section 7.1).

Two wells were installed and test pumped (by constant rate pumping and recovery tests) in turn, with the other well being used as an observation well. Although these wells were intended as trial or test wells they were located carefully so that, if the trial was successful, the wells could be incorporated in

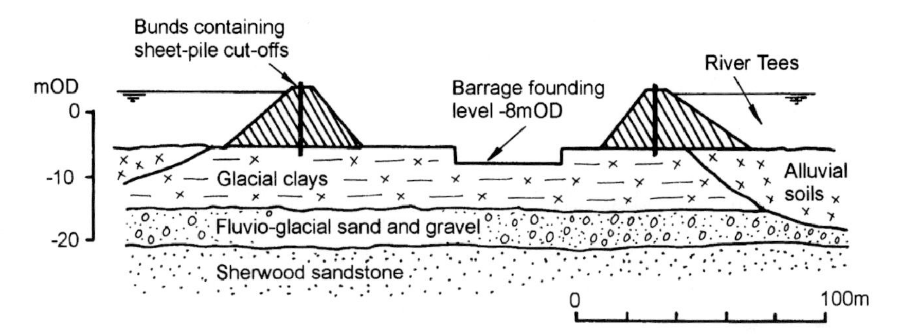

Figure 10.11 Section through construction basin for Tees Barrage showing ground conditions (after Leiper and Capps 1993).

the final deep well system. The trial allowed flow rate and drawdown data to be gathered, and the system was designed using the cumulative drawdown method (see Section 7.4). A particular feature of the test data was that, when pumping was interrupted, piezometric pressures recovered rapidly to close to their original levels. This is a characteristic feature of confined aquifers, and means that any design should strive to ensure that any breakdowns or interruptions in pumping are minimized.

The final pumped system employed had the following key elements, several of which were specified based on the expectation of the rapid recovery of piezometric levels:

i A system of sixteen deep wells was installed to abstract from the sand and gravel aquifer. This included an additional two wells over the minimum number required to allow for maintenance or individual pump failure. The wells were drilled by cable percussion methods at 300 mm boring diameter, to allow installation of 200 mm nominal diameter well screen and casing, which in turn allowed installation of electro-submersible pumps of 10 l/s nominal capacity.

ii The power supply was split into two separate systems, with each part feeding half the pumps. This was to reduce the risk of a power system failure knocking the whole system out.

iii Standby generators were permanently connected into the system ready for immediate start up, and were connected to alarm systems.

iv 24 h supervision was provided by the dewatering subcontractor.

v Passive relief wells were installed in the base of the excavation, to provide additional pressure relief capacity in the event of a total system failure. In such an event the wells would have overflowed and flooded the excavation in a controlled manner, but without the risk of heave at formation level.

vi All piezometers, deep wells and pressure relief wells were monitored regularly to ensure the system was operating satisfactorily. A positive management system was established for the monitoring, with a proper inspection record kept by an approved and suitably experienced and qualified individual.

The system produced a total yield of 95 l/s in steady state pumping (compared with a nominal installed pumping capacity of sixteen wells at 10 l/s each), and operated for the eleven month life of the basin. On completion the pumps were removed and the deep wells and relief wells backfilled with gravel, bentonite and concrete.

References

Banks, D., Cosgrove, T., Harker, D., Howsam, P. J. and Thatcher, J. P. (1993). Acidisation: borehole development and rehabilitation. *Quarterly Journal of Engineering Geology*, **26**, 109–125.

Clark, L. J. (1988). *The Field Guide to Water Wells and Boreholes*. Open University Press, Milton Keynes.

Cruse, P. K. (1986). Drilling and construction methods. *Groundwater, Occurrence, Development and Protection* (Brandon, T. W., ed.). Institution of Water Engineers and Scientists, Water Practice Manual No. 5, London, pp 437–484.

Driscoll, F. G. (1986). *Groundwater and Wells*. Johnson Division. Saint Paul, Minnesota.

Franklin, J. B. and Capps, C. T. F. (1995). Tees Barrage – construction. *Proceedings of the Institution of Civil Engineers, Municipal Engineering*, **109**, 196–211.

Leiper, Q. J. and Capps, C. T. F. (1993). Temporary works bund design and construction for the Tees Barrage. *Engineered Fills* (Clarke, B. G., Jones, C. J. F. P. and Moffat, A. I. B., eds). Thomas Telford, London, pp 482–491.

Parsons, S. B. (1994). A re-evaluation of well design procedures. *Quarterly Journal of Engineering Geology*, **27**, S31–S40.

Preene, M., Roberts, T. O. L., Powrie, W. and Dyer, M. R. (2000). *Groundwater Control – Design and Practice*. Construction Industry Research and Information Association, CIRIA Report C515, London.

Rowles, R. (1995). *Drilling for Water: A Practical Manual*, 2nd edition. Avebury, Aldershot.

Stow, G. R. S. (1962). Modern water well drilling techniques in use in the United Kingdom. *Proceedings of the Institution of Civil Engineers*, **23**, 1–14.

Stow, G. R. S. (1963). Discussion of modern water well drilling techniques in use in the United Kingdom. *Proceedings of the Institution of Civil Engineers*, **25**, 219–241.

Chapter 11

Other dewatering systems

11.0 Introduction

This chapter describes some other less commonly used dewatering and groundwater control systems, and outlines conditions appropriate to their use.

The systems described in this chapter are:

- i Ejector systems, appropriate to low permeability soils such as very silty sands or silts.
- ii Horizontal wellpointing, mainly suitable for dealing with large trenching or pipeline projects of limited depth.
- iii Pressure relief wells.
- iv Collector wells.
- v Electro-osmosis, applicable to very low permeability soils, or for increasing the shear strength of very soft soils.
- vi Dewatering systems used in conjunction with exclusion systems.

Another specialized technique, artificial recharge of water back into the ground is described in Chapter 13 in relation to control of the side effects of groundwater lowering.

Some of the techniques described are specialized, and perhaps rather esoteric in nature. An engineer working in groundworks and excavations might spend a whole career in the field and never have to apply any of these techniques. Nevertheless it is important to be aware of the specialist methods that may be of help when faced with difficult conditions.

11.1 Ejectors

The ejector system (also known as the eductor or jet-eductor system) is suitable for pore water pressure reduction projects in low permeability soils such as very silty sands, silts, or clays with permeable fabric. In such soils the total flow rate will be small and some form of vacuum-assistance to aid drainage is beneficial (see Section 5.4); the characteristics of ejectors are an ideal match for these requirements.

Essentially, the ejector system involves an array of wells (which may be closely spaced like wellpoints or widely spaced like deep wells), with each well pumped by a jet pump known as an ejector. Ejector dewatering was developed in North America in the 1950s and 1960s, when jet pumps used in domestic supply wells were first applied to groundwater lowering problems. Since then the technique has been applied in Europe, the Far East and the former Soviet Union. Ejectors were not widely employed in the United Kingdom before the late 1980s; the A55 Conwy Crossing project was one of the first UK projects to make large-scale use of ejectors (Powrie and Roberts 1990). Some applications of the ejector method in the United Kingdom are described by Preene and Powrie (1994) and Preene (1996).

11.1.1 Merits of ejector systems

The ejector system works by circulating high pressure water (from a tank and supply pumps at ground level) down riser pipes and through a small-diameter nozzle and venturi located in the ejector in each well. The water passes through the nozzle at high velocity, thereby creating a zone of low pressure and generating a vacuum of up to 9.5 m of water at the level of the ejector. The vacuum draws groundwater into the well through the well screen, where it joins the water passing through the nozzle and is piped back to ground level via a return riser pipe and thence back to the supply pump for recirculation. A schematic ejector well system is shown in Fig. 11.1. Two header mains are needed. A supply main feeds high pressure water to each ejector well and a return main collects the water coming out of the ejectors (consisting of the supply water plus the groundwater drawn into the well). This large amount of pipework is needed to allow the recirculation process to continue.

The most obvious advantage of an ejector system is that ejectors will pump both air and water; as a result if the ejectors are installed in a sealed well in low permeability soil a vacuum will be developed in the well. This is one of the main reasons why ejectors are suited to use in low permeability soils, where the vacuum is needed to enhance drainage of soils into the wells. Another advantage is that the method is not constrained by the same suction lift limit as a wellpoint system (see Chapter 9). Drawdowns of 20–30 m below pump level can be achieved with commonly available equipment, and drawdowns in excess of 50 m have been achieved with systems capable of operating at higher supply pressures.

These characteristics mean that ejector systems are generally applied in one of two ways:

i As a vacuum-assisted pore water pressure control method in low permeability soils.
ii As a form of 'deep wellpoint' in soils of moderate permeability as an alternative to a two-stage wellpoint system or a low flow rate deep well system.

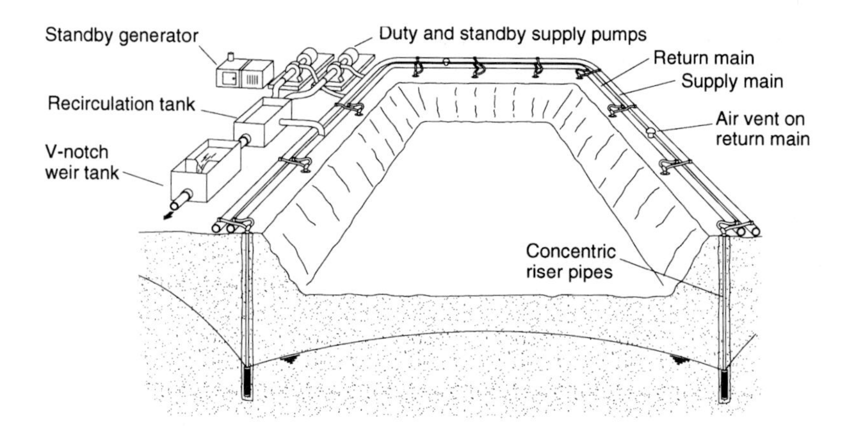

Figure 11.1 Ejector system components (from Preene *et al.* 2000; reproduced by kind permission of CIRIA).

It is also important to be aware of some of the practical limitations and drawbacks of the ejector system. Perhaps the most significant drawback is the low mechanical (or energy) efficiency of ejector systems. In low to moderate permeability soils, where flow rates are small this may not be a major issue, but in higher permeability soils the power consumption and energy costs may be huge in comparison to other methods. This is probably the main reason why the ejector system is rarely used in soils of high permeability. Another potential problem is that, due to the high water velocities through the nozzle, ejector systems may be prone to gradual loss of performance due to nozzle wear or clogging. This can often be mitigated by regular monitoring and maintenance, but it may make long-term operation less straightforward.

11.1.2 Types of ejectors

An ejector is a hydraulic device which, despite having no moving parts, acts as a pump. Several different designs of ejectors are available, each having different characteristics. These different ejector designs can be categorized into two types, based on the arrangement of the supply and return riser pipes:

i Single (or concentric) pipe (Fig. 11.2). This design has the supply and return risers arranged concentrically, with the return riser inside the supply riser. The supply flow passes down the annulus between the pipes, through the ejector and then returns up the central pipe.
ii Twin (or dual) pipe (Fig. 11.3). Here the supply and return risers are separate, typically being installed parallel to each other.

(a)

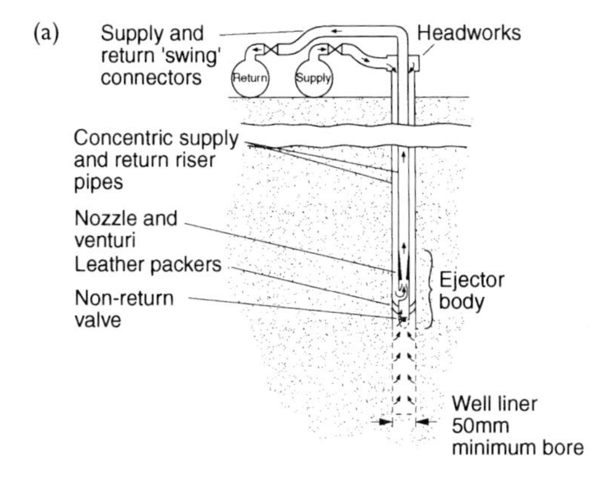

(b)

Figure 11.2 Single-pipe ejectors. (a) Schematic view (from Preene *et al.* 2000; reproduced by kind permission of CIRIA). (b) Ejectors attached to central riser pipes ready for installation into well (courtesy of WJ Groundwater Limited).

Both types are used in dewatering systems. The single-pipe ejector has the advantage in that the outer pipe can also be used as the well casing, provided it has sufficient pressure rating; this allows ejectors to be installed in well casings of 50 mm internal diameter. Twin-pipe ejectors need to be accommodated in rather larger well casings (approximately 100 mm internal diameter). The installation and connection of twin-pipe ejectors involves rather simpler plumbing than for single-pipe designs; this can make the twin-pipe

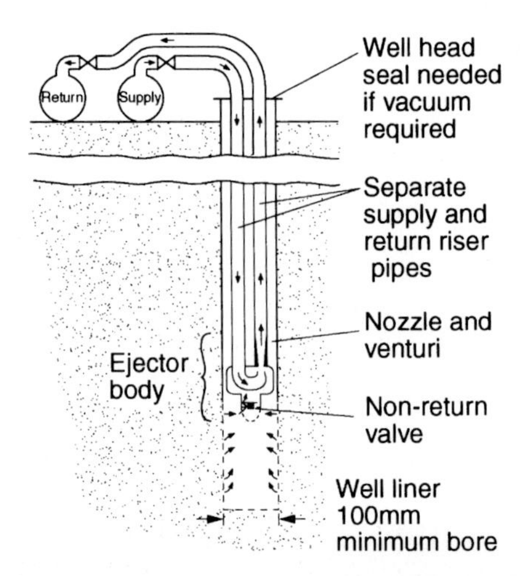

Figure 11.3 Twin-pipe ejectors – schematic view (from Preene *et al.* 2000; reproduced by kind permission of CIRIA).

type more suitable for use in localities where skilled labour is scarce, and it is desired to keep dewatering equipment as simple as is practicable.

11.1.3 Installation techniques

The ejectors themselves (the jet pumps) are installed in wells, which are generally installed by similar methods to deep wells: cable percussion boring, rotary drilling, jetting or auger boring (see Chapter 10). Well casings and screens and filter packs are then installed in the borehole. If single-pipe ejectors are to be used, the smaller diameter casings are sometimes installed by methods more akin to wellpointing (see Chapter 9) than deep wells. Ejector wells will normally need to be developed following installation and prior to placement of the ejectors and risers.

The ejectors are connected to the supply and return riser pipes (twin or single-pipe depending on design) and are lowered down to the intended level, typically near the base of the well. The headworks are fitted to seal the top of the riser pipes to the well liner, and the flexible connections are made to the supply and return mains (Fig. 11.1). This is needed to connect each ejector to the pumping station(s) which supply the high pressure water which is the driving force behind the pumping system.

11.1.4 Ejector pumping equipment

The pumping equipment making up an ejector system consists of three main parts: the ejector (and associated riser pipes and headworks); the supply pumps; and the supply and return pipework.

Ejectors are a form of pump, and like any pump, their output will vary with discharge head. An additional complication is that, in order to function, the ejector must be supplied with sufficient supply of water at adequate pressure (typically 750 to 1500 kPa measured at ground level). Each design of ejector will have different operational characteristics, so performance curves will be needed for the model being used (Fig. 11.4). It is important that the performance curves are representative of the conditions in a well (see Miller (1988) and Powrie and Preene (1994) for further details). At a given supply pressure and ejector setting depth, the pumping capacity (the induced flow rate) can be estimated from the performance curves. If this is sufficient to deal with the anticipated well inflow, then the required supply flow per ejector (at the specified pressure) and the number of ejectors to be fed by each pump can be used to select the supply pumps.

If the ejector induced flow rate is not large enough to deal with the predicted well inflow, it may be possible to increase the capacity of the ejector by increasing the supply pressure. However, Fig. 11.4 shows that, as the supply pressure is increased, the induced flow rate of the ejector tends to plateau. This is caused by cavitation in the ejector, and additional increases in supply pressure beyond that point do not give a corresponding increase in ejector capacity. With some ejector designs it is possible to increase capacity by fitting a larger diameter nozzle and venturi. This gives an increased induced flow rate at a given supply pressure, at the expense of an increase in the required supply flow per ejector.

The supply pumps should be chosen to be able to supply the required total supply flow at the necessary pressure, taking into account friction losses in the system. Adequate standby pumping capacity should be provided. Supply pumps are typically high-speed single or multi-stage centrifugal pumps; the pump and motor may be either horizontally or vertically-coupled (Figure 11.5). Pumps are typically electrically powered, but units with diesel prime movers can be used in remote locations or for emergency projects. If the system consists of relatively few ejectors, one pump may be used to supply the whole system. However, large systems (more than fifteen to twenty-five ejectors) could be supplied by one large pump (Fig. 11.5(a)) or a bank of several smaller pumps connected in parallel (Fig. 11.5(b)). The use of several smaller pumps can allow a more flexible approach with additional pumping capacity being easy to add as needed, and allows more economical use of standby pumps.

The supply pipework is normally made of steel with the pipes and joints rated to withstand the supply pressure. The return pipework does not need so large a pressure rating, but may still be under considerable positive pressure, especially if the wells are drawing air into the system. Supply header pipes are typically 100 or 150 mm in diameter. Return header pipes are typically 150 mm in diameter, although bigger diameters may be appropriate if large

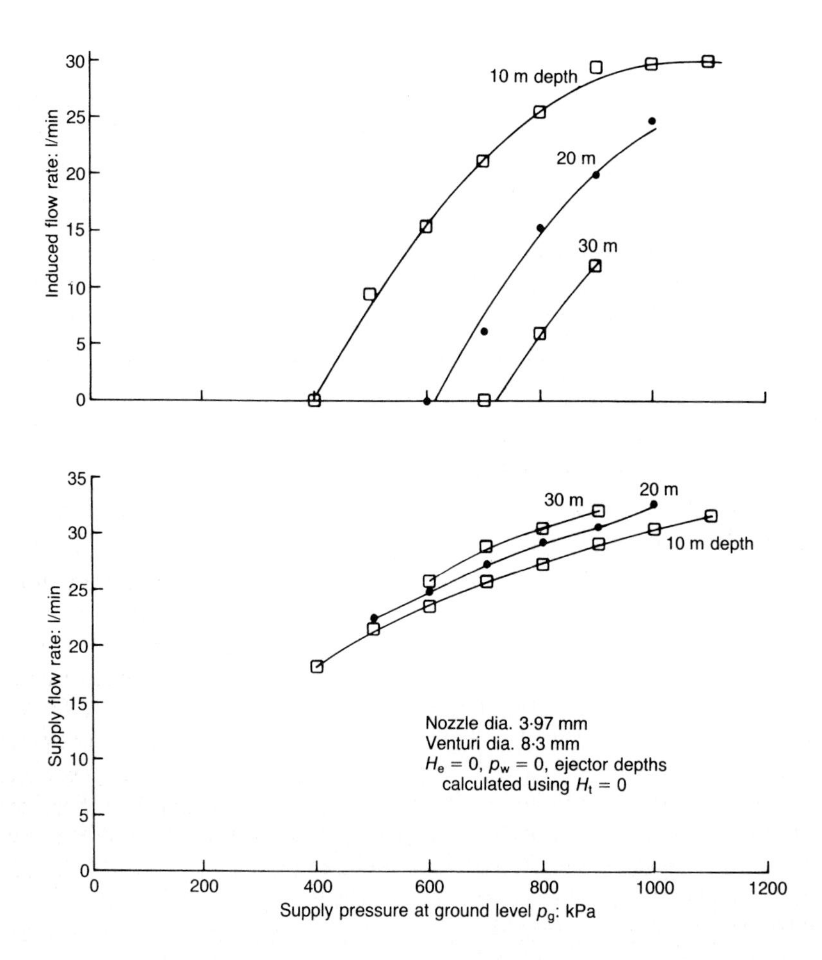

Figure 11.4 Example ejector performance curves (from Powrie and Preene 1994, with permission). The upper graph is used to determine the supply pressure necessary to achieve the desired induced flow rate per ejector at the specified ejector setting depth (10, 20 and 30 m shown). The lower graph is then used to determine the supply flow rate per ejector necessary to maintain the specified supply pressure.

numbers of ejectors are in use; air elimination valves may need to be incorporated in the return header to prevent airlocking. Figure 11.6 shows the pipework arrangement for a large-scale ejector system.

The components of an ejector set-up form a complex hydraulic system. For all but the smallest of systems, the design of the pumping systems should be carried out with care. Work by Miller (1988), Powrie and Preene

(a)

(b)

Figure 11.5 Ejector supply pumps. (a) Horizontally-coupled pump, (b) vertically-coupled pumps (courtesy of WJ Groundwater Limited).

Figure 11.6 Ejector pipework (courtesy of WJ Groundwater Limited). An ejector headworks is shown in the foreground. The large diameter supply and return header pipes can be seen in the background, linked to each ejector by flexible hoses.

(1994) and Powers (1992) is recommended as further reading to those faced with such a problem.

11.1.5 Operation of an ejector system and potential imperfections

Operation of an ejector system is relatively straightforward. In general, individual ejectors do not need adjusting or trimming, as might be required for wellpoints. The system should be very stable in operation and the supply pressure (displayed on gauges at the supply pump) should hardly vary once the system is primed and running. Any changes in supply pressure may indicate a problem with the system and should be investigated.

Successful long-term operation of ejector systems relies on ensuring the circulation water is not contaminated with suspended fine particles of soil or other detritus. Significant amounts of suspended solids will damage the supply pumps (which are generally intended for clean water only), and may build up in the ejector risers and bodies, restricting flow. However, the most

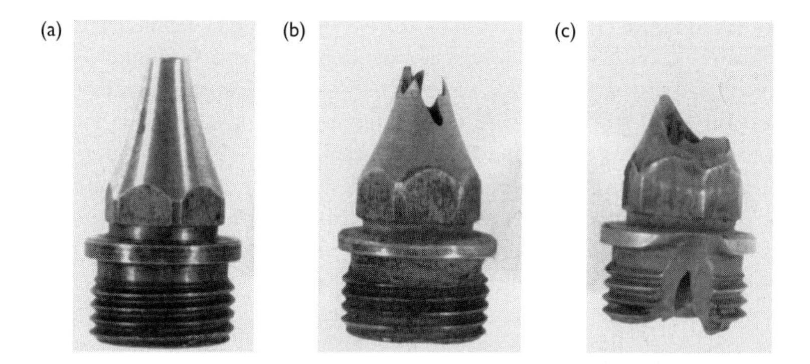

Figure 11.7 Examples of ejector nozzle wear (courtesy of W. Powrie). (a) New nozzle (b) After 6 months in slightly silty system (c) After 1 month in a moderately silty system.

serious effect of suspended solids is excessive wear of the nozzle and venturi in the ejector due to the abrasive action of the particles as these pass through the nozzle at high velocity. Even low levels of suspended solids can cause nozzle wear over weeks or months of pumping. As the nozzle wears, its opening enlarges, the supply pressure falls and system performance deteriorates. Figure 11.7 shows examples of pristine, moderately worn and severely worn nozzles.

Suspended solids may enter the circulation water in a number of ways:

(1) The material may have been present when the system was assembled. If appropriate supervision and workmanship are not employed it is not unknown for tanks and pipework to contain sand, dead leaves and other extraneous material left over from previous use or storage. If these are not cleaned out prior to commissioning the system, the system will probably either clog up after a few minutes or suffer severe nozzle wear over the next few days – neither is particularly satisfactory. When installing a system the pipes should be as clean as is practicable. Prior to start up the system must be primed with *clean* water. The water must be run to waste to flush the system clean; flushing should only stop when the water runs clear. It is important to allow for this flushing out when estimating the volume of water needed for priming.
(2) Silt or sand particles may be drawn from the wells into the system, either during initial stages of pumping (due to inadequate development) or continually (due to ineffective well filters). These problems should be avoided by ensuring appropriate well filters are in place, and that development is not neglected.

(3) Silt or sand may be drawn in from an individual well that may have a poorly installed filter or where the wellscreen has been fractured by ground movement. If this well is not identified and switched off it will continually feed particles into the system. The circulating action means that the sand from one well can damage all the ejectors in a system.
(4) Soil or debris may have inadvertently been added to the circulation tank as a result of construction operations. This has occurred where spoil skips have been craned over the tank location and small amounts of soil have fallen into the tank. This can be avoided by fitting a lid to the tank.
(5) The growth of biofouling bacteria (see Section 14.8) can generate suspended solids in the form of iron-related compounds associated with the bacteria's life cycle. This is less straightforward to deal with, but one of the simplest solutions is to periodically dispose of the circulation water and flush out the system with clean water.

If significant nozzle wear does occur, once the cause of the problem has been identified and dealt with, the ejectors will need to be removed from the wells and the worn nozzles and venturi replaced with pristine items. If there are many ejectors in the system or they are particularly deep, this can be quite an undertaking. Nozzle wear is definitely one case where prevention is better than cure.

Biofouling (item 5 above) may also cause problems by allowing material to build up in and around the ejectors, causing clogging rather than wear. In general, if the level of dissolved iron in the groundwater is more than a few mg/l, the potential for clogging should be considered (see Section 14.8).

11.2 Horizontal wellpoints

The horizontal wellpoint method uses a horizontal flexible perforated pipe, pumped by a suction pump, to effect lowering of water levels. Typically, the perforated pipe (the horizontal wellpoint) is installed by a special land-drain trenching machine (Fig. 11.8). One end of the pipe is unperforated and is brought to the surface and connected to a wellpoint suction pump. A horizontal wellpoint is very efficient hydraulically because it has a very large screen area, and horizontal flow will be plane to the sides of the perforated pipe. This contrasts with flow local to vertical wellpoint systems, where flow lines converge radially to each wellpoint, and where the screened area is limited by the short length of the wellpoint screens.

The main use of the method is for large-scale shallow cross-country pipelines as an alternative to single-stage wellpointing, when rapid rates of installation and progression are required. The principal restriction on the use of the method is local availability of the specialist trenching machines to lay the perforated pipe at adequate depth. Trenching machines capable

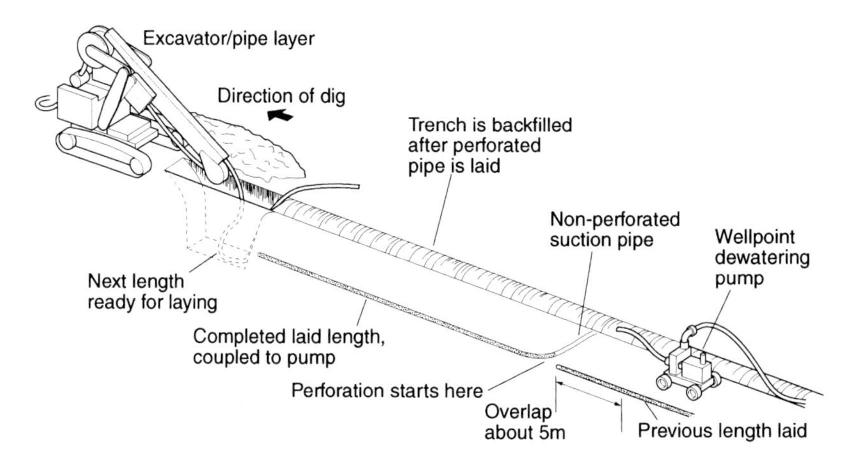

Figure 11.8 Horizontal wellpoint installation using a land drain trenching machine (from Preene *et al.* 2000; reproduced by kind permission of CIRIA).

of installing drains to 6 m depth are relatively commonly available in Holland and North America, but are less readily available in the United Kingdom. Large trenching machines were used on British pipeline and motorway cutting projects in the 1960s and 1970s, but nowadays most of the trenching machines in use in the United Kingdom are limited to installation depths of 3–4 m.

In addition to pipeline work, horizontal wellpoints are occasionally used instead of vertical wellpoints to form perimeter dewatering systems around large excavations for dry docks and the like (Anon 1976). If large drawdowns are required, multiple stages of horizontal wellpoints can be installed in a similar way to vertical wellpoint systems.

Although the horizontal method is best suited to the long straight runs associated with pipelines or the perimeter of large excavations, it can also be used to form a grid or herringbone pattern to dewater large areas to shallow depth. The method has also been used to consolidate areas of soft soils in conjunction with vacuum pumping systems (Anon 1998).

11.2.1 Merits of horizontal wellpoint systems

There are two principal practical advantages to the use of the horizontal wellpoint method. First, very rapid rates of installation can be achieved by specialist trenching machines (up to 1000 m per day in favourable conditions); this can be vital when trying to keep ahead of the installation of cross-country pipelines. Second, the absence of vertical wellpoints and surface header pipes alongside the trench allows unencumbered access for the pipe-laying operations. This has the additional benefit that there are fewer above-ground dewatering installations which might be damaged by the contractor's plant, with the associated risk of interruption of pumping.

Other practical advantages are: a supply of jetting water is not necessary for installation; and once the drainage pipe has been laid, installation and dismantling is simple and rapid, because only the pumps and discharge pipes are involved, without the need for header pipes.

From the cost point of view, although the horizontal drain cannot be recovered and is written off on the job, the cost of disposable vertical wellpoints and the hire cost of header pipes are saved. This can make the installation rate per metre of trench very cost-effective. However, overall costs should include for mobilization and demobilization costs, which can be high (in comparison to conventional wellpoint equipment) for large trenching machines and supporting equipment. This may be one of the reasons why this method tends not to be used on smaller contracts.

11.2.2 Installation techniques

Horizontal drains can be installed using conventional trench excavation techniques (e.g. by hydraulic excavator). However, these methods tend to be slow, and may have problems in maintaining stability of the trench while the drain is placed; such methods will only be cost-effective on relatively small projects or where rapid progression is not desired.

Horizontal wellpoint systems are more commonly installed using crawler-mounted trenching machines (Fig. 11.9) equipped with a continuous digging chain, which typically cuts a vertical sided trench of 225 mm width as the machine tracks forward. Typical depths of installation are between 2 and 6 m. A reel of flexible perforated drainage pipe feeds through the boom supporting the digging chain and is laid in the base of the trench. As the machine tracks forward either the spoil is allowed to fall back into the trench, or the trench is backfilled with filter media.

The perforated pipe used as the drain is typically uPVC land drain of 80–100 mm diameter (although 150 mm pipe is sometimes used); the pipe is generally wrapped in a filter of geotextile mesh, coco matting or equivalent. The pipe comes in continuous reels, perhaps 100 m in length. One end of the pipe is sealed with an end plug, and the other end is unperforated for the first 5–10 m. When the machine starts to cut the trench the unperforated end is fed out first and is left protruding from ground level. The machine tracks away, cutting the trench and laying the drain almost simultaneously. When the reel of drain runs out the sealed end is left in the base of the trench. A new reel of drain is fitted to the machine, and the next section is laid in a similar way, with around a 5 m overlap between sections (Fig. 11.8). Some trenching machines have the facility for addition of filter media above the drain to improve vertical drainage in stratified soils (Fig. 11.10). Filter media should be selected on the same basis as for vertical wellpoint systems (see Section 9.3).

When installing a horizontal drain for pipeline dewatering, topsoil is normally stripped off prior to installation. This is normal practice to allow the site to be reinstated at the end of the project, but has the added advantage in allowing the trenching machine to track on the firmer sub-soil. Some of the larger machines weigh up to 32 tonnes and can be difficult to operate on soft soils. It may be necessary to fit wider crawler tracks to reduce ground pressures.

For pipeline works the drain is typically installed along the centreline of the proposed pipeline, at a depth below the pipeline formation level (Fig. 11.11). When pumped, this allows the drain to directly dewater the area beneath the proposed pipeline. When construction is complete, the horizontal drain is left in place and abandoned. Occasionally, it is necessary to grout up the drain at the end of the project to prevent any influence on long-term groundwater conditions.

The specialist trenching machines can be effective tools, but are not without their problems. Firstly, it can be difficult to detect unexpected ground conditions through which the drain has been laid. If coarse gravel, cobbles or boulders are present, progress may be slowed and wear to the trenching machine may be excessive; in extreme cases there is a risk of the digging chain breaking. If layers of soft clay are present in conjunction with a high

(a)

Figure 11.9 Specialist trenching machine for installation of horizontal drains (courtesy of T. O. L. Roberts). (a) Trenching machine laying drains, (b) detail of continuous cutting chain.

(b)

Figure 11.9 Continued.

water table, the clay may 'slurry up' and coat the perforated pipe, clogging it as it is laid. If such ground conditions are anticipated, judgement should be applied before committing to the horizontal wellpoint method.

11.2.3 Pumping equipment

Horizontal wellpoints are pumped by connecting a conventional wellpoint pump to the unperforated section of the drainage pipe where it emerges from the ground. As the horizontal system is pumped on the suction principle, it is subject to drawdown limitations for similar reasons to the vertical wellpoint system. In general, the maximum achievable drawdown will be

Figure 11.10 Trencher equipped for addition of filter media to drain trench (courtesy of T. O. L. Roberts). A spoil deflector ensures the arisings do not fall back into trench. A special chute allows filter media to be tipped into the trench.

limited to between 4.5 and 6 m depending on ground conditions and the type of pump used (see Chapter 12).

11.3 Pressure relief wells

When an excavation is made into a low permeability layer above a confined aquifer, there is a risk that pore water pressures in the confined aquifer may

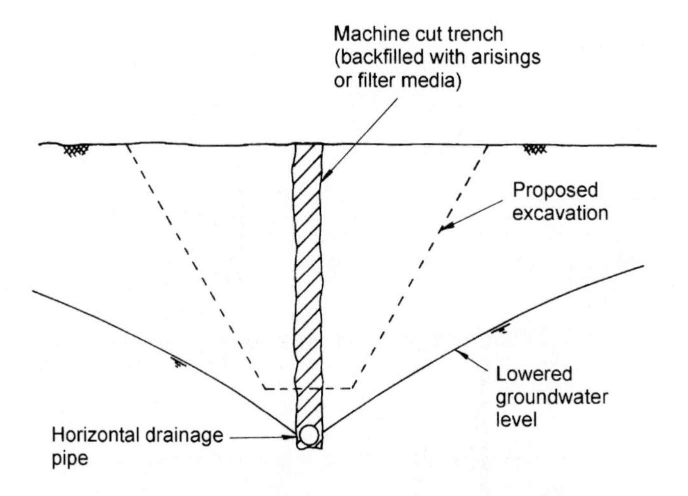

Figure 11.11 Installation of horizontal drains for pipeline trench.

cause the base of the excavation to become unstable. The base of the excavation may 'heave' because the weight of soil remaining beneath the excavation is insufficient to balance the uplift force from the aquifer pore water pressure (see Section 4.3). One way of avoiding this potential instability is to reduce pore water pressures in the aquifer by pumping from an array of deep wells (or, for shallow excavations, wellpoints).

Such pumped well systems are termed 'active' because pumps are used. The system of pressure relief (or bleed) wells offer an alternative method of reducing pore water pressures in confined aquifers. These systems are 'passive'; this means they are not directly pumped but merely provide preferential pathways for water from the aquifer to 'bleed' away, driven by the existing groundwater heads.

A schematic view through a typical relief well system is shown in Fig. 11.12. The wells are normally drilled prior to commencement of excavation, or at least before the excavation has progressed below the piezometric level in the aquifer. As excavation continues, the wells will begin to overflow, relieving pore water pressures in the aquifer and ensuring stability. The water flowing from the relief wells is typically disposed of by sump pumping (Chapter 8).

The discussion of relief wells described in this section will be mainly restricted to wells of relatively large diameter (greater than 100 mm) formed by drilling or jetting, and backfilled with sand or gravel. This type of wells are distinct from the smaller diameter vertical drains installed for soil

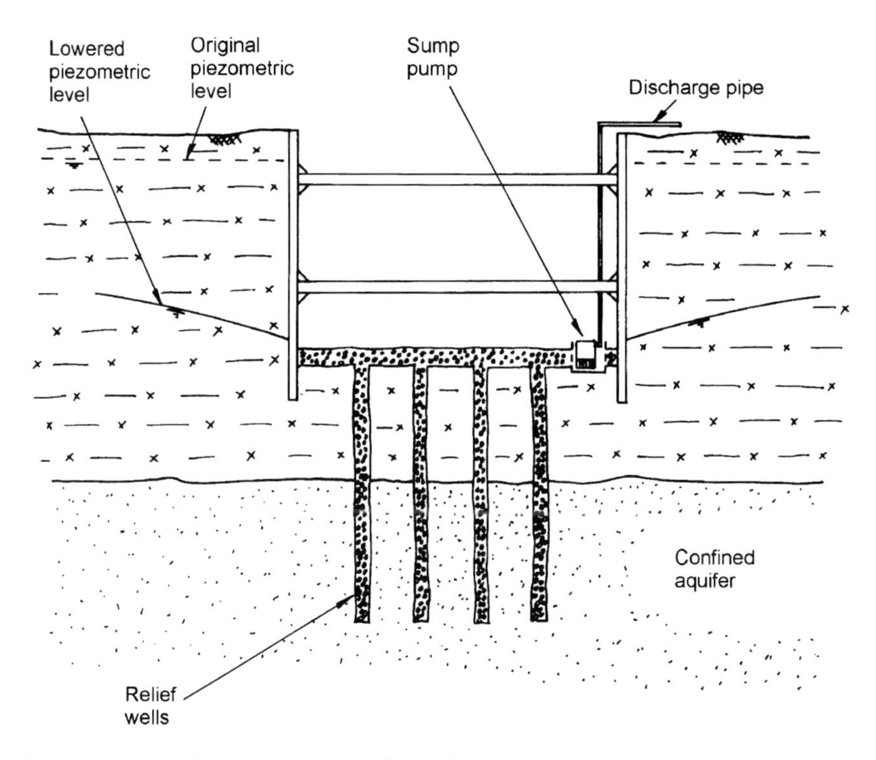

Figure 11.12 Relief well system. The relief wells overflow into a granular drainage blanket in the base of the excavation. The water is pumped away from a sump in the drainage blanket.

consolidation purposes, often formed using a mandrel to push plastic drainage wicks into soft clay and silt soils (Institution of Civil Engineers 1982).

11.3.1 Merits of pressure relief well systems

When used in appropriate conditions the principal advantages of relief well systems are cost and simplicity. Typical relief wells consist of a simple gravel-filled borehole; since the wells do not need to accommodate pumps or well screens they can be of modest diameter, reducing drilling and installation costs. The water flowing from the wells is removed by conventional sump pumps, which are more readily available and more robust in use than borehole electro-submersible pumps used to pump from deep wells.

Relief wells are best employed in shafts or deep cofferdams where the sides of the excavation are supported and the stability of the excavation base is the primary concern. The method is most appropriate to use where the excavation base is in stiff clay or weak rock (such as chalk, soft sandstones or fractured mudstones). Because the water from the relief wells overflows onto the excavation formation, there may be a risk of the water causing softening of exposed soils (especially in clays that have a permeable fabric). This can lead to difficult working conditions. It may be possible to avoid this problem by installing a granular drainage blanket and network of collector drains to direct water to the sumps and prevent ponding in the excavation.

A key question when considering a relief well system is: how many wells of a given diameter will be required? An initial stage is obviously to estimate the rate at which groundwater must be removed by the wells to achieve lowering to formation level – typically this is estimated by treating the excavation as an equivalent well (see Section 7.3).

Powers (1992) suggests that the capacity of a gravel-filled vertical drain can be estimated by direct application of Darcy's law (see Section 3.2).

$$Q = kiA \qquad (11.1)$$

where Q is the vertical flow rate along a relief well (m^3/s), k the permeability of the gravel backfill (m/s), A the cross-sectional area of the well (m^2) and i the vertical hydraulic gradient along the well.

Powers suggests that the vertical hydraulic gradient in the relief well be taken as unity. Table 11.1 presents theoretical maximum capacities of relief wells from equation (11.1) assuming fully saturated conditions and unit hydraulic gradient.

Powers states that although many engineers will find these values surprisingly low, field experience suggests that these values are theoretical maximums, rarely achieved in practice. In fact, the actual capacity may be significantly less than the theoretical capacity for a variety of reasons including smearing or clogging of the borehole wall during drilling, or segregation of the filter material during placement.

While the number of relief wells required must consider the well capacity, some thought must also be given to the spacing between wells. If small

Table 11.1 Maximum theoretical capacity of sand or gravel-filled relief wells

Permeability of filter backfill (m/s)	*100 mm borehole* (l/min)	*150 mm borehole* (l/min)	*300 mm borehole* (l/min)
1×10^{-4}	0.05	0.11	0.42
5×10^{-4}	0.24	0.53	2.1
1×10^{-3}	0.47	1.1	4.2
5×10^{-3}	2.4	5.3	21
1×10^{-2}	4.7	11	42
5×10^{-2}	24	53	212

flows are predicted only a few wells may be necessary to deal with the volume of water. However, it may be prudent to install additional relief wells to ensure the distance between wells is not excessive (say greater than 5–10 m). If the wells are widely spaced and the ground conditions may be variable (especially if the wells are installed into fissured rock), there is a danger that the wells may not adequately intercept sufficient permeable zones or fissures. This could lead to unrelieved pressures remaining, with the possibility of local base heave in the areas between the wells.

11.3.2 Installation techniques

Relief wells are typically drilled by similar methods to deep wells: cable percussion boring, rotary drilling, jetting or auger boring (see Chapter 10). The borehole is drilled to full depth, any drilling fluids used are flushed clear and then the filter media (sand or gravel) are added to backfill the bore up to the required level. The diameter of drilling is typically between 100 and 450 mm.

The filter media may be placed in the well via a tremie pipe or may be simply poured in from ground level. This latter approach is acceptable provided that the filter media has a very uniform grading so that there is little risk of segregation of the filter particles as they settle to the bottom of the bore. In practice, many relief wells installed in soft rocks, where the performance of the filter is not critical, are filled with uniform coarse gravel. Ideally, to permit maximum transmission of water the gravel should be of the rounded pea shingle type of 10–20 mm nominal size. However, on remote sites it may be necessary to use locally available material (which may consist of angular crushed aggregates), and accept some reduction in well efficiency. In certain cases, where the long-term performance of the relief wells is critical, the gravel used may need to be designed to match soil conditions in a similar way to deep wells (see Section 10.2). If no well screen is installed, it is not normally possible to develop the relief well.

11.3.3 Relief wells – is a well casing and screen needed?

Previous sections have mainly discussed gravel-filled relief wells, but there are cases when it is appropriate to install casings and screens (surrounded by a gravel pack) in relief wells. Although introducing additional cost and complexity, relief wells with screens can be used in the following cases:

(i) If the confined aquifer is of high permeability, a screened well will have a greater vertical flow capacity than a purely gravel filled well. Use of screened wells may reduce the number of wells required.

(ii) If a pumping test is needed to confirm aquifer permeability and flow rate the casing and screen will allow the test to be carried out using a submersible pump. This may be appropriate for large excavations where a widely spaced grid of relief wells is installed, based on an initial assessment of permeability. Pumping tests are then carried out on some of the wells to estimate the actual permeability and determine whether additional relief wells are needed to fill in the gaps between the original wells.

(iii) If during critical stages of construction (e.g. during casting of concrete structures) overflowing water would be an inconvenience, it may be possible to install submersible pumps in the wells and temporarily lower water levels below formation level. This would allow the critical works to be completed in more workable conditions.

Relief wells are sometime also used as part of the permanent works, to provide long-term pressure relief for deep structures (e.g. basements or deep railway cuttings) constructed above confined aquifers. Permanent relief wells are often installed with screen and casings to allow the wells to be cleaned out and re-developed if their performance deteriorates after several years of service.

11.4 Collector wells

A collector well (Todd 1980) consists of a vertical shaft typically 5 m in diameter, sunk as a caisson, from which laterals (horizontal or sub-horizontal screened wells, typically of 200–250 mm diameter) are jacked radially outward (Fig. 11.13). Collector wells are sometimes known as Ranney wells

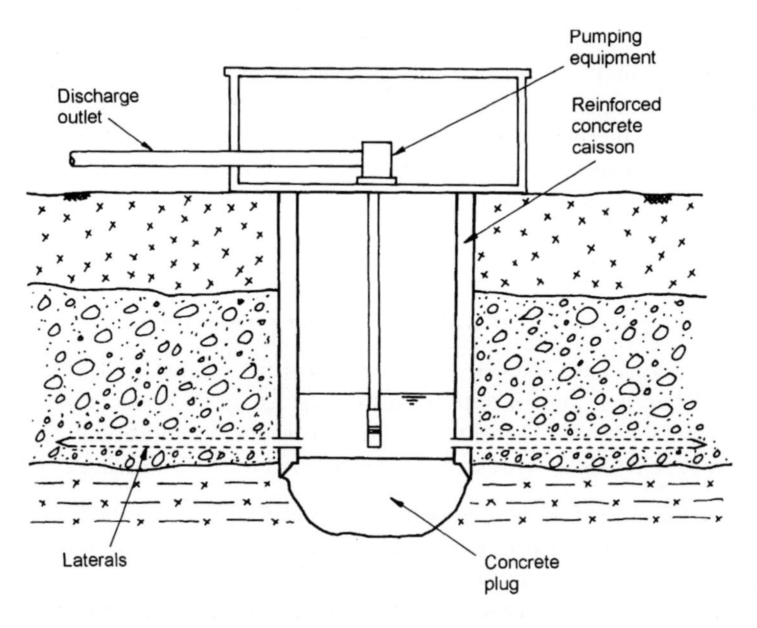

Figure 11.13 Collector well.

after a proprietary system (first developed by Leo Ranney in 1933) used to form this type of well (Drinkwater 1967).

A collector well is pumped by a submersible or lineshaft pump located in the shaft, which lowers the water level in the well. This creates a pressure gradient along the laterals causing them to flow freely into the main shaft. This method is particularly suited to large capacity permanent water supply installations in shallow sand and gravel aquifers, but is rarely used for groundwater control operations. In general, the cost of constructing the central shaft and forming the laterals is likely to be prohibitive for temporary works applications. Nevertheless, the method has occasionally been used, for example to aid dewatering of tunnel crossings beneath roads and railways where access to install conventional wells was restricted (Harding 1947).

11.5 Electro-osmosis

Electro-osmosis is suitable for use in very low permeability soils such as silts or clays where groundwater movement under the influence of pumping would be excessively slow. Electro-osmosis causes groundwater movements in such soils using electrical potential gradients, rather than hydraulic gradients. A direct current is passed through the soil between an array of anodes and cathodes installed in the ground. The potential gradient causes positively charged ions and pore water around the soil particles to migrate from the anode to the cathode, where the small volumes of water generated can be pumped away by wellpoints or ejectors (Fig. 11.14). The method can reduce the moisture content of the soil, thereby increasing its strength. In many ways, electro-osmosis is not so much a groundwater control method as a ground improvement technique. One of the drawbacks of the method is that it is a decelerating process, becoming slower as the moisture content decreases.

Electro-osmosis is a very specialized technique, and is used rarely, mainly when very soft clays or silts are required to be increased in strength. Casagrande (1952) describes the development of the method. Some relatively recent applications are given by Casagrande *et al.* (1981) and Doran *et al.* (1995). In some applications, electro-osmosis is used in conjunction with electro-chemical stabilization (see Bell and Cashman 1986), when chemical stabilizers are added at the anodes to permanently increase the strength of the soil.

In application the electrode arrangements are straightforward, typically being installed in lines, with a spacing of 3–5 m between electrodes. Anodes and cathodes are placed in the same line, in an alternating anode–cathode–anode sequence (Fig. 11.14). Water is to be pumped from the cathodes, so these can be formed from steel wellpoints or steel well liners.

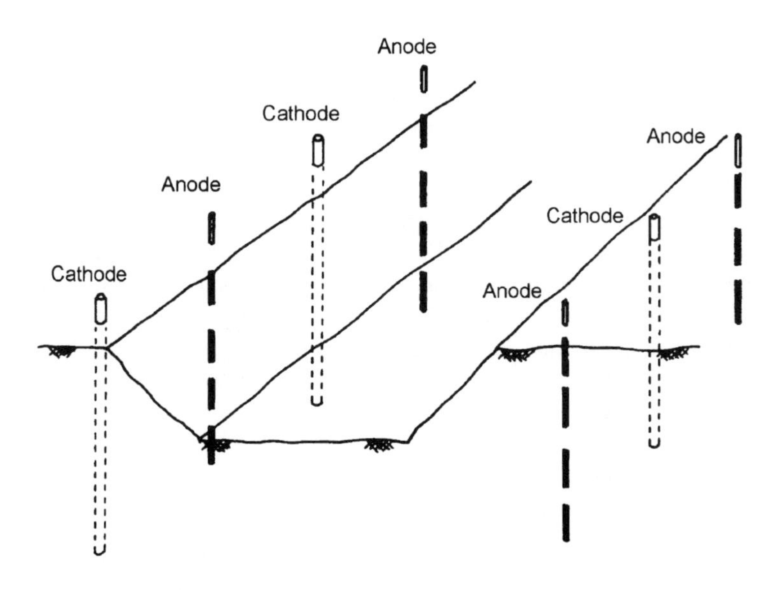

Figure 11.14 Electro-osmosis system. A direct current is applied to the electrodes. Water is drawn to the cathodes. The anodes can be simple metal stakes, but the cathodes are formed as wells and are pumped by wellpoints or ejectors.

If it is desired to use plastic well liners at the cathodes a metal bar or pipe (to form the electrode) will need to be installed in the sand filter around the well (Fig. 11.15(a)). The anodes are essentially metal stakes (Fig. 11.15(b)); gas pipe, steel reinforcing bar, old railway lines or scrap sheet-piles can be used.

Applied voltages are generally in the range 30–100 V. Effectiveness can be improved if the potential gradient can be in the same direction as the hydraulic gradient. Casagrande (1952) states that the potential gradient should not exceed 50 V/m to avoid excessive energy losses due to heating of the ground. However, it might be advantageous to operate at 100–200 V/m during the first few hours to give a faster build up of groundwater flow. Reduction in power consumption may be possible if the system can be operated on an intermittent basis.

11.6 Use of dewatering and exclusion in combination

Pumped groundwater control systems are frequently used in combination with exclusion methods (see Section 5.3). The methods might be combined for a number of reasons.

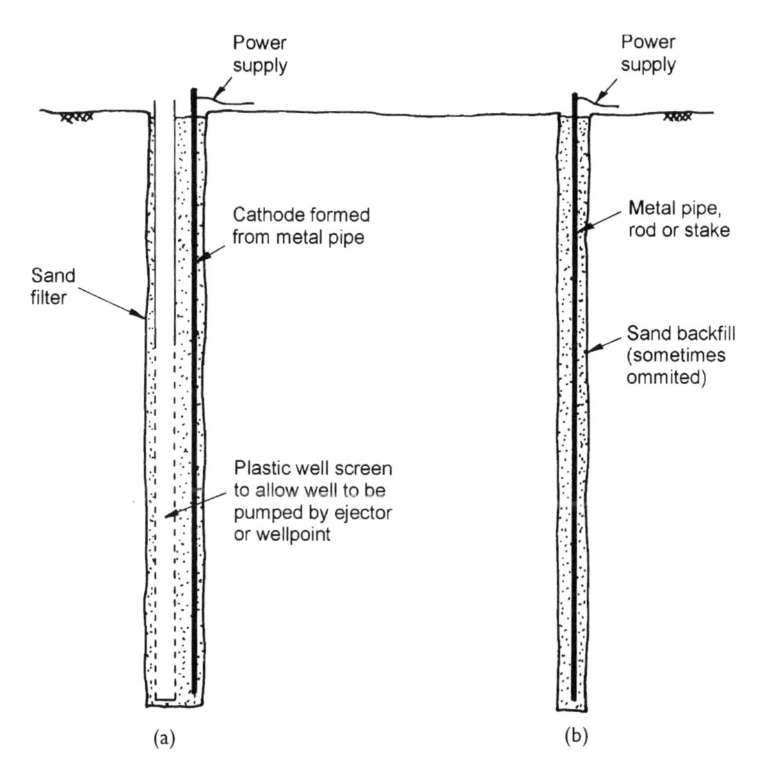

Figure 11.15 Typical electrode details. (a) Cathode well (b) Anode.

11.6.1 Exclusion methods to reduce pumped flows in high permeability soils

In high permeability soils, where the pumped flow will be large, cut-off methods (e.g. sheet-pile walls, slurry walls, grout curtains, etc.) can be used to reduce the potential inflows to an excavation. If the abstraction points (sumps or wells) are located within the cut-off walls the pumped flow rate will be reduced to a lesser or greater degree depending on the effectiveness of the exclusion method.

11.6.2 Groundwater pumping to dewater cofferdams

If the soil stratification allows the cut-off wall to form a complete cofferdam (e.g. if sheet piles are driven down to an impermeable layer) the excavation will be isolated from the surrounding groundwater regime. However, some groundwater will remain trapped inside the cofferdam, and will need to be pumped away. This is sometimes done by crude sump pumping as excavation proceeds. However, this may result in loss of fines or loosening of the soils that lie beneath the formation of the structure under construction. Sometimes wells or wellpoints are installed inside the cofferdam to pre-drain water levels prior to excavation, thus reducing detrimental effects on the soil formation.

11.6.3 Groundwater lowering used to reduce loading on cut-off structures

If the sides of an excavation are temporarily supported by sheet-pile or concrete pile walls, the stresses on the walls will arise partly from soil and partly from groundwater loading. In some circumstances groundwater lowering can be used to reduce the groundwater loading on a wall, to minimize temporary propping requirements. This can be a highly effective expedient; fewer props in the construction area can allow work to proceed quicker and with less obstruction.

However, a note of caution should be sounded. If pumping is interrupted (e.g. because of power or pump failure) the recovery of groundwater levels will increase loading on the walls and props, leading to overstressing, distortion and, in the worst case, collapse. Adequate standby facilities (perhaps arranged for automatic start-up) are essential. Alternatively, pressure relief holes could be formed through the wall above formation level; if water levels rose these would relieve the pressure behind the wall, albeit at the inconvenience of allowing the excavation to be flooded.

References

Anon. (1976). Dewatering for Andoc's Hunterson platform site. *Ground Engineering*, **9**(7), 42–43.

Anon. (1998). Vacuum packed: a new rapid and cost effective method of pre-consolidation is being used to combat the rapid settlement of soft, compressible soils in the Netherlands. *Ground Engineering*, **31**(2), 18–19.

Bell, F. G. and Cashman, P. M. (1986). Groundwater control by groundwater lowering. *Groundwater in Engineering Geology* (Cripps, J. C., Bell, F. G. and Culshaw, M. G., eds), Geological Society Engineering Geology Special Publication No. 3, London, pp 471–486.

Casagrande, L. (1952). Electro-osmotic stabilisation of soils. *Journal of the Boston Society of Civil Engineers*, **39**, pp 51–83.

Casagrande, L., Wade, N., Wakely, M. and Loughney, R. (1981). Electro-osmosis projects, British Columbia, Canada. *Proceedings of the 10th International Conference on Soil Mechanics and Foundation Engineering*, Stockholm, Sweden, pp 607–610.

Doran, S. R., Hartwell, D. J., Roberti, P., Kofoed, N. and Warren, S. (1995). Storebælt Railway tunnel – Denmark: implementation of cross passage ground treatment. *Proceedings of the 11th European Conference on Soil Mechanics and Foundation Engineering*, Copenhagen, Denmark.

Drinkwater, J. S. (1967). The Ranney method of abstracting water from aquifers. *Water and Water Engineering*, pp 267–274.

Harding, H. J. B. (1947). The choice of expedients in civil engineering construction. *Journal of the Institution of Civil Engineers*, Works Construction Paper No. 6.

Institution of Civil Engineers. (1982). *Vertical Drains: Géotechnique Symposium in Print*. Thomas Telford, London.

Miller, E. (1988). The eductor dewatering system. *Ground Engineering*, **21**(1), 29–34.

Powers, J. P. (1992). *Construction Dewatering: New Methods and Applications*, 2nd edition. Wiley, New York.

Powrie, W. and Preene, M. (1994). Performance of ejectors in construction dewatering systems. *Proceedings of the Institution of Civil Engineers, Geotechnical Engineering*, **107**, July, 143–154.

Powrie, W. and Roberts, T. O. L. (1990). Field trial of an ejector well dewatering system at Conwy, North Wales. *Quarterly Journal of Engineering Geology*, **23**, 169–185.

Preene, M. (1996). Ejector feat. *Ground Engineering*, **29**(4), 16.

Preene, M. and Powrie, W. (1994). Construction dewatering in low permeability soils: some problems and solutions. *Proceedings of the Institution of Civil Engineers, Geotechnical Engineering*, **107**, January, 17–26.

Preene, M., Roberts, T. O. L., Powrie, W. and Dyer, M. R. (2000). *Groundwater Control – Design and Practice*. Construction Industry Research and Information Association, CIRIA Report C515, London.

Todd, D. K. (1980). *Groundwater Hydrology*, 2nd edition. Wiley, New York, 213–214.

Chapter 12

Pumps for groundwater lowering duties

12.0 Introduction

There is a wide range of pumps and pumpsets available for pumping water. However, many pumps are not suitable for temporary works dewatering installations, so equipment must be selected with care. The categories of pumpsets appropriate to a particular groundwater lowering site requirement will depend on the technique in use.

In this chapter, the categories appropriate to the most common groundwater lowering techniques (wellpoints, deep wells and sump pumping) are described together with some discussion of their strengths and weaknesses.

12.1 Units for wellpoint pumping

The wellpoint pump operates on the suction principle and is positioned at some elevation above the level of the installed wellpoints (see Chapter 9) – generally at the level of the header main. It is required to suck the groundwater into the perforated wellpoints, up the individual unperforated riser pipes, into the suction header main to the intake of the wellpoint pumpset and thence discharge the pumped groundwater, via a discharge manifold to a disposal point. It follows, therefore that an efficient wellpoint pumpset must develop:

(a) Adequate vacuum to lift the groundwater from the soil at the level of the wellpoints and deliver the groundwater to the pump intake.
(b) Sufficient residual power to discharge the pumped water to the disposal area.

All efficient wellpoint pumpsets are designed to pump only clean water, with minimal suspended solids. There are three categories of pump types commonly used for wellpoint applications:

i Double acting piston pumps
ii Self-priming centrifugal pumps
iii Vacuum tank units.

12.1.1 Double acting piston pump

The Callans type reciprocating double acting piston pump (Fig. 12.1) is not widely employed on wellpoint installations in the United Kingdom but in continental Europe it is much used. Its energy consumption is significantly less than that of a vacuum assisted self-priming centrifugal pumpset of comparable output. If there are suspended fine particles in the abstracted groundwater (perhaps because the sanding-in of wellpoints is inadequate) these will act as an abrasive that will cause wear of the pistons and piston liners. In time this will lead to a reduced vacuum and associated deterioration of pumping efficiency.

Units are available with diesel or electric prime movers. A typical 100 mm unit (which refers to the nominal size of the discharge outlet) has a power requirement of 5.5 kW and can pump up to 18 l/s at 10–15 m head. The larger 125 mm unit (7.5 kW) can pump up to 26 l/s at 10–15 m head. Comparison with the capacities of self-priming centrifugal units shows that piston pumps are generally suited to lower flow rate situations, and that if high flows are anticipated a centrifugal unit may be a better choice.

These pumps are nominally self-priming, but even for a pump in good condition priming may be slow. Accordingly, initial priming (by filling the pump and header main with clean water) should be considered. The amount of vacuum that it can generate when primed is slightly less than

Figure 12.1 Reciprocating piston pumpset used for wellpointing (courtesy of WJ Groundwater Limited).

that generated by a comparable self-priming vacuum assisted centrifugal pumpset. As a rule of thumb, a piston pump is not appropriate if a suction lift of more than 4.5–5 m is required.

12.1.2 Vacuum assisted self-priming centrifugal pumpset

The conventional vacuum assisted self-priming centrifugal wellpoint pumpset has four separate components:

(1) An enclosed chamber (the float chamber) complete with an internal baffle and a float valve. This chamber serves to separate the air and water drawn into the pump. The inlet side of the float chamber is connected to the header main and the outlet is connected the pump unit (3). The volume of this chamber should be generous so as to ensure adequate separation of the air from the water. This is especially important when the rate of water flow from the header main is substantial. The air should be extracted by the vacuum pump so that the water fed to the eye of the pump impeller, has little or no air; otherwise there will be a risk of cavitation of the pump when operating. The purpose of the float valve is to shut off the vacuum when the water level in the float chamber reaches a predetermined level, thereby preventing carry-over of droplets of water to the vacuum pump; and to re-establish vacuum when the water level in the float chamber has fallen.

(2) A vacuum pump connected to the float chamber (1), to augment significantly the small amount of vacuum generated by the centrifugal pump unit itself so as to suck the groundwater continuously into the float chamber and to remove air from the water.

(3) A 'clean water' centrifugal pump (i.e. with a minimal clearance between the impellers and the inside of the volute casing) to discharge the water from the float chamber to the discharge main. Generous clearance between impeller and volute is only appropriate to usage as a contractor's 'all purpose' pump (see Section 12.3) for dealing with solids in suspension, but the extra clearance needed for the all purpose duty impairs efficient performance for a wellpointing duty.

(4) The motor (prime mover) will generally be either diesel or electric. Its rating must be adequate to drive both the vacuum pump and the water pump to produce the expected flow with adequate allowance to cope with total head from all causes. If it is anticipated that the pump will be required for duty at a ground elevation significantly above mean sea level, and/or the ambient temperature will be above normal, a larger engine will be needed to produce the required output.

A centrifugal wellpoint pump is operated continuously at a high vacuum (i.e. low positive pressure at the eye of the impeller) and so is liable to cavitation. A cavitating pump exhibits a rattling sound similar to a great snoring noise and the unit tends to vibrate. If the condition persists the surface of the impeller will become pitted and the bearings damaged. The pump shaft may be fractured.

The most commonly used size of centrifugal pump for wellpoint installations is the 150 mm water pump size with a 10 l/s vacuum pump (Fig. 12.2). Such pumps typically have water flow capacities of up to 55 l/s at 10 m head. The 100 mm size is also used on some small wellpoint installations (capacity up to 40 l/s at 10 m head). Larger units are available in the 200 mm, 250 mm and 300 mm sizes but are not often required, it generally being preferable to use two or more 150 mm pumpsets in place of larger units.

Wellpoint pumps are most commonly powered by diesel prime movers. These units are relatively noisy, and there is growing pressure for the use of 'silenced' units in built-up areas. In fact, silenced units are far from silent; if noise is a major concern then electrically powered pumps (running from a mains supply) should be considered, as these are the quietest units available.

Figure 12.2 Vacuum assisted self-priming centrifugal pumpset used for wellpointing (courtesy of Sykes Pumps).

The vacuum assisted self-priming centrifugal pump may also be used for sump pumping of clean, filtered groundwater (see Chapter 8) and for operating a bored shallow well system (see Chapter 10).

12.1.3 *Vacuum pumps*

The two types of rotary vacuum pumps most commonly fitted to centrifugal wellpoint pumpsets to assist priming are:

i The liquid ring or water recirculating vacuum pump.
ii The flood lubricated or oil sealed vacuum pump.

The vacuum units for the wellpoint application are required to operate continuously over a wide range of air flow rates from the very high to the very low. The service duty to be fulfilled is onerous. Generally, the vacuum pump is belt driven off the drive shaft of the prime mover. The energy consumption of the vacuum pump is of the order of 15 per cent of the total of the motor output. The heat that may be generated can be considerable so the cooling arrangements must be reliable; this requirement is particularly pertinent to the oil sealed vacuum unit.

The liquid ring vacuum pump is better able to cope with low water flow conditions because the liquid ring pump recirculates some of the pumped groundwater, which is generally cool. Thus, the cooling requirements are less demanding than for the flood lubricated vacuum pump. However, if it is expected that the pumped water will have a detrimental effect (e.g. if the groundwater is corrosive) special cooling modifications may be desirable, depending on the anticipated length of the pumping period and the chemical characteristics of the water to be pumped.

The liquid ring type is simpler than the flood lubricated vacuum pump but the degree of vacuum that it can generate is slightly less (about 5 per cent lower). It is more robust and can be made capable of handling high air flow rates. The air handling capacity of the liquid ring pump is in the range 1.4–14 m^3/min as compared to a capacity 1–3 m^3/min of the flood lubricated pump – both at 0.85 bars and ambient temperatures and pressures.

Adequate separation of air from water in the float chamber is very important when a flood lubricated pump is used. If the air is not adequately separated from the water, some water will be 'carried over' to the vacuum pump and cause emulsification of the oil that forms the vacuum seal; this will cause damage to the vacuum pump.

12.1.4 *Vacuum tank unit*

The vacuum tank unit (Fig. 12.3) is a variation on the float chamber but of considerably greater volume (of the order of 5 m^3) and therefore more efficient at separating the air from the water. A continuous vacuum can be applied to

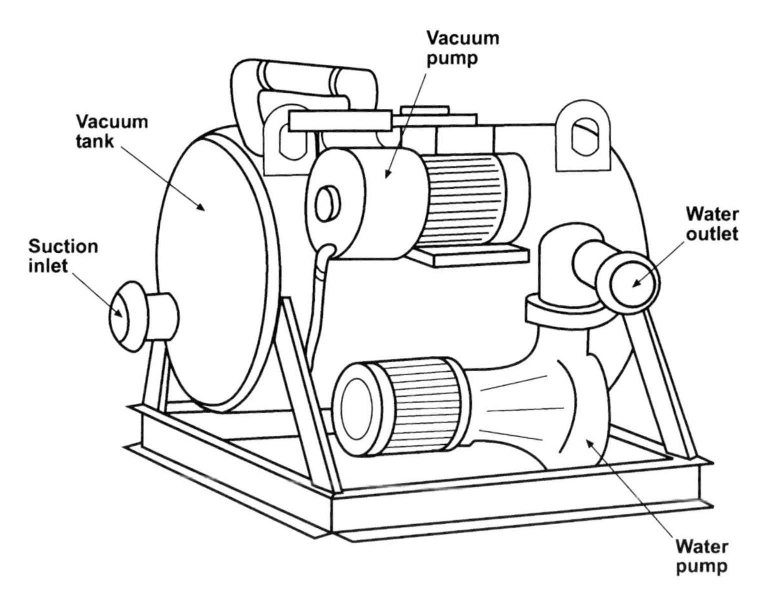

Figure 12.3 Vacuum tank unit for wellpointing.

the tank by one or more vacuum pumps; this draws water to the vacuum tank, which gradually fills with water. The tank is fitted with one or more water pump(s) to discharge the water. Level sensors are used to switch the water pumps on or off depending on the level of the water in the tank. Typically, the whole unit is electrically operated. The vacuum tank unit is efficient, especially for dealing with low rates of flow; because while the application of vacuum is continuous (and in most units the amount of applied vacuum can be varied depending on the duty required) the energy for pumping the water is only mobilized as and when required to empty the tank. However, the sophistication of the total unit is considerable. These units are used fairly frequently in continental Europe but are little used in the United Kingdom.

The vacuum tank unit is also useful for ground improvement of very low permeability soils since, by reducing the moisture content, the shear strength of the soil can be increased. The rate of increase in shear strength will be slow. The rate of pumping will tend to be very low, perhaps of the order of 100–200 l/hr. A self-priming centrifugal pump could not cope with such a duty; it would grossly overheat within a relatively short period of time.

12.2 Jetting pumps

Wellpoint jetting units are high-pressure centrifugal clean water pumps (often not self-priming) used for the installation of wellpoints and their risers, placing tubes, etc. Their outputs are usually in the range of 12–200 l/s at pressures of 4–23 bars. The most commonly used jetting pump for the installation of wellpoints is rated to deliver about 20 l/s at a pressure of about 6–8 bars.

12.3 Units for sump pumping

Sump pumping (Chapter 8) is one of the most common pumping applications on construction sites. Many types of pumps are used for sump pumping; some are suitable but some others are not. The sump pumping duty is most frequently needed to deal with surface water run-off. Unfortunately, usually little or no consideration is given to filtering of the water to be pumped. Thus the units used for sump pumping must be able to cope with some suspended solids in the pumped water.

12.3.1 Contractor's submersible pump

A pump type commonly used for sump pumping is known as the contractor's submersible pump (Fig. 12.4). It is an electric submersible unit with a sealed motor that usually runs in oil. The pump is of the bottom intake type (i.e. the water intake is beneath the motor). The pump is installed in a sump (Fig. 8.2); the pumped water flows around the outside of the motor casing and thereby helps to keep the motor cool. Most units can operate on intermittent 'snore' where the water level is drawn down to the pump invert and the pump draws both air and water. It is prudent to suspend the unit so that the bottom of the pump intake is about 300 mm above the bottom of the sump; this allows for some accumulation of sediment in the sump. These units are less mechanically efficient than some other types of pump. When high flows are involved the energy costs are significant.

The submersible pump does not operate on the suction principle, so there is no limit to the possible drawdown, provided pumps of sufficient power and head rating are used. Commonly used pumps range from 4.5 kW units with 75 mm outlets (capable of pumping around 10 l/s at 10 m head) up to 40 kW units with 200 mm outlets (of capacity 100 l/s at 25 m head). Hydraulic submersible pumps are also available, where the submersible pump is driven by a diesel hydraulic power pack located outside the sump at ground level.

12.3.2 All purpose self-priming centrifugal pump

These units are very similar to pumpsets used for wellpointing duties (see Section 12.1), but adapted to be tolerant of some suspended solids in the pumped water. This requires the pump to be manufactured with generous clearance between the impeller and the volute casing, to allow fines to pass. However, by allowing for extra clearance in the pump internals, some

Figure 12.4 Contractor's electric submersible pump (courtesy of Sovereign Pumps Limited).

hydraulic efficiency is lost, and 'all purpose' pumps normally have a reduced performance compared with dedicated wellpoint units. As with units for wellpointing duties, pumps may be driven either by diesel or electrical power.

12.4 Pumps for deep wells

There are two types of pumps commonly used to pump from deep wells:

1 The borehole electro-submersible turbine pump.
2 The vertical lineshaft turbine pump.

The pump end is similar in configuration in both types but the drive is different. The submersible pump is driven by a submersible electric motor incorporated in a common casing with the pump (the wet end). The complete unit is positioned near the bottom of the well. The lineshaft pump unit (the wet end) is likewise positioned near the bottom of the well but is driven via a lineshaft, by a motor (either electric or diesel) mounted at the surface.

12.4.1 Borehole electric submersible turbine pumps

Generally, in developed countries, deep well pumps for temporary works projects are of the borehole electro-submersible turbine type. These are much used by the water supply organizations. Hence, a great variety of these are readily available; almost off the shelf in many instances. The submersible turbine pump has a high mechanical efficiency – 70 to 80 per cent is common. Figure. 12.5 is a photograph of a typical submersible turbine pump.

The borehole submersible units are slim and so economize on the borehole and well screen diameter necessary to accommodate the pump. The smallest electro-submersible pumps in common use have a capacity of around 3 l/s and can be installed inside well screen and casing of 110–125 mm internal diameter. Pumps of 10 l/s capacity can normally be installed inside 152 mm internal diameter (the old imperial six inch size) well screen and casing. Pumps of up to 40–80 l/s capacity are available, and require well liner and casing of 250–300 mm or larger internal diameter.

The pump manufacturer's specification will normally state the minimum internal well diameter into which a given model of pump can be installed and operated. This should be used as a guide, but remember the manufacturer's recommendation will assume the well is straight and true. In reality, the well casing may have been installed with slight twists or deviations from plumb, and these may make the pump a tight fit when, in theory, it should pass freely. Pumps do sometimes become stuck and have to be abandoned down the well (resulting not only in the need for a new pump, but also for a new well), much to the chagrin of all concerned. If cost and available drilling methods allow, it is good practice to slightly increase the diameter of the well liner and casing above the minimum required for the pump.

For each pump capacity there is a range of pumps offering the same flow rate but at increasing head. The additional head is achieved by adding additional stages of impellers to the wet end, and correspondingly increasing the power requirements of the electric motor. Pumps are typically constructed largely from stainless steel, but some plastic, cast iron or bronze components may also be used.

Most borehole submersible turbine pumps have their electric motor located in the lower part of the pump body casing with the wet end above. The wet end consists of the water intake and stator (these are both integral parts of the casing); and the impellers which are fixed to the drive shaft from the motor.

There is a waterproof seal on the pump shaft between the motor and the pump unit. Beneath the motor there is a bottom bearing which is designed

Figure 12.5 Borehole electro-submersible turbine pump (courtesy of Grundfos A/S). The pump shown is a 'cut-away' demonstration version to show the shaft and impellers inside the wet end at the top part of the pump.

to take the weight and end thrust of the whole pumpset. Most submersible turbine pump units supplied for deep wells have three phase motors. However, single-phase units are available for low duties and outputs. They have motors up to about 5 kW.

The pump is installed centrally in the well on the end of its riser pipe (Fig. 12.6) after the well has been developed so the pumped groundwater has no fines in suspension (see Section 10.6).

Figure 12.6 Lowering borehole electro-submersible pump and riser into well (courtesy of Grundfos A/S).

In order to prevent water in the riser pipe running back through the impellers into the well when the power is switched off, there should be a non-return valve immediately above the pump outlet. Such a back flow would be harmful if an attempt were made to restart the pump while the water in the riser pipe was still running back; under such circumstances the additional start up load could cause overloading. However, a non-return valve will mean that the riser pipe will remain full of water when the pump is switched off. This increases the weight to be lifted when the pump is removed from the well. It is acceptable to drill a 5 mm diameter hole in the non-return valve to allow water to drain from the riser pipe sufficiently slowly.

12.4.2 Borehole electro-submersible pumps – operational problems

There are three possible causes of motor failure:

(a) Wear of the seal between the motor and the impellers. In time wear would allow water to enter the sealed motor thus causing failure of the motor windings.
(b) Uneven wear of the bottom thrust bearing. This will lead to uneven wear of the upper seal also – result as (a) above.
(c) Overheating of the motor. This will damage the windings of the motor.

Seal wear
In order to avoid failure due to seal wear, it is necessary to ensure that the water pumped is clean. However, if the grading of the filter media around the well screen is too coarse to retain the soil particles (or the well has not been adequately developed, see Section 10.6) the pumped water may contain fine particles in suspension. In time this will cause wear of the impellers and stator leading to reduced output. Also the seal between the impellers and the motor will be worn, eventually causing failure of the motor windings. The rewinding of a pump motor is both costly and time consuming.

Bearing failure
It is important that the pumpset hangs vertically. Otherwise the thrust on the bottom bearing pad will be uneven, leading to uneven wear of the seal between the motor and the impellers; eventually water will penetrate the windings of the electric motor, causing it to fail.

Motor overheating
In operation the pump casing is immersed in the groundwater and also the motor windings are surrounded by a jacket filled with oil or water based emulsion coolant fluid. Overheating of the motor windings is unlikely, provided the flow to the well is significant compared with the volume of water surrounding the pump motor casing. However, if the rate of pumping of the well is low (e.g. in a low permeability soil), the pump motor casing may be in 'dead water'. The heat generated by the motor will gradually raise the temperature of the dead water in the well sump above the generally cool

temperature of the inflowing groundwater. Eventually, the motor windings will overheat and burn out.

The risk of encountering dead water in the sump is more prevalent where the sump length of a well is formed in the top of an underlying clay layer. Often, it is necessary to form the sump section to penetrate into the clay to achieve maximum drawdown by positioning the pump intake beneath the level of the base of the aquifer. This is particularly relevant where the proposed formation level is close to an aquifer/clay interface and the wetted screen length per well is limited.

Apart from changing to a smaller output pump – technically the most satisfactory course of action – there are two expedients that can be used where there is risk of overheating of the motor windings. Fitting a shroud over the pump intake and motor casing will ensure that water from the cool groundwater source flows upwards over the motor casing before reaching the intake and ensures that the motor is cooled (Fig. 12.7(a)). An alternative is to fit a small bore (3 mm or 6 mm) by-pass pipe tapped into the riser pipe

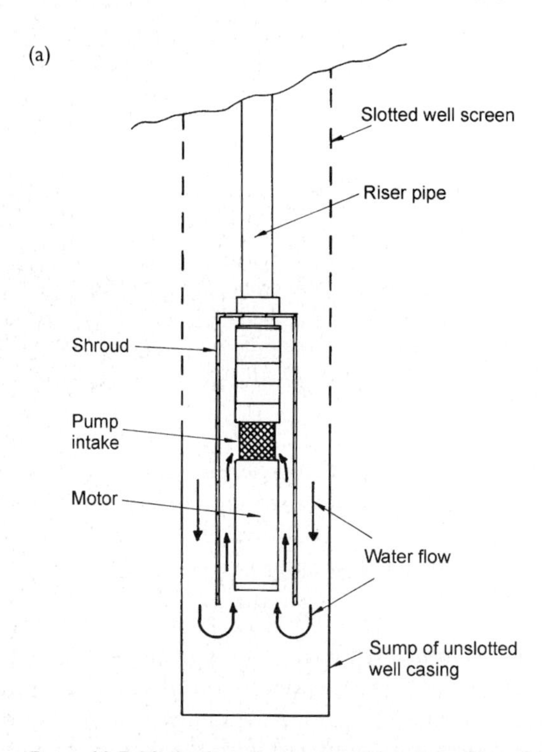

Figure 12.7 Methods to prevent overheating of borehole electro-submersible pump motors. (a) Motor shroud: The shroud is a tube, open only at its base, enclosing the electro-submersible pump. Water can only reach the pump intake by flowing over the motor, helping to keep it cool. (b) By-pass pipe: The small flow diverted from the riser pipe via a by-pass pipe is discharged near the base of the motor to provide a cooling flow.

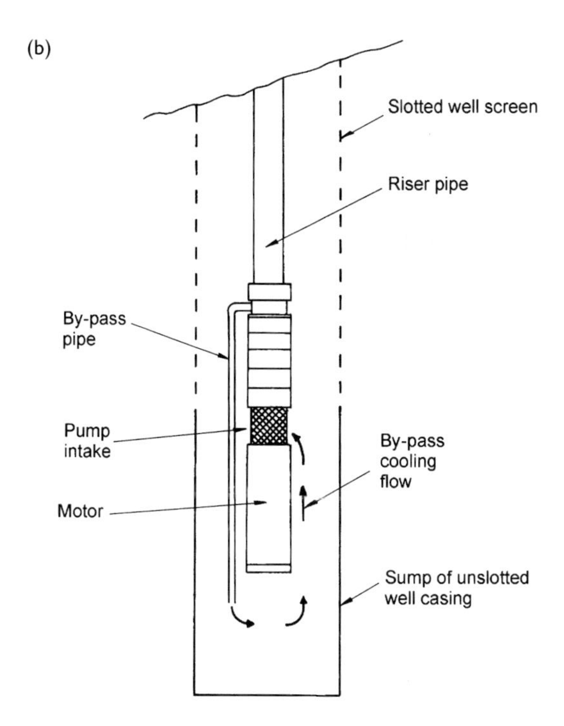

Figure 12.7 Continued.

above the pump outlet but below the non-return valve, to divert some of the cool intake flow into the sump around and below the motor casing (see Fig. 12.7(b)). This small and continuous flow of intake water will maintain cool water around the pump motor casing.

There is an insidious variation on motor overheating – though the discharge water may appear clear to the naked eye, this is no guarantee that there are no fines in suspension. When operating a low yield well there is a risk that the sump water is dead. Small fine particles, not visible to the naked eye, may settle in the dead water and gradually build up around the outside of the motor casing with the result that there will be no sump water to cool the motor casing. In time the windings will fail due to overheating. The fitment of either a shroud or a small bore bypass pipe, as described above, will prevent sediment build-up in the sump.

12.4.3 Vertical lineshaft turbine pumps

Lineshaft pumps (Fig. 12.8) are particularly suitable for high volume outputs at low heads and for high horsepower duties. This type of pump is commonly used in developing countries, probably due to its greater simplicity in operation.

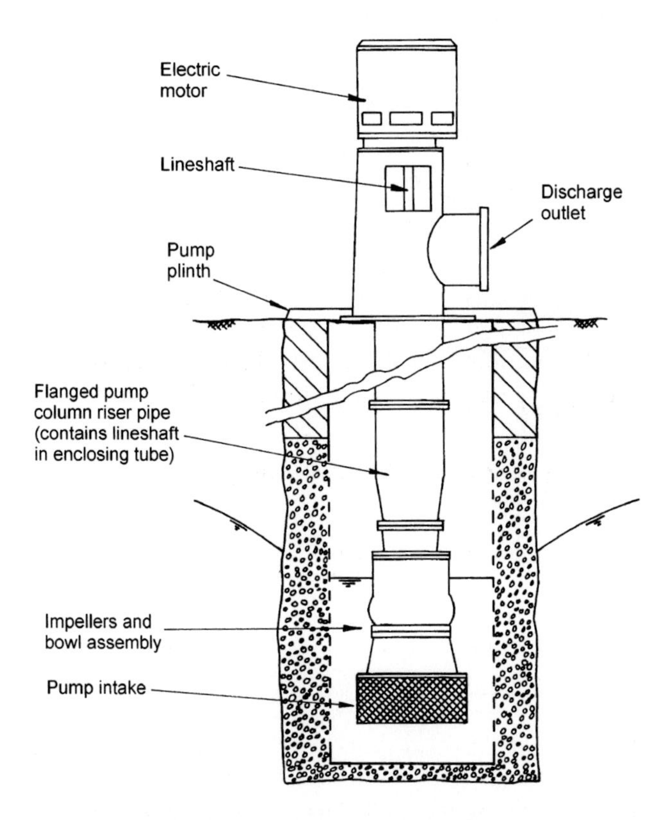

Figure 12.8 Vertical lineshaft turbine pump. The pump is powered by an electric motor mounted above the pump column.

The wet-end unit of a lineshaft pumpset (which consists of one or more impellers in a bowl assembly), like that of the submersible pump, is located near the bottom of the well. However, unlike the submersible pumpset, the prime mover unit is located on top of or adjacent to the wellhead. The prime mover can be either petrol driven (small output pumps only) or diesel driven, with a separate energy source for each well. However, the prime movers may be electrically driven from a common power source.

The impellers of the pump unit are powered via a vertical lineshaft drive (with bearing assemblies) inside the pump column riser pipe which is connected to the prime mover unit located at ground level. The verticality and straightness of the well is as important to the trouble free operation of the lineshaft pump as for the submersible pump. The connection from the prime mover to the drive shaft depends on the type of prime mover and may be a direct coupling, a belt drive, or right angle gear. As with a borehole electro-submersible pump, it is prudent to incorporate a non-return valve at the pump outlet.

12.4.4 Lineshaft pumps – operational problems

If the pumped water contains fines in suspension (either because the well was not developed adequately or the grading of the filter media is too coarse) in time the impeller stages and casing will wear and the pump output will decline. Otherwise, the lineshaft unit is generally trouble free and reliable in use provided that the pump unit with its lineshaft assembly has been properly installed in the well, though prime mover troubles there may be. All the foregoing assumes that the pump unit sizing is appropriate to the actual well yield.

12.4.5 Comparison of merits of lineshaft pumps vs electro-submersible pumps

The mechanical efficiency of lineshaft pumps tends to be slightly greater than that of submersible pumps. The initial cost of small output lineshaft pumps tends to be greater than a comparable sized submersible pump but this is reversed for the large output units.

The installation procedures for lineshaft pumps are more onerous. The well must be plumb. Also, skilled personnel are required for installation. Separate connections have to be made to each drive shaft length and pump column riser pipe joint assemblies, as these are being installed in the well. The standard length of shaft components may not be the same as those of the riser pipes; this can cause tedium in installation of the riser pipe and drive shaft in the well. In contrast, the installation of an electro-submersible pump unit entails only the connections of the riser pipe; the electric power cable will be in one continuous length and is simply paid out as the pump is lowered into the well. However, a considerably lesser standard of skills are required for the maintenance and repair of lineshaft pumps as compared to the submersible unit. Motor or prime mover repairs are particularly easy for lineshaft pumps because of the above ground installation.

Each individual wellhead prime mover unit of a lineshaft pump installation should to visited at regular intervals and performance and fuel requirements checked. However, for a submersible pump installation, all the control switch gear, pump starters and associated process timers (for automatically restarting an individual pump if it trips out) and the change-over switchgear from mains power to standby generators (in case there is a failure of the mains supply, for whatever cause) etc. can all be located inside a single switch house. This makes supervision of the running of a deep well submersible pump installation easier and more straightforward as well as less labour intensive.

12.5 Sizing of pumps and pipework

On most groundwater lowering projects, standard or 'off the shelf' pumpsets will be used, either bought new, hired-in or re-used from a previous project. The pumps must be selected so they can achieve the anticipated flow rate both when drawdown is established, and also during the initial period of pumping, when flow rates may be higher.

The flow rate produced by a given type of pumpset varies (to a greater or lesser degree) with the 'total head' against which the pump must act. In general, the greater the total head the lower the output from a given pumpset. Different types of pumps will respond to changes in head in different ways; pump manufacturers produce performance charts describing the idealized performance of each model of pumpset.

Knowing the total head against which a pumpset must work, the manufacturer's performance charts allow the pump output under field conditions to be estimated. If the predicted pump output is less than that required, larger pumpsets could be used or (if the system design permits) multiple units could be employed.

The total head of pumping consists of three components: suction lift, discharge head and friction losses. The relationship between these components for submersible and suction pumps is shown in Fig. 12.9.

Friction losses are often small in comparison with the other components of total head and are rarely a major issue in the design of dewatering systems

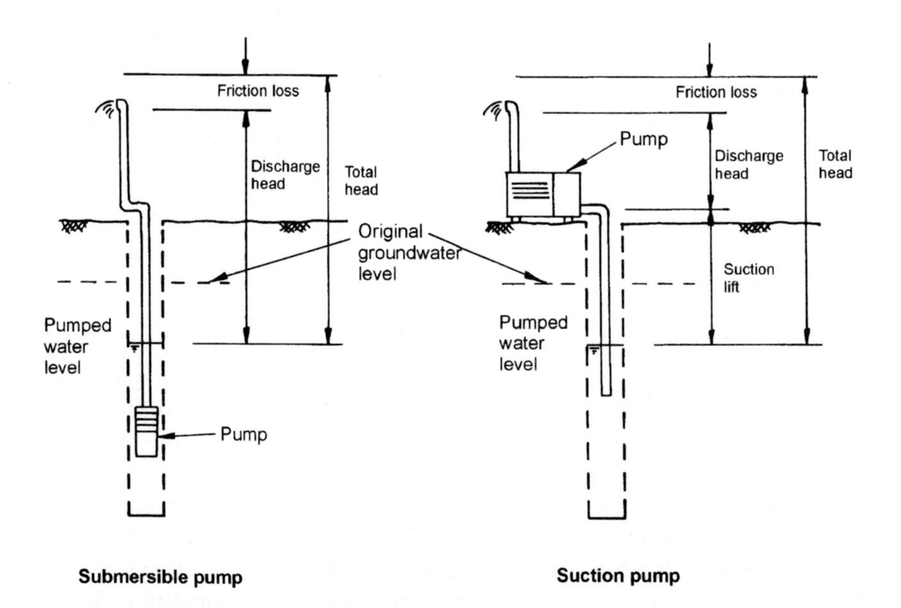

Figure 12.9 Discharge heads for pump sizing.

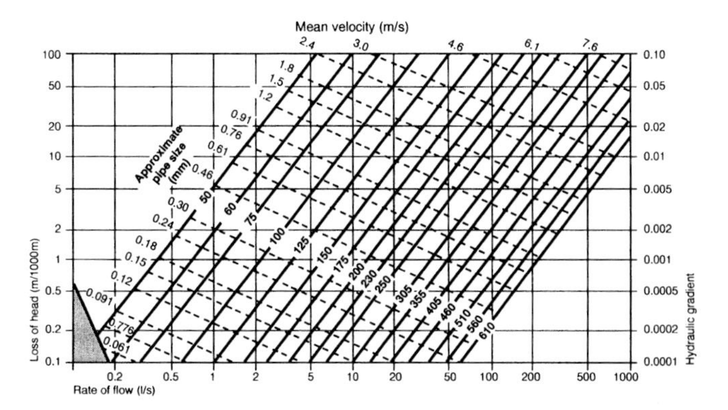

Friction losses in valves and fittings as an equivalent length of straight pipe in metres

Type of Fitting		Nominal pipe diameter (mm) 150	200	250	300	450	600	750	1065	1200
Gate valve	Open	1.1	1.4	1.7	2.0	2.8	4.3	5.2	6.4	7.6
	¼ closed	6.1	7.9	10.1	12.2	18.3	24.4	30.5	41.2	48.8
	½ closed	30.5	39.6	51.8	59.5	91.5	122.0	152.0	213.0	244.0
	¾ closed	122.0	159.0	213.0	244.0	366.0	488.0	610.0	854.0	976.0
Standard Tee	Flow in line	2.9	4.3	5.0	5.9	9.1	11.9	15.1	22.0	24.7
	Flow to/from branch	9.8	12.8	16.8	19.8	30.5	39.6	50.3	73.2	82.3
Medium Sweep 90° elbow		4.3	5.5	6.7	7.9	12.2	15.9	21.3	28.0	32.0
Long sweep 90° elbow		3.2	4.3	5.3	6.1	9.1	12.2	15.2	21.3	24.4
Square 90° elbow		9.8	12.8	16.8	19.8	30.5	39.6	50.3	73.2	82.3
45° elbow		2.3	3.1	3.7	4.6	6.4	8.5	10.7	15.2	10.3
Sudden enlargement	d/D = ¼	4.9	6.4	8.4	9.9	15.2	19.8	25.2	36.6	41.2
	d/D = ½	3.2	4.3	5.3	6.1	9.1	12.2	15.2	21.3	24.4
	d/D = ¾	2.9	3.7	4.9	5.6	8.4	11.0	13.7	19.8	22.9
Sudden contraction	d/D = ¼	2.3	3.1	3.7	4.6	6.4	8.5	10.7	15.2	18.3
	d/D = ½	1.7	2.3	2.9	3.4	4.9	6.4	8.2	11.3	12.8
	d/D = ¾	1.1	1.4	1.7	2.0	2.8	4.3	5.2	6.4	7.6

Figure 12.10 Friction losses in pipework (from Preene *et al.* 2000; reproduced by kind permission of CIRIA).

Note: Friction head loss may be estimated by assuming that the total output from the wellpoints flows the full length of the header pipe

using standard pump and pipework sizes. The main exception is in high flow rate systems (greater than around 50 l/s) where the water has to be discharged to considerable distances (greater than 100 m). In such cases the friction losses may be significant and may reduce the output from the pumps. Figure 12.10 shows charts and tables (from Preene *et al.* 2000) that allow friction losses to be estimated for commonly used pipe sizes and fittings.

If friction losses are perceived to be large enough to detrimentally effect pump outputs, then a number of remedial measures are possible.

i Replace the pump sets with units rated at higher heads.
ii Provide additional pumpsets so that a greater number of pumps producing a lower output can achieve the required total flow rate.
iii Modify the discharge pipework by either: planning the pipework layout carefully to avoid unnecessary bends, junctions or constrictions; using discharge pipework of larger diameter, if available; or, if larger pipework is not available, one or more additional lines of discharge pipework could be laid, reducing the flow rate taken by a individual pipe.
iv Reduce the distance that the discharge water must be pumped (e.g. by locating an alternative discharge point).
v Provide 'booster pumps' between the dewatering pumps and the discharge point. This will reduce the total head on the dewatering pumps, increasing their output.

References

Preene, M., Roberts, T. O. L., Powrie, W. and Dyer, M. R. (2000). *Groundwater Control – Design and Practice*. Construction Industry Research and Information Association, CIRIA Report C515, London.

Chapter 13

Side effects of groundwater lowering

13.0 Introduction

Groundwater lowering operations may affect water levels over a wide area, which can extend for some distance beyond the construction site itself. In many cases this will not cause problems, but there are some circumstances when undesirable side effects may arise. This chapter outlines some potential side effects of groundwater lowering, the conditions in which they may occur and possible mitigation measures, including artificial recharge systems.

The effects discussed in this chapter are:

- i Settlement resulting from the instability of excavations when groundwater is not adequately controlled.
- ii Ground settlements caused by loss of fines.
- iii Ground settlements induced by increases in effective stress, and associated structural damage or distress.
- iv Derogation or depletion of groundwater sources.
- v Changes in groundwater quality, including movement of contamination plumes and saline intrusion.
- vi The impact of discharge flows on the surface water environment.
- vii Other less common effects, including the drying out of timber piles and the desiccation of wetlands and vegetation.

The use of artificial recharge systems to mitigate some of the side effects of groundwater lowering is described. Practical problems that may occur during operation of recharge systems are discussed, and a case history is presented.

13.1 Settlement due to groundwater lowering

Ground settlements are an inevitable consequence of every groundwater lowering exercise. In the great majority of cases the settlements are so small that no distortion or damage is apparent in nearby buildings. However, occasionally settlements may be large enough to cause damaging distortion or distress of structures, which can range from minor cracking of architectural finishes to major structural damage. In extreme cases these effects have extended several hundred metres from the construction site itself, and have affected large numbers of structures.

If there is any concern that groundwater lowering (or any other construction operation) may result in ground settlements beneath existing structures it is essential that a pre-construction building condition survey be carried out. This exercise (sometimes known as a dilapidation survey) involves recording the current state of any structures that may be affected by settlement. This should provide a detailed record of any pre-existing defects, so that if damage is alleged, there is a basis for judging the veracity of the claims. Unfortunately, for groundwater lowering projects, where the influence of the works may extend for several hundred metres from the site, the building condition survey area selected is often too small and does not cover all the structures that may be significantly affected. Ideally, the extent of the survey should be finalized following a risk assessment exercise.

Settlements caused by groundwater lowering may be generated by a number of different mechanisms, some easily avoidable, some less so:

1 Settlement resulting from the instability of excavations when groundwater is not adequately controlled.
2 Settlement caused by loss of fines.
3 Settlement induced by increases in effective stress.

13.1.1 Settlement due to poorly controlled groundwater

This book describes the philosophy and methods whereby groundwater can be controlled to provide stable excavations for construction works. However, sometimes groundwater is not adequately controlled, leading to instability of excavation, uncontrolled seepages and perhaps a groundwater 'blow' (see Chapter 4). These problems may result from several causes, such as: failure to appreciate the need for groundwater control; a mis-directed desire to reduce costs by scaling down or deleting groundwater control from the temporary works; inadequate standby or back up facilities to prevent interruption in pumping; and ground or groundwater conditions not anticipated by the site investigation or design, and not identified by construction monitoring.

Ideally, with adequate investigation and planning, most of these issues can normally be avoided, especially the last one. Sadly, these problems still occur, leading to the failure of excavations and both significant additional costs and delays to the construction project (see Bauer *et al.* 1980 and Greenwood 1984 for case histories). If there is a sudden 'blow' or failure of an excavation, soil material will be washed into the excavation. This can create large and unpredictable settlements around the excavation, much larger than the effective stress settlements associated with groundwater control. Any buildings in the area where the uncontrolled settlements occur are likely to be severely damaged.

13.1.2 Settlement due to loss of fines

Settlement can also occur if a groundwater lowering system continually pumps 'fines' (silt and sand sized particles) in the discharge water – a problem known as 'loss of fines'. Most dewatering systems will pump fines in the initial stages of pumping, as a more permeable zone is developed around the well or sump (see well development, Section 10.6). However, if the pumping of fines continues for extended periods, the removal of particles will loosen the soil and may create sub-surface erosion channels (sometimes known as 'pipes'). Compaction of the loosened soil or collapse of such erosion channels may lead to ground movements and settlement.

Continuous pumping of fines is not normally a problem with wellpoints, deep wells or ejectors, provided that adequate filter packs have been installed and monitored for fines in their discharge. Occasionally, a sand pumping well may be encountered, perhaps caused by a cracked screen or poor installation techniques. Such wells should be taken out of service immediately.

The method which most commonly causes loss of fines is sump pumping (see Chapter 8). This is because the installation of adequate filters around sumps is often neglected, allowing fine particles in the soil to become mobile as groundwater is drawn towards the pump. Section 8.6 shows an example of the problems this can cause. Powers (1985) lists soil types where the use of sump pumping is fraught with risk. These include:

- Uniform fine sands
- Soft non-cohesive silts and soft clays
- Soft rocks where fissures can erode and enlarge due to high water velocities
- Rocks where fissures are filled with silt, sand or soft clay, which may be eroded
- Sandstone with uncemented layers that may be washed out.

In these soil types, even the best-engineered sump pumping systems may encounter problems. Serious consideration should be given to carrying out groundwater lowering by a method using wells (wellpoints, deep wells or ejectors) with correctly designed and installed filters.

13.1.3 Settlement due to increases in effective stress

Lowering of groundwater levels will naturally reduce pore water pressures, and hence increase effective stress (see Section 4.2). This will cause the soil layer to compress, leading to ground settlements. In practice however, for the great majority of cases the effective stress settlements are so small that no damage to nearby structures results.

The magnitude of effective stress settlements will depend on a number of factors:

i The presence and thickness of a highly compressible layer of soil below the groundwater level, which will be affected by the pore water pressure reduction. Examples include soft alluvial silts and clays or peat deposits. The softer a soil layer (and the thicker it is), the greater the potential settlement.
ii The amount of drawdown. The greater the drawdown of the groundwater level, the greater the resulting settlement.
iii The period of pumping. In general, at a given site, the longer the pumping is continued, the greater the settlement.

Powers (1992) states that the most significant of these factors is the presence of a layer of highly compressible soil. It is certainly a truism that damaging settlements are unlikely to result from groundwater lowering on sites where highly compressible soils are absent. The corollary of this is that the potential for damaging settlements should be investigated carefully on any site where there is a significant thickness of highly compressible soils.

Effective stress settlements can be calculated using basic soil mechanics theory. The final (or ultimate) compression ρ_{ult} of a soil layer of thickness D is:

$$\rho_{ult} = \frac{\Delta u D}{E'_0} \tag{13.1}$$

where Δu is the reduction in pore water pressure, and E'_0 is the stiffness of the soil in one-dimensional compression (which is equal to $1/m_v$, where m_v is the coefficient of volume compressibility). E'_0 can be estimated using several techniques; values used in calculations should be selected with care (see Preene *et al.* 2000).

Equation (13.1) is based on the assumption that total stress remains constant as the groundwater level is lowered. This is a reasonable simplifying assumption, since the difference between the unit weight of most soils in saturated and unsaturated conditions is generally small and, given the uncertainties in other parameters, can be neglected without significant error.

To be useful in practice, equation (13.1) needs to be written in terms of drawdown s. This will be different for aquicludes, aquifers and aquitards (see Section 3.3).

1 Aquicludes. This type of strata is of very low permeability. During the period of pumping from an adjacent aquifer, no significant pore water pressure reduction (and hence compression) will be generated.
2 Aquifers. This type of strata is normally of significant permeability and is pumped directly by the wells or wellpoints. Pore water pressure reductions will occur effectively at the same time as the drawdown. In

practice, this means compression and settlement of aquifers occurs instantaneously once drawdown occurs. The rate at which drawdown occurs, and the resulting pattern of drawdown around a groundwater lowering system can be estimated using the methods given in Section 7.6.

3 Aquitards. These strata are at least one to two orders of magnitude less permeable than aquifers, and are not generally pumped directly by wells or wellpoints. Aquitards drain vertically into the aquifer, at a rate controlled by the vertical permeability of the aquitard. In practice this means that, even if the drawdown and compression of the aquifer has stabilized, water may still be draining slowly out of the aquitard, and will continue to do so until the pore water pressures equilibrate with those in the aquifer. The pore water pressure reductions and compressions of aquitard layers will tend to lag behind the aquifer, and may take weeks, months or even years to reach their ultimate value. Of course if the groundwater lowering system operates for only a short period of time, the ultimate compression will not develop fully.

Figure 13.1 shows that, where aquifers and aquitards are present, the distribution of pore water pressure reduction with depth will depend on the nature of the strata at the site, and will change with time. This should be taken into account when estimating ground settlements.

Settlements caused by groundwater lowering will generally increase with time, and will be greatest at the end of the period of pumping. There are two separate time-dependent effects at work:

i Increasing aquifer drawdown with time. When groundwater is pumped from an aquifer by a series of wells, a zone of drawdown propagates away from the system at a rate controlled by the pumping rate and aquifer properties. This means that at a given point, some distance from the system, drawdown in the aquifer will increase with time, as will compression of the aquifer. For all practical purposes compression of an aquifer occurs contemporaneously with drawdown.

ii Slow drainage from aquitards. When drawdown occurs in an aquifer, any adjacent aquitard will begin to drain vertically into the aquifer. This drainage may occur quite slowly. It follows that the compression of aquitards will lag behind the drawdown in the aquifer.

For a simple case (Fig. 13.2) with wells fully penetrating a confined or semi-confined aquifer, groundwater flow will be horizontal, Δu will be constant with depth and $\Delta u = \gamma_w s$, where γ_w is the unit weight of water and s is the drawdown of the piezometric layer, giving the ultimate compression of a soil layer as:

$$\rho_{ult} = \frac{\gamma_w s D}{E_0'} \tag{13.2}$$

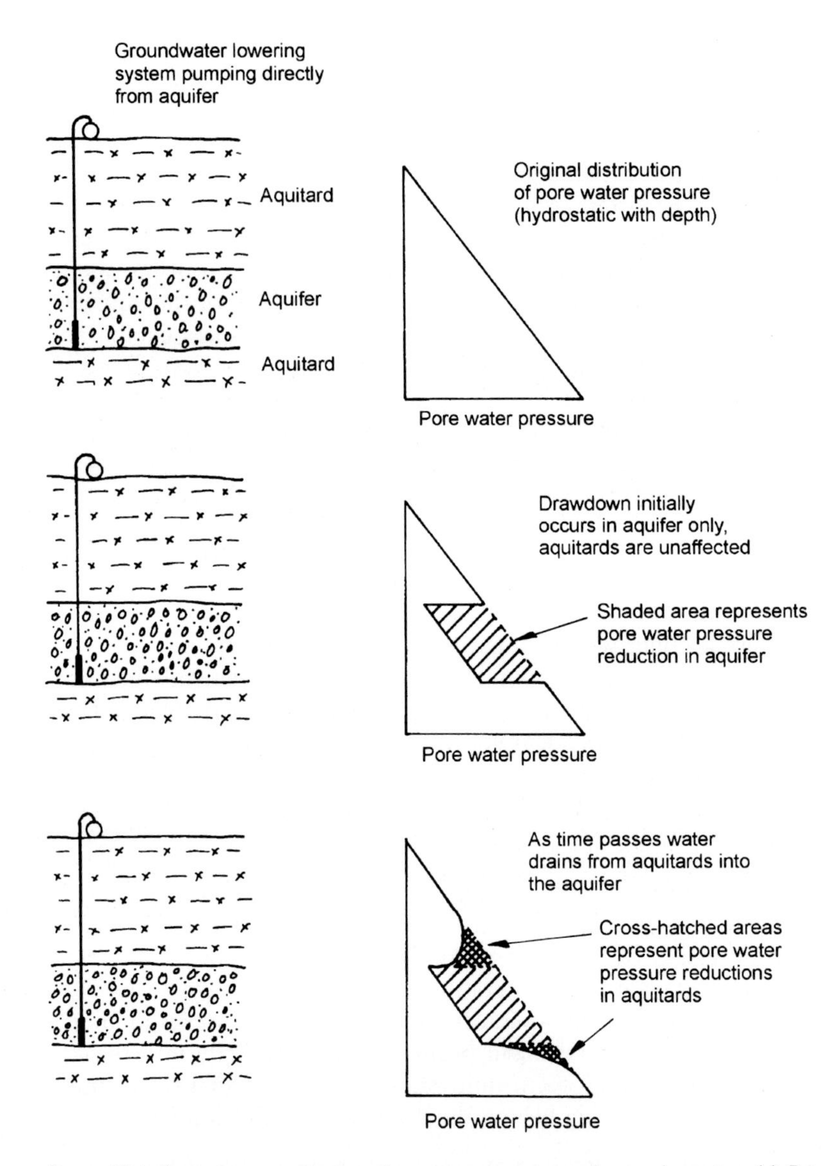

Figure 13.1 Pore water reductions in response to groundwater lowering. (a) Prior to pumping – Hydrostatic conditions prevail in the aquifer and aquitards (b) After short-term pumping – Pore water pressure reductions have not yet occurred in the aquitards (c) After long-term pumping – Drainage from aquitards is occurring – as pumping continues the pore water pressure reduction will propagate further from the aquifer into the aquitards.

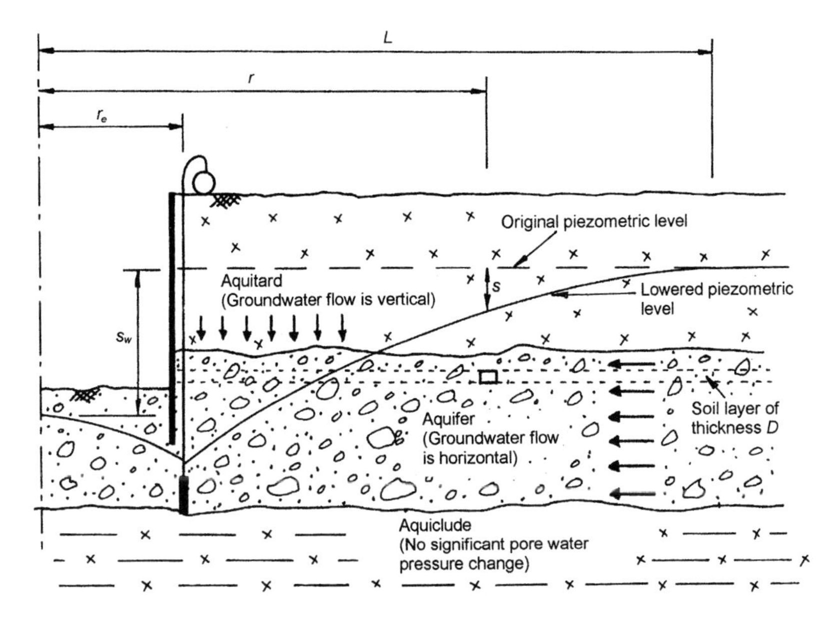

Figure 13.2 Groundwater flow to fully penetrating wells.

At the location under consideration, drawdown s can be estimated *approximately* using standard solutions for the shape of drawdown curves, such as those given in Section 7.6.

Equation (13.2) shows the ultimate compression. However, because compression may lag behind aquifer drawdown, settlement calculations should concentrate on the effective compression ρ_t at a time t after pumping commenced. For an aquifer, compression occurs rapidly, so the effective compression is the ultimate compression:

$$(\rho_t)_{\text{aquifer}} = \rho_{\text{ult}} \tag{13.3}$$

However, the effective compression of an aquitard layer may be less than the ultimate compression

$$(\rho_t)_{\text{aquitard}} = R\rho_{\text{ult}} \tag{13.4}$$

where R is the average degree of consolidation (with a value of between zero and one) of the aquitard layer, determined in terms of a non-dimensional time factor T_v as shown in Fig. 13.3.

The ultimate compression calculated in equation (13.2) is based on pore water pressure reductions consistent with a confined aquifer, where the piezometric head is not drawndown below the top of the aquifer. However, for greater drawdowns in confined aquifers, for unconfined aquifers, and

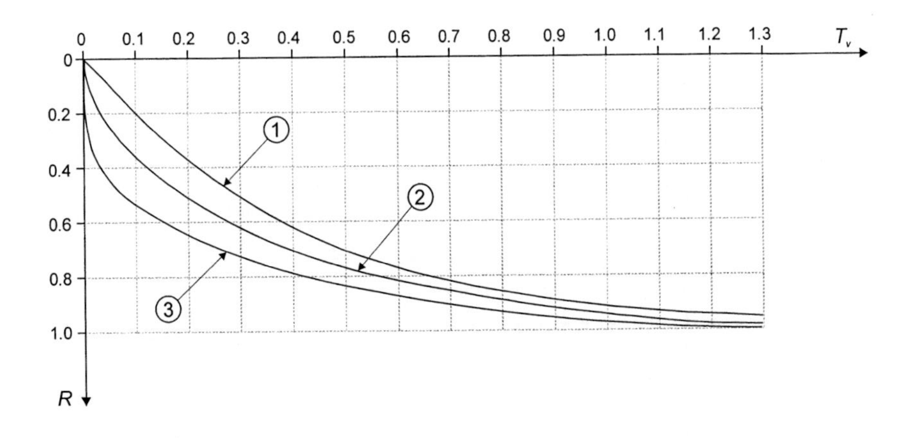

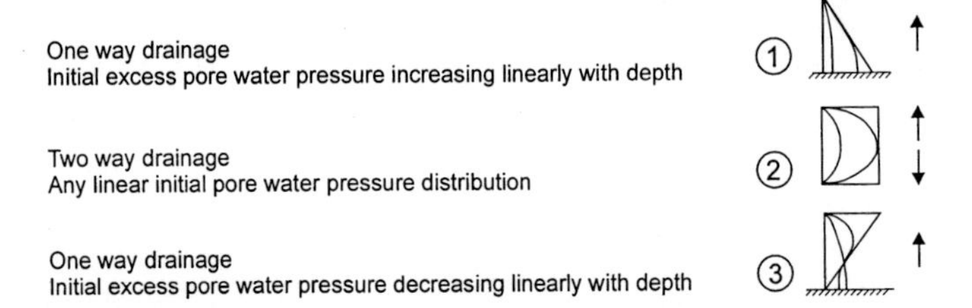

Figure 13.3 Average degree of consolidation of a soil layer vs time (after Powrie 1997).

Note:
$T_v = c_v t/h^2$ using aquitard parameters, where c_v = coefficient of consolidation, t = time since pumping began, h = length of drainage path.

for aquitards the pore water pressure distribution will be different. Therefore the effective compression estimated from equations (13.2) to (13.4) needs to be converted to corrected compression ρ_{corr}

$$\rho_{\text{corr}} = C_{\text{d}}\rho_t \tag{13.5}$$

where C_{d} is a correction factor for effective stress (with values of between 0.5 and 1.0) shown in Fig. 13.4. Note that this correction will tend to reduce the compression compared to the uncorrected values.

The above calculations are for the compression of individual soil layers. The result of interest to the designer is the resulting ground settlement ρ_{total}. This is determined by summing the corrected compressions of all aquifer and aquitard layers affected by the groundwater lowering.

$$\rho_{\text{total}} = (\rho_{\text{corr}})_{\text{all aquifers}} + (\rho_{\text{corr}})_{\text{all aquitards}} \tag{13.6}$$

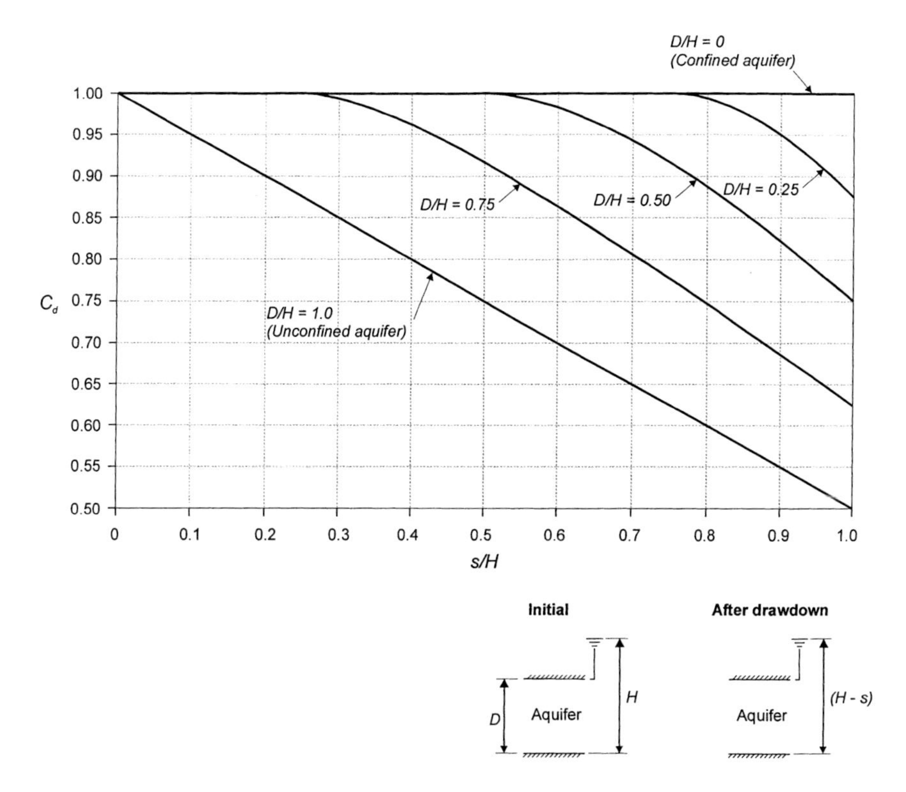

Figure 13.4 Correction factor for effective stress (after Preene 2000).

13.1.4 Settlement damage to structures

From a practical point of view, it is necessary to consider the magnitude of settlement that will result in varying degrees of damage – predicted settlements will be of less concern if they will not damage nearby structures. Considerable work has considered the damage to structures due to self weight settlements (Burland and Wroth 1975) or due to tunnelling-induced settlements (Lake *et al.* 1996). There is a paucity of data on damage that may result from groundwater lowering induced settlements. This may be because pre-construction building condition surveys are often not carried out over a wide enough area around a groundwater lowering system – damaging settlements may occur up to several hundred metres away. If damage occurs (or is alleged by property owners) the actual damage caused by groundwater lowering can be difficult to assess, since the original condition of the property is not known. This is in contrast to settlements from tunnelling or deep excavation, which rarely extend more than a few times the excavation depth; building condition surveys normally cover almost all of the structures at risk.

Structures are not generally damaged by settlements *per se*, rather by differential settlement or distortion across the structure. In uniform soil conditions the typical convex upward drawdown curve will create a settlement profile that will distort structures in hogging (Fig. 13.5). The drawdown curve will propagate away from the groundwater lowering system with time; at a given location the slope of the drawdown curve (and hence the differential settlements and distortion) will increase while pumping continues.

In fact, most groundwater lowering operations in uniform soils do not cause damaging settlements (unless the soils are very compressible). This is because the distortions and ground slopes resulting purely from the drawdown curve are generally slight. Variations in soil conditions or foundation type can allow more severe differential settlements or distortions to occur (Fig. 13.6).

Damage risk assessment exercises are often carried out for tunnel projects by assessing the maximum settlement and tilt that a structure will experience (Lake *et al.* 1996). A similar approach can be applied to groundwater lowering projects (Preene 2000). Table 13.1 shows tentative values to be used in initial damage risk assessments for settlements caused by groundwater lowering.

Table 13.1 Tentative limits of building settlement and tilt for damage risk assessment (from Preene 2000)

Risk category	*Maximum settlement (mm)*	*Building tilt*	*Anticipated effects*
Negligible	<10	<1/500	Superficial damage unlikely.
Slight	10–50	1/500–1/200	Possible superficial damage, unlikely to have structural significance.
Moderate	50–75	1/200–1/50	Expected superficial damage and possible structural damage to buildings; possible damage to rigid pipelines.
Severe	75	>1/50	Expected structural damage to buildings and expected damage to rigid pipelines or possible damage to other pipelines.

Notes:
1 Maximum settlement is based on the nearest edge of the structure to the groundwater control system.
2 Tilt is based on rigid body rotation, assuming that all of the maximum settlement occurs as differential settlement across the width of the structure, or across an element of the structure.
3 The risk category is to be based on the more severe of the settlement or tilt criteria.

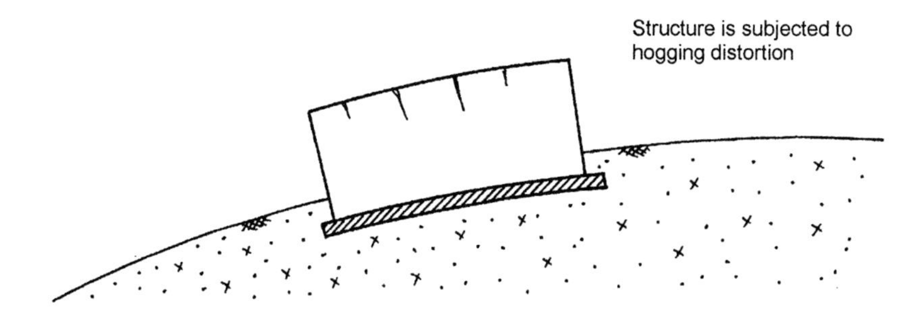

Figure 13.5 Deformation of structure due to settlement profile in uniform conditions (from Preene 2000).

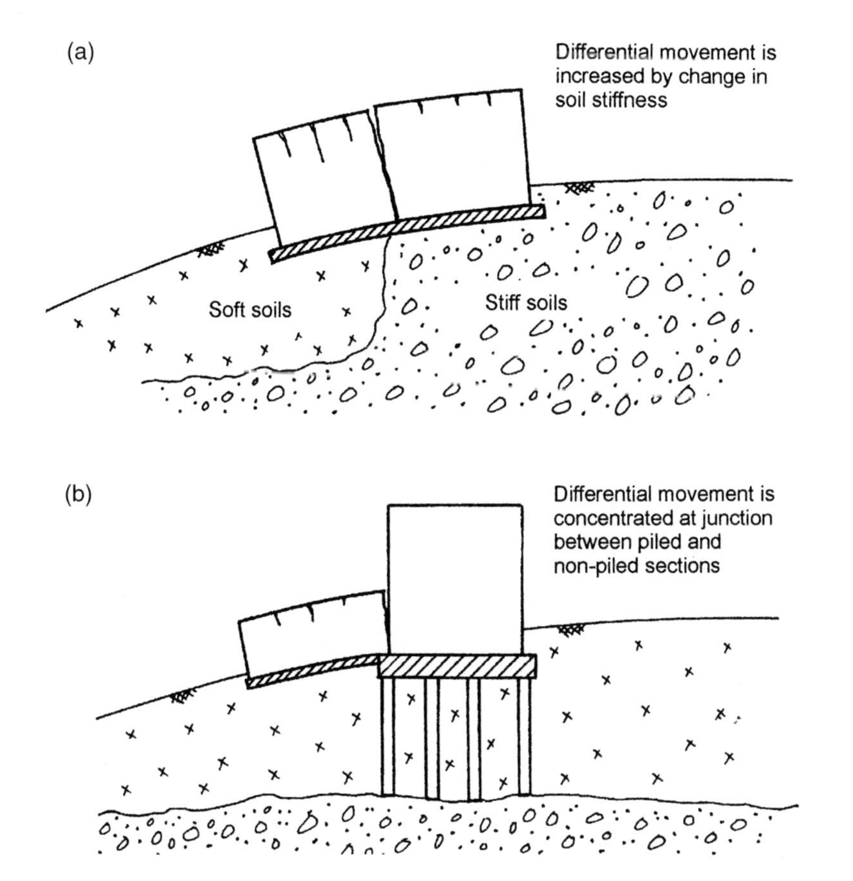

Figure 13.6 Deformation of structure due to settlement profile in non-uniform conditions (after Preene 2000). (a) Change in soil conditions (b) Change in foundation type.

13.1.5 Risk assessment of settlement damage from groundwater lowering

Once the settlement at various distances from the dewatering system has been estimated (using the equations outlined earlier, and drawdown curves from Section 7.6) the values shown in Table 13.1 can be used to delineate risk zones. These are defined as areas where structures may experience particular levels of settlement, and hence degrees of damage. The simplest form of risk zones assume soil conditions that do not vary with distance. For radial flow the risk zones will be a series of concentric circles centred on the groundwater lowering system (Fig. 13.7(a)); for plane flow to pipeline trenches the risk zones will be parallel lines either side of the trench. The

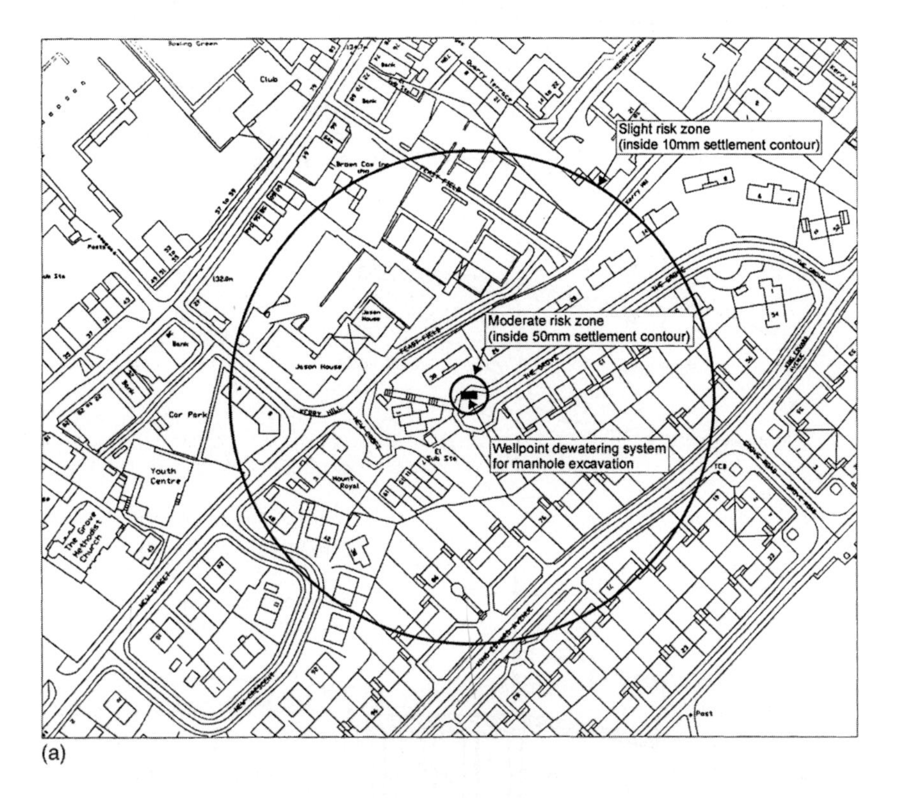

Figure 13.7(a) Settlement risk zones (after Preene 2000). Idealized settlement risk zones for radial flow to a groundwater lowering system.

Notes:

1 Predicted settlement at excavation is less than 75 mm, so no severe risk zone is generated.
2 No buildings lie within the moderate risk zone.
3 Numerous buildings lie within the slight risk zone. Building condition surveys should be considered for this zone. Any sensitive structures should be identified.
4 Settlement assessment assumes thickness of Alluvium is constant. Thickness based on borehole at manhole location.

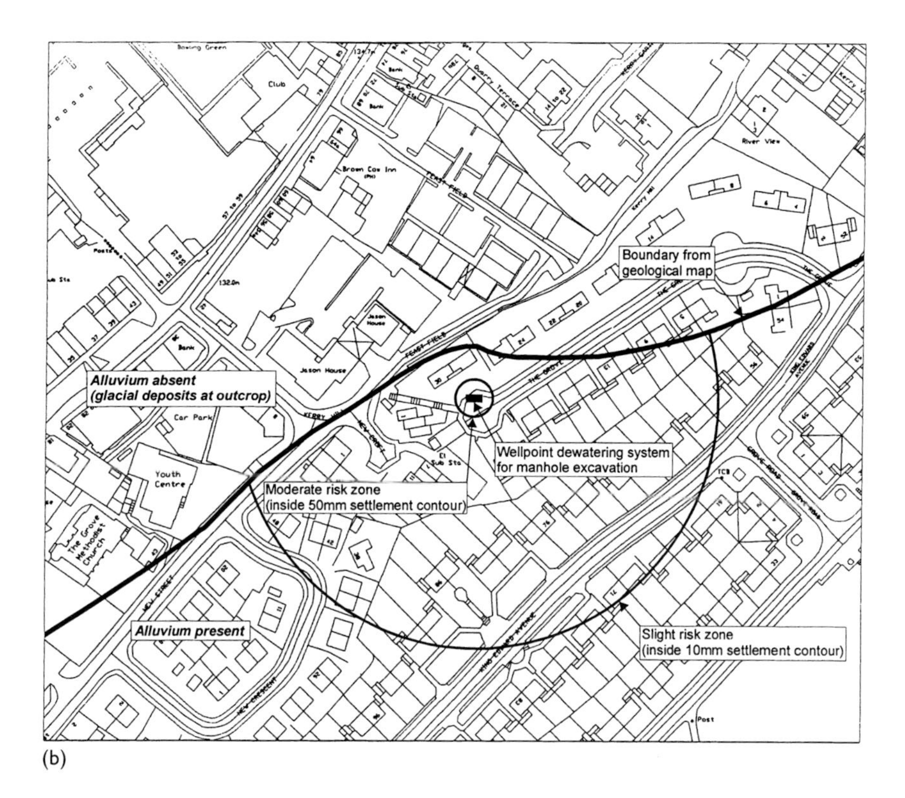

(b)

Figure 13.7(b) Settlement risk zones for radial flow to a groundwater lowering system based on variation in soil type from geological mapping.

Notes
1 Geological boundary taken from published geological mapping, confirmed by nearby boreholes.
2 Predicted settlement in area where Alluvium is absent is less than 10 mm.
3 Settlement assessment assumes thickness of Alluvium is constant when present. Thickness based on borehole at manhole location.

risk zones can be refined by including any known changes in ground conditions with distance. If settlements are significant it is likely that compressible alluvial or post glacial soils are present. Geological mapping (from published data, or from site investigation) can help determine the extent of these soils; damaging settlements are much less likely where these soils are absent. Application of geological mapping data will tend to produce non-circular risk zones for radial flow (Fig. 13.7(b)), and may reduce the extent of the zones and the number of structures at risk.

When the risk zones have been determined, and the number and type of structures at risk identified, appropriate action is required. Initial actions are summarized in Table 13.2. Additional, more detailed, assessments may be

Table 13.2 Suggested actions for settlement risk categories (from Preene 2000)

Risk category	*Description of likely damage*	*Actions required*
Negligible	Superficial damage unlikely	None, except for any buildings identified as being sensitive, for which a detailed assessment should be made.
Slight	Possible superficial damage, unlikely to have structure significance	Building condition survey, to identify any pre-existing cracks or distortions. Identify any buildings or pipelines which may be sensitive, and carry out detailed assessment. Determine whether mitigation or avoidance measures are required locally.
Moderate	Expected superficial damage and possible structural damage to buildings; possible damage to rigid pipelines	Building condition survey and structural assessment. Assess buried pipelines and services. Determine whether anticipated damage is acceptable or whether mitigation or avoidance measures are required. (applies to Moderate and Severe)
Severe	Expected structural damage to buildings and expected damage to rigid pipelines or possible damage to other pipelines	

needed for structures in the moderate and severe risk zones, and for sensitive structures in the slight risk zone.

13.1.6 Mitigation and avoidance of settlement

Dependent on the type of project and the number and nature of structures at risk, some (or perhaps all) of the anticipated damage may be deemed unacceptable. Table 13.3 lists settlement mitigation or avoidance measures, including the use of artificial recharge (see Section 13.6).

In addition to mitigation or avoidance there is a third option, rarely considered explicitly – that is the acceptance of settlement. Powers (1992) has suggested that, if damage risk is no more than slight, it may be more economical to accept third party claims, rather than deploying large-scale mitigation measures. This might be quite a controversial approach on many projects, and would present a public relations challenge, but could be appropriate where relatively few structures were classified as being slightly at risk. A pre-construction building condition survey would be essential for this approach.

Table 13.3 Measures to mitigate or avoid groundwater lowering induced settlement damage (from Preene 2000)

Mitigation of settlement	*Possible measures*
Protect individual structures	Prior to the works, underpin the foundations of some or all of the structures at risk
Reduce the number of structures at risk	Reduce drawdowns by reducing the depth of excavation below groundwater level Reduce the extent of the risk zones by minimizing the period of pumping Use cut-off walls to reduce external drawdowns Use an artificial recharge system to minimize external drawdowns
Avoidance of settlement	Relocate excavation away from vulnerable structures Re-design project to avoid excavation below groundwater level, or excavate underwater Carry out excavation within a notionally impermeable cut-off structure[a]

Notes:
a If a cut-off structure is used to avoid external drawndowns it is essential that a groundwater monitoring regime is in place to allow any leaks in the cut-off to be identified before significant settlements can occur.

Powers (1985) suggests that a pre-construction building condition survey should include a photographic and narrative report on the interior and external condition of buildings. Particular attention should be paid to the condition of concrete foundations, structural connections, brickwork, and the condition of plasterwork or other architectural finishes that are particularly susceptible to cracking. Additionally, the condition of other structures must also be documented – examples include bridges, utility enclosures and historic monuments. Paved surfaces (roads, pavements, hardstandings) should also be examined and their condition recorded. Ideally, the survey should be carried out by an independent organization (to avoid later charges of bias in the event of claims).

The legal position under United Kingdom law related to settlement damage resulting from the abstraction of groundwater is outlined in Section 15.3.

13.2 Effect on groundwater supplies

This book deals mainly with groundwater as a problem, needing to be controlled to allow construction excavations to proceed, but groundwater is also a resource used by many. Groundwater is obtained from wells and springs as part of public potable water supplies and for private supplies for domestic dwellings and industrial users such as breweries, paper mills, etc. If temporary works groundwater lowering is carried out in the vicinity of existing well or spring abstractions, there is a risk that the abstractions will be 'derogated' – in other words it will be harder for the user to abstract water, and in extreme cases the source may even dry up completely. The interaction between groundwater supplies and civil engineering works is discussed further by Brassington (1986).

Occasionally water quality may deteriorate as a result of changes in the groundwater flow direction. The effect is often temporary, and may cease soon after the end of temporary works pumping, but can cause considerable inconvenience and cost to groundwater users. The legal issues must also be considered, since in England and Wales licensed groundwater abstractors have a legal right to continue to obtain water (see Section 15.3).

13.2.1 Derogation of groundwater supplies

The primary effect on nearby abstractions is a general lowering of water levels, which will affect operating water levels in existing wells, with a corresponding reduction in output. The magnitude of the reduction in output will depend on factors including:

1 Aquifer characteristics, including permeability and storage coefficient.
2 Distance between groundwater lowering wells and supply wells, and their location in relation to any existing hydraulic gradients in the aquifer.
3 The dewatering pumping rate and period of pumping (low flow rate and short duration pumping systems will have less of an effect on supply wells).
4 The depth, design and condition of the supply wells.

Any rational assessment of the effect of groundwater lowering on supply wells will require some form of conceptual groundwater model to be developed. This could then be used as the basis for a numerical model, or an initial assessment could be made using the methods of Chapter 7, treating the groundwater lowering system as an equivalent well.

If the estimated effects on the supply wells are small, this may be deemed acceptable with no further mitigation measures. However, if the effects are more severe, Powers (1985) suggests the following mitigation measures:

1 If only a few low volume users are affected, and the dewatering period is short, the lost supply might be replaced by a temporary tanker supply.
2 If the supply well is deep but with the pump set at a fairly high level, it may be possible to install higher head pumps at greater depth in the

supply well. This would allow abstraction to continue even with the additional drawdown generated by groundwater lowering.

3 If the supply wells are shallow, it may be necessary to deepen the wells, perhaps into another aquifer. Alternatively, for small diameter shallow wells it may be more economic to simply drill a new, deeper well.
4 A portion of the dewatering discharge may be piped to the affected user by temporary pipeline. Dependent on water quality, point of use treatment may need to be provided to ensure the water is suitable for use.
5 Public water mains may be extended into the affected area, giving a permanent benefit for the money spent.

Some of these measures have huge cost, time and public relation implications, and clearly need to be compared with the alternative of constructing the project without groundwater lowering, or even of relocating the project away from the supply wells.

If the effect on nearby groundwater abstractions are of real concern, it is essential that they are addressed early in the planning of a project, because it is unrealistic to expect the contractor to bear all the costs and risk of some of these measures. The project client will have to face up to the potential need for some of these measures, and perhaps allow for them when negotiating with landowners for wayleaves, etc. The mitigation measures might be included at the very start of site works as part of the enabling works. Alternatively, the supply wells may be monitored during the works, with a contingency in place that the mitigation measures will be applied if the well is affected beyond a certain pre-defined level.

13.3 Effect on groundwater quality

Groundwater quality (i.e. the chemical composition of the water) varies from place to place and aquifer to aquifer (see Section 3.8). In some cases groundwater is almost pure enough to be potable with only minimal treatment in the form of chlorination to destroy any harmful bacteria. In other locations the groundwater may contain considerable impurities which could be naturally occurring, or man-made. It is important to realize that pumping from groundwater lowering systems changes natural groundwater flow in aquifers, and may cause existing contamination plumes or zones to migrate. Two of the most important cases to be considered are contaminated groundwater, often left over from industrial land use, or intrusion of saline water in coastal areas.

13.3.1 Movement of contaminated groundwater

The study of groundwater contamination is a major field in itself, and the reader is commended to texts such as the one by Fetter (1993) to obtain the full background on the subject. Contaminants interacting with groundwater flow exist in one of three forms (or phases):

1 Dissolved (or aqueous) phase. A wide range of substances are soluble in water, and so become part of the water itself, travelling with it.
2 Non-aqueous phase. This describes liquids which are immiscible with water. They may have densities less than water (light non-aqueous phase liquids or LNAPLs) and will float on top of the water table (examples are petrol and diesel compounds). Dense non-aqueous phase liquids (DNAPLs) also exist, which are denser than water and tend to sink below the water table until they meet a low permeability layer (examples include chlorinated hydrocarbons such as trichloroethylene).
3 Vapour phase. Volatile compounds in the contaminant can move in gaseous form in the unsaturated zone above the water table.

Some contaminants (e.g. hydrocarbons such as petroleum products) may create all three phases when they reach groundwater.

Contamination may be caused by a variety of land uses:

- Industrial processes – mainly from spillages or leakages from stored materials. Sites of concern include manufacturing and chemical production sites, but also vehicle storage areas, airports and other locations where fuel and detergents are used.
- Landfilling and waste disposal – both from official and unofficial disposals.
- Agricultural practices – such as fertilizer or pesticide use.
- Urban use – including leaking sewers, fuel spills, etc.

The absence of active use of a site does not provide assurance that the site is uncontaminated. A legacy of contamination may exist in the ground and groundwater for years or decades after the pollution is stopped. A guide to the types of pollution that can be expected from former industrial sites can be found in CIRIA reports on contaminated land (Harris *et al.* 1995). A desk study (see Chapter 6) should investigate former uses of a site to determine the risk of contaminants being present at problematic levels.

As described in Chapter 3, groundwater is constantly in motion, and where contamination exists that will tend to move too, gradually forming a plume stretching away from the original source of contamination. The rate and direction of movement of contamination depends on many factors, including hydraulic gradients, the geological structure of the aquifer, the nature of the contaminant and any chemical changes in the contaminant with time. Detailed consideration of these factors is beyond the scope of this book, and the references cited earlier are recommended for further study. However, it is vital that anyone designing or carrying out a groundwater lowering understands that pumping may change considerably the existing groundwater gradients and velocities, affecting both the magnitude (generally increasing flow velocities) and direction. This means groundwater lowering can cause the extent of a contamination plume to change, perhaps much more rapidly than previously. If movement of contamination is of real concern, a thorough site investigation followed by development of a groundwater flow and contaminant transport model is essential, and specialist advice should be obtained at an early stage.

When planning groundwater lowering on or near a contaminated site, there are two important issues to be addressed, in addition to the dewatering design itself:

1 How can the influence of groundwater lowering on the contamination plume be minimized or controlled? The effect on contamination plumes is similar to settlement in that it is largely controlled by drawdown; the methods described in Section 13.1 can be used to minimize effects on existing contamination. The use of physical cut-off barriers (Privett *et al.* 1996) to hydraulically separate the groundwater lowering system from adjacent contaminated sites is a method often used.
2 How can the discharge be disposed of? The water pumped from the wells may contain problematic levels of contamination, preventing direct discharge to watercourses or sewers. On occasion it has been necessary to establish a temporary water treatment plant on site to clean up the discharge water quality (see Section 13.4).

Because pumping of groundwater affects the extent of contaminant plumes, and can remove some of the contamination with the discharge water, methods based on groundwater lowering technology can be used to help clean up sites. Pumping of groundwater and treatment of discharge prior to disposal, with the aim of reducing contamination levels, is known as the 'pump and treat' method. This is a specialist method, and its effectiveness should be compared to other competing clean up techniques (see Holden *et al.* 1996).

13.3.2 Saline intrusion

Saline intrusion describes the way more mineralized water is drawn into fresh water aquifers under the influence of groundwater pumping. This is a particular problem where large volumes of groundwater are abstracted for potable supply, because if saline water reaches the well, it may have to be abandoned. Saline intrusion principally affects coastal aquifers, but saline water can sometimes be found in inland aquifers, where the water has become highly mineralized at depth.

Saline intrusion is a complex process affected by aquifer permeability, rate of recharge, natural groundwater gradients and the effect of any existing pumping wells. Any significant groundwater lowering operations will affect the boundary between fresh and saline water. If saline water is drawn to the groundwater lowering system that may not be a problem in itself (provided that the water can be disposed of), but any saline water drawn toward nearby supply wells is of much greater concern. The risk of saline intrusion may need to be investigated using hydrogeological modelling to assess the effect on local and regional water resources; the reader is referred to hydrogeological texts such as Todd (1980) for further details.

13.3.3 Contamination via boreholes

One potential effect on groundwater quality that is sometimes overlooked is the conduit for contamination formed by wells, wellpoints, etc. (and to a lesser extent by piles and some forms of ground improvement). It is now recognized that the vulnerability of aquifers to pollution resulting from surface sources (e.g. surface run-off, fuel spills, etc.) will depend on the aquifer type and structure. For example, a high permeability unconfined aquifer will be much more vulnerable to contamination than a deep confined aquifer overlain by a thick clay layer, which can act as a barrier to pollution. In the United Kingdom, aquifer vulnerability maps have been produced for many areas (Palmer and Lewis 1998).

The installation of wells (and possibly the construction excavation itself) may puncture low permeability layers, increasing the risk of surface pollution finding its way down into the aquifer. If changes in aquifer vulnerability are of concern a number of measures should be considered:

1. The well design should include appropriate grout seals to prevent vertical seepage around the outside of the well casing.
2. The well casing should stand sufficiently proud of ground level to prevent surface waters being able to pass into the well casing (and thence into the aquifer) in the event of localized flooding around the well.
3. The top of the well casing should be capped when not in use, and sealed around the pumping equipment when in use. This will reduce the chances of noxious substances being dropped or poured down the well, either maliciously or by accident.
4. The wells should be adequately capped or sealed on completion (see Section 14.7).

13.4 The impact of discharge flows on the surface water environment

Any system that lowers groundwater by pumping will produce a discharge flow of water that must be disposed of. Poorly managed discharges may have adverse impacts on the environment. This section describes good practice for discharges to the surface water environment. Disposal of discharge by artificial recharge is described in Section 13.6. Legal permissions necessary for the discharge of groundwater are outlined in Section 15.3.

13.4.1 Erosion caused by discharge flows

Poorly located discharges can cause erosion of river banks or watercourses if the flow is concentrated in one location – particularly if the flow rate is large. A scour hollow will form under the discharge point, possibly undermining the bank. Problems may be created downstream, as the scoured material is re-deposited, blocking or changing the flow in the watercourse.

This problem can be avoided by designing the discharge system to reduce the potential for scour. Materials such as gabion baskets, stone or geotextile mattresses, or even straw bales can be placed at the discharge point to dissipate the energy of the water, before it passes into the watercourse proper.

13.4.2 Suspended solids

Suspended solids, in the form of clay, silt and occasionally sand-sized particles, are a common problem resulting from dewatering discharges. Silt discharges are a highly visible aesthetic problem (Fig. 13.8), but silt also

Figure 13.8 Silt plume in a watercourse resulting from the discharge of water from poorly controlled sump pumping (courtesy of T. O. L. Roberts).

harms aquatic plant, fish and insect life and can build up in watercourses, blocking flow.

The suspended solids in sediment-laden discharges can be difficult to deal with economically, so the best approach is to tackle the problem at source and avoid silt being drawn into the discharge water. This requires appropriately designed and installed filters to be included in the system design. This is the norm for the wellpoint, deep well and ejector methods and such systems do not commonly produce discharges with high sediment loads, except for a short period during initial well development.

Most problems with suspended solids in discharge water arise from sump pumping operations. In many cases adequate filters are not installed around the sumps and fine particles can be drawn out of the soil and entrained in the discharge water. Where sump pumping is carried out in soils containing a significant proportion of fine particles, the discharge will need to be treated to reduce any suspended solids to acceptable levels prior to discharge. If treatment is not possible, a change of dewatering method (e.g. by using wellpoints with adequate filters) should be considered.

The most common form of treatment for suspended solids is by passing them through settlement tanks or lagoons. Small portable steel tanks (typically 3 m by 1.5 m in plan and up to 1.5 m deep) are often used to reduce suspended solids. Purpose-built tanks are available, or improvised tanks can be made from waste skips available on site (Fig. 13.9(a)). Such tanks can be effective in removing sand-size particles, but silt and clay-size particles settle slowly and will pass through small tanks.

Lagoons will be necessary to provide sufficient retention time for silt and clay size particles to settle out. A typical lagoon is shown in Fig. 13.9(b), and consists of an earthwork bund or pit with some form of waterproof lining on the base and sides; edge protection is necessary to reduce hazards to personnel. It may be necessary to operate two or more lagoons in parallel. When one lagoon is full the other receives the discharge, allowing more time for settlement in the first. Water from each lagoon is decanted and disposed of in turn; sediment needs to be removed periodically from the lagoons. If large lagoons are not feasible, the addition of chemical flocculants might be considered to allow smaller tanks to be used (Smethurst 1988). Such treatment can be complex, and specialist advice should be obtained.

13.4.3 Oil and petroleum products

Oil and petroleum based products may find their way into dewatering discharges as a result of spills or leaks from plant or fuel storage areas. This is a particular risk with sump pumping because any spills or leaks in the base of the excavation will be carried to the pumps by surface water. Petroleum products may also occur in discharges if contaminated groundwater is pumped.

(a)

(b)

Figure 13.9 Treatment of discharges by settlement. (a) Small settlement tanks improvised from waste skips (b) Large settlement lagoons.

Petroleum products are generally of lower density than water and do not mix well with it. They are known as light non-aqueous phase liquids (LNAPLs) and will tend to appear as floating films or layers on top of ponds or tanks of water. The petroleum products must be separated from the discharge before disposal. This can be accomplished by passing the discharge through proprietary 'petrol interceptors' or 'phase separators'. The water is then discharged as normal, and the product (which collects in the interceptor) is disposed of (e.g. to a waste oil company) at appropriate intervals. Very thin oil layers may be removed by the use of floating skimmer pumps or sorbent booms or pillows placed in discharge tanks or lagoons.

13.4.4 Contaminated groundwater

If groundwater is pumped from a contaminated site, or from adjacent to such a site, the resulting dewatering discharge may itself be contaminated. Unless the flow is discharged to a sewage treatment works capable of dealing with the contaminants, some form of treatment will be required prior to discharge.

A wide range of treatment methods is available. Often, a given contaminant could be treated by several quite different methods; the choice of method will depend on the concentration of contaminants, the discharge flow rate, the duration of pumping and the availability of treatment equipment and technologies. Some of the available technologies are described by Nyer (1992) and Holden *et al.* (1996).

The scale of treatments used in practice varies greatly. Occasionally, on very heavily contaminated sites with low pumped flow rates the discharge may be pumped directly to special road tankers, for off-site disposal at a licensed facility. In other cases on-site treatment may be feasible. This ranges from simple dosing with caustic soda for pH adjustment of acidic groundwater, through to the construction of modular treatment plants on heavily contaminated sites.

A programme of chemical testing will be needed to monitor discharge water quality, and the environmental authorities (see Section 15.3) will need to be kept fully informed. Often the targets for treatment of the discharge will be set by the environmental regulators, based on a consideration of the environment impact of the discharge.

13.5 Other effects

Occasionally, other, less common, side effects may be of concern.

13.5.1 Damage to timber piles

It is sometimes considered that timber piles supporting older structures may be detrimentally affected by drawdown of water levels. Powers (1985) states that the damage may result from fungi present in the timber thriving in an aerobic environment created if water levels are drawn down, exposing the tops of the piles to air. However, Powers also states that the most severe cases of aerobic attack have been when piles were exposed in excavations, and that observed decay due to drawdown has been less severe. This is probably because the oxygen supply to the timber surface is not increased substantially when the piles are in dense or fine-grained soils.

Nevertheless, a sensible approach is to proceed cautiously when working in areas where older structures are founded on timber piles. Even if aerobic attack does not compromise pile stiffness, soil consolidation and pile downdrag due to negative skin friction should be considered. When groundwater lowering has been used in Scandinavian cities where timber piling is commonplace, artificial recharge systems have been used to reduce drawdowns, and minimize the effect on piles.

13.5.2 Vegetation and wetlands

It is rare for groundwater lowering systems to have a noticeable effect on vegetation. This is mainly due to the short-term nature of pumping, and the fact that plants generally draw their water from immediately below the surface, above the water table. This zone is much more likely to be affected by changes in precipitation and infiltration, than by deeper pumping. Longer term pumping (for certain types of quarries or open cast mines) may need to consider this issue further, and the services of an experienced ecologist can be very useful in that regard.

Wetlands (areas of marsh, fen or peatland, or areas covered with shallow water or poorly-drained areas subject to intermittent flooding) may sometimes be affected by longer-term groundwater pumping. To assess each case the interaction between the surface water and groundwater will need to be assessed. Some wetlands are directly supported by groundwater seepages, while others (if the soil is of low permeability) may receive little contribution from groundwater. Wetlands tend to vary with the seasons (and also from year to year), so the additional influence of groundwater pumping may or may not be significant in comparison. Merritt (1994) gives a thorough background to the creation and management of wetlands.

When assessing the potential impact on wetlands, specialist ecological and hydrological advice is likely to be required. It may be possible to reduce effects on wetlands by constructing a groundwater cut-off barrier between the dewatering system and the wetland. An alternative approach would be to use artificial recharge of groundwater or surface water or to pipe a portion of the discharge water directly to the wetland. If the latter approach is adopted, the temperature, chemistry and sediment content of the discharge water must be assessed to ensure that there will not be adverse reactions with the water and ecology of the wetlands; the risk of additional erosion must also be considered. It is worth noting that in the United Kingdom many wetlands are protected as Sites of Special Scientific Interest or other similar designations, and as such are protected by law. Negotiations will have to be opened with the appropriate authorities (such as the Environment Agency and English Nature) when planning work near such sites. Similar protected environments exist in other parts of the world.

13.6 Artificial recharge systems

Artificial recharge is the process of injecting (or recharging) water into the ground in a controlled way, typically by means of special recharge wells or trenches (Fig. 13.10). For groundwater lowering applications the recharge water is generally the discharge from the abstraction system, injected into the ground to control or reduce drawdowns around the main excavation area, or simply as a means to dispose of the discharge water. Occasionally, water other than groundwater (e.g. from mains water supply) is used for recharge.

Artificial recharge systems are not straightforward in planning or operation and are carried out on only a small minority of groundwater lowering projects.

13.6.1 Applications of artificial recharge systems

When used in conjunction with groundwater lowering systems, artificial recharge can be used as a mitigation measure to control or reduce drawdowns

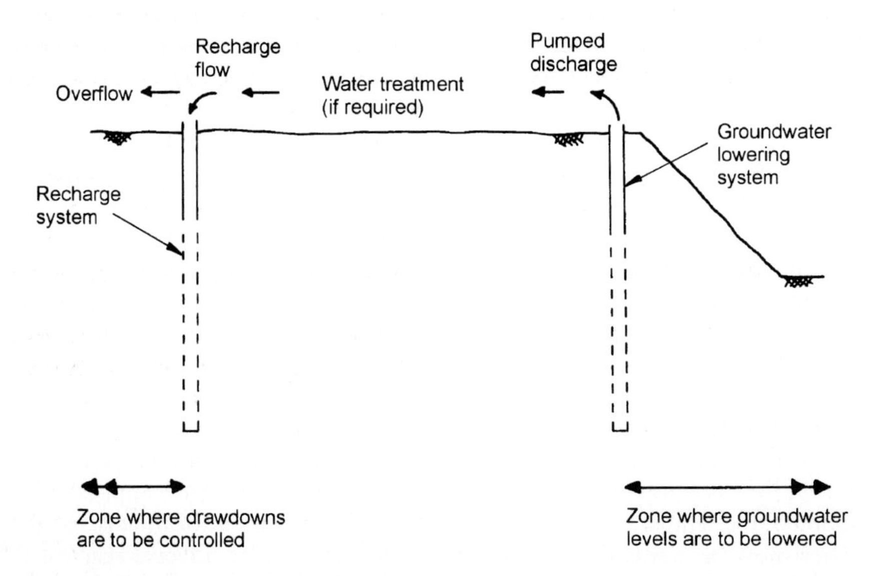

Figure 13.10 Artificial recharge to control drawdowns around a groundwater lowering system.

away from the main area of groundwater lowering, in order to minimize the side effects described earlier in this chapter.

Artificial recharge is also occasionally used as a means to dispose of some or all of the discharge water, if other disposal routes are not practicable. If this option is being considered, it must be noted that in the United Kingdom, a discharge consent (see Section 15.3) must be obtained from the regulatory authorities to allow artificial recharge – there is no automatic right to recharge groundwater back into the ground. Similar regulations exist in several countries.

Recharge wells or trenches must be located with care. If the system is intended to reduce drawdowns at specific locations then the recharge points will generally be between the groundwater lowering system and the areas at risk. The recharge locations may be quite close to the pumping system and much of the recharge water may recirculate back to the abstraction wells, leading to an increase in the pumping and recharge rate. Unless the soil stratification can be used to reduce the connection between the pumping and recharge system, a physical cut-off could be used to minimize recirculation of water through the ground. If the water is being recharged to prevent derogation of water supplies, the recharge should be carried out further away to avoid excessive recirculation.

It is worth noting that artificial recharge is being increasingly applied in the water supply field. Artificial recharge in that sense involves injecting water (often treated to potable standards) into the aquifer at a location where there is a supply at ground level (e.g. from a river). This injection effectively provides more water to the aquifer that can be abstracted from the aquifer down gradient, where it is to be used. A variant on artificial recharge, used increasingly in the United States and Europe is aquifer storage recovery (ASR). ASR involves storage of potable water in an aquifer at one location. This is done by injecting water at times when supply is available, and later, when it is needed to be put back into supply, recovering the water from the same wells. This interesting application of groundwater technology is described in detail by Pyne (1994).

13.6.2 What is the aim of artificial recharge?

If an artificial recharge system is being planned, it is vital that the aim of the system is clear at the start. It is sometimes thought that an artificial recharge system should prevent any drawdown or lowering of groundwater levels around a site. However, Powers (1985) suggests that the aim of a system should not be to maintain groundwater levels *per se*, but to prevent side effects (such as settlement or derogation of water supplies) reaching unacceptable levels.

Groundwater levels vary naturally in response to seasonal recharge and under the influence of pumping. Other natural effects include tidal groundwater responses in coastal areas. A specification for an artificial recharge system that required 'zero drawdown' would be not practicable. It may be more appropriate to set a target such that groundwater levels in selected observation piezometers shall not fall below defined levels (which are likely to be different in different parts of the site).

Assuming that groundwater levels around a site vary naturally with time, and that this variation is not causing detrimental side effects, the lowest acceptable groundwater level in a monitoring well is often set as the seasonal minimum. Occasionally, if the groundwater lowering system is intended to generate very large drawdowns in relatively stiff aquifers, the lowest acceptable groundwater level is set rather lower, at a few metres below the seasonal minimum.

It is sometimes useful to specify maximum groundwater levels in selected piezometers. This reduces the temptation for overzealous recharging raising the groundwater levels above the seasonal maximum, which could lead to problems with flooded basements and the like. By defining an allowable minimum and maximum groundwater level in each observation well (with at least 0.5 m between maximum and minimum levels), the system operator has a little leeway within which to adjust the system.

13.6.3 Recharge trenches

A simple method of recharge for shallow applications is the use of recharge trenches (Fig. 13.11). These are excavated from the surface, and the dewatering discharge is fed into them from where the water infiltrates into the ground.

Recharge trenches have been employed to good effect when pumping from mineral workings (Cliff and Smart 1998), but are less effective in construction situations and are rarely used. Clogging of the trenches (by the growth and decay of vegetation or the build up of sediment) reducing infiltration rates is often a problem. Trenches typically require periodic cleaning out by excavator. It is difficult to control and measure the volumes entering the trenches, making this method less easy to adjust than a system

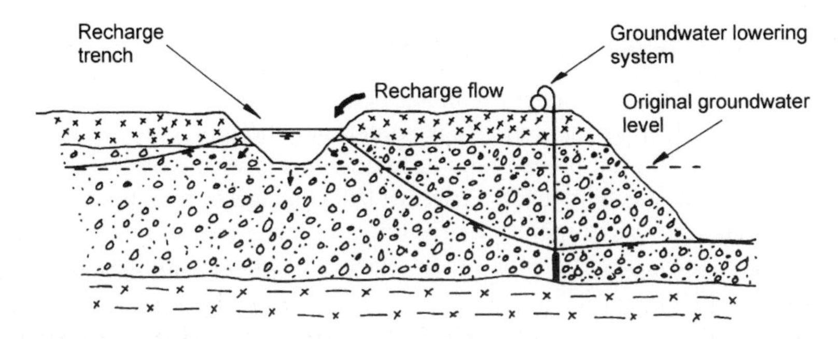

Figure 13.11 Recharge trench.

of recharge wells. An overflow channel is normally required to prevent overtopping of the trenches.

The recharge trench method is best suited to unconfined aquifers with water tables near ground level, allowing shallow trenches to be used. The method is less practicable for recharging water where the water table is relatively deep, or if the aquifer is confined by overlying clay layers.

13.6.4 Recharge wells

Recharge wells are generally superior to recharge trenches because the wells can be designed to inject water into specific aquifers beneath a site; this can allow the stratification of soils at the site to be used to advantage in controlling drawdowns. Recharge wells also allow better control and monitoring of injection heads and flow rates than do trench systems. A recharge well operates in a similar way to a pumping well, except that the direction of flow is reversed. A recharge well should allow water to flow into the aquifer with as little restriction as possible. However, recharge of water into the ground is more difficult than pumping. A pumping well is effectively self-cleaning, in that any loose particles or debris will be removed from the well by the flow. In contrast, a recharge well is effectively self-clogging – even if the water being recharged is of high quality, any suspended particles or gas bubbles will be trapped in the well (or the aquifer immediately outside) leading to clogging and loss of efficiency.

In broad terms, a recharge well should be designed, installed and developed in the same way as a pumping well (see Chapter 10). The two key differences are:

1. To maximize hydraulic efficiency, the well filter media and slot size (see Section 10.2) should be as coarse as possible, while still allowing the well to be pumped without continuous removal of fines during re-development.
2. Because a recharge well does not generally have to accommodate a submersible pump, the well casing and screen can be of smaller diameter than a pumping well. However, the well casing and screen must be large enough to allow the well to be re-developed if necessary.

Figure 13.12 shows a typical recharge well. Key features are:

i Well casing and screen, surrounded by filter media, with a grout or concrete seal around the well casing to prevent water short-circuiting up the filter pack to ground level. If the top of the well casing is also sealed around the recharge pipework, this can allow recharge flows to be fed by header tanks raised above ground level, to increase the injection rate in each well.

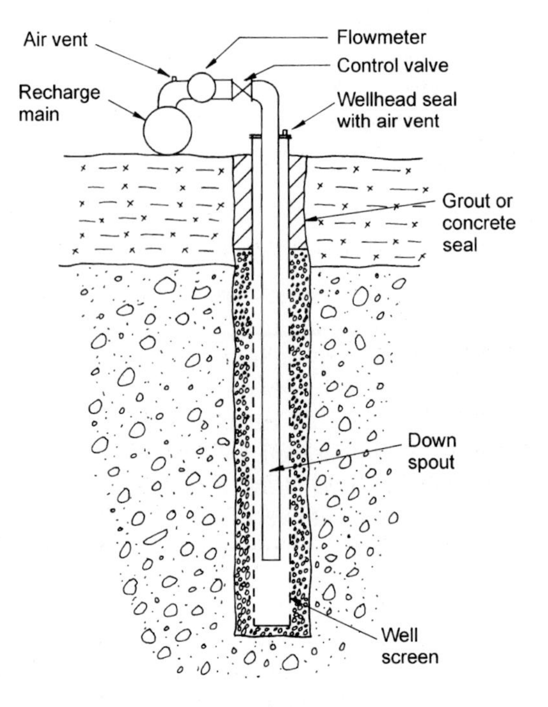

Figure 13.12 Recharge well.

ii A down spout to prevent the recharge water from cascading into the well and becoming aerated. Aeration of the water may promote bio-fouling and other clogging processes.
iii Air vents at the top of the well and pipework to purge air from the system when recharging commences, and to prevent airlocks in the system.
iv Control valve and flowmeter to allow monitoring and adjustment of flow to the well.

If water is to be recharged into a shallow aquifer, a system of recharge wellpoints may be considered. The recharge wellpoints are installed at close spacings using similar methods to conventional wellpointing (see Chapter 9).

Almost all recharge wells suffer from clogging to some degree, and it is vital that the design used allows for appropriate re-development. This is discussed further in the following sections.

13.6.5 Water quality problems and clogging

Experience has shown that it is much harder to artificially recharge water into the ground than it is to abstract it. There are various rules of thumb which say that, to recharge water back into the aquifer from which it came two or three recharge wells will be needed for every abstraction well. Many of the practical difficulties with artificial recharge arise from water quality problems leading to clogging. The designer and operator are being unrealistic if they do not expect recharge wells or trenches to clog (to a lesser or greater degree) in operation.

It is worthwhile considering the mechanisms that can lead to the clogging of recharge wells (similar processes affect recharge trenches). Pyne (1994) has outlined five clogging processes:

(1) Entrained air and gas binding. Bubbles of vapour (such as air or methane) present in the recharge water build up in the well, inhibiting flow of water through the filter pack into the aquifer. The gas bubbles may result from the release of dissolved gases from the recharge water; from air drawn into the recharge pipework where changes in flow generate negative pressures; or, if water is allowed to cascade into the recharge well, from air entrained into the water. Careful sealing of pipework joints, and minimizing aeration of the recharge water can help avoid this problem. Degassing equipment can be installed onto the pipework to bleed off any gas before it reaches the recharge wells.
(2) Deposition of suspended solids from recharge water. Particles (colloidal or silt and sand-sized soil particles, biofouling detritus, algal matter, loose rust or scale from pipework, etc.) carried with the recharge water will build up in the well and filter pack, blocking flow. Control of this problem requires effective design and development of the abstraction wells to minimize suspended solids in the water; the discharge from sump pumping is rarely suitable for recharge. For low flow rate systems the use of sand filter systems to clean up the water might be considered.
(3) Biological growth. Bacterial action in the recharge wells can result in clogging of the wells themselves. Additionally, biofouling of the abstraction wells and pipework can release colloidal detritus (the result of the bacteria life cycle) into the water. The flow of water will carry these particles inexorably into the recharge wells, leading to further clogging. The severity of any clogging will depend on several factors (see Section 14.8). Since some of the problem bacteria are aerobic, minimizing any aeration of the recharge water is advisable. Periodic dosing of the wells with a dilute chlorine solution has been used to inhibit bacterial growth.
(4) Geochemical reactions. The recharge water can react with the natural groundwater or with the aquifer material. Such reactions are most likely if the recharge water is not groundwater from the same aquifer (e.g. if mains water is used for recharge, or water from one aquifer is recharged back to another aquifer) or if the pumped groundwater is allowed to change chemically prior to re-injection. Typical reactions include the

deposition of calcium carbonate or iron/manganese oxide hydrates. The potential significance of these reactions can only be assessed following study of the aquifer and groundwater chemistry, but are generally reduced in severity if water pressure and temperature changes during recharge are minimized. Chemical dosing of the recharge water might be considered to inhibit particular reactions.

(5) Particle re-arrangement in the aquifer. Although not usually a significant effect, the permeability around the well may reduce due to loose particles around the well being re-arranged by the flow of recharge water out of the well. This effect can be minimized by effective development on completion and periodic re-development.

These clogging effects should be considered when designing and operating an artificial recharge system. After all, a recharge system is not going to achieve its aims if it is unable to continue to inject water into the ground.

13.6.6 Operation of recharge systems

An artificial recharge scheme broadly consists of an abstraction system, a recharge system, and a transfer system (between the abstraction and recharge points). The abstraction system should be straightforward in operation and maintenance (see Chapter 14), but the transfer and recharge systems may be more problematic. The crux of the issue is to manage the recharge water quality to limit the clogging of recharge wells to acceptable levels that do not prevent the system from achieving its targets.

As has been stated previously, most artificial recharge systems will suffer from clogging. There are two approaches to dealing with this, often used in combination:

1. Prevention (or reduction) of clogging – principally by treatment of the discharge water to remove suspended solids and retard clogging processes such as bacterial growth.
2. Mitigation of clogging – accepting that clogging will occur, and then planning for it. This may involve providing spare recharge wells, to be used when efficiency is impaired, and implementing a programme of regular re-development of clogged wells.

Measures to prevent clogging must focus on ensuring the recharge water is of high quality and does not promote clogging processes. Possible measures include:

- Removal of suspended solids by filtration or sedimentation.
- Chemical dosing to precipitate out problematic carbonates or iron and manganese compounds.
- Intermittent or continuous chlorine dosing of the recharge water to reduce bacterial action.

- Prevention of aeration of water to reduce the risk of gas binding.
- Use of mains water if the pumped groundwater cannot be rendered suitable.
- Careful management of the system on start-up to avoid slugs of sediment-laden water entering the wells (the initial surge of water may pick up sediment deposited in the pipework and should be directed to waste, not to the recharge wells).

Measures to mitigate clogging include:

- Designing recharge well screens to have ample open area to accept flow. Driscoll (1986) recommends that the average screen entrance velocity (recharge flow divided by total area of screen apertures) is less than 0.15 m/s, approximately half the maximum velocity recommended for abstraction wells.
- Re-developing the recharge wells on a regular basis to rehabilitate the wells. This is most commonly done by airlift development (see Section 10.6) that dislodges any sediment, loose particles, precipitates and bacterial residues. Depending on the severity of clogging (see Section 14.8), wells in a system may be re-developed on a rolling programme, with an individual well being re-developed every few weeks or months. Even after re-development wells may not return to their pre-clogging performance – if the system operates for very long periods, additional recharge wells may need to be installed after a few years to prevent the overall system performance from dropping to unacceptable levels.
- Providing more than the minimum number of recharge wells necessary to accept the recharge flow. In this way the flow can still be accepted if some of the wells are badly clogged, allowing more time for redevelopment.

Monitoring of artificial recharge systems follows good practice for abstraction systems (see Chapter 14). For recharge wells the water level below ground level (by dipping) or pressure head above ground level (by pressure gauge) should be monitored. Ideally, flowmeters should be installed on each well to record recharge flow rates, but some designs may be affected by clogging. At the very least the overall recharge rate should be monitored, perhaps by V-notch weir.

13.6.7 Case history: Misr Bank, Cairo

An interesting application of recharge was carried out in the centre of Cairo, Egypt, described by Troughton (1987) and Cashman (1987). A two-storey basement was to be constructed for the new Bank Misr headquarters immediately adjacent to the existing bank building. The basement was to be formed inside a sheet-piled cofferdam to a depth of 7.5 m below ground level, and approximately 4.5 m below original groundwater level.

Although the sides of the excavation were supported by sheet-piles, and a stratum of fissured clay was expected at final formation level, the presence of permeable sand layers below the excavation required reduction of pore water pressures at depth to prevent base heave. In planning the groundwater lowering operations, two issues were of concern:

i To control drawdowns beneath adjacent structures (which were founded on compressible alluvial soils) to prevent damaging settlements.
ii To dispose of the discharge water, which could not be accepted by the local sewer network, which was heavily overloaded.

A system of recharge wells, used in combination with wellpoints pumping from within the excavation, was used to solve the problem. Without such a system it is unlikely that the Cairo authorities would have allowed the work to proceed.

The sequence of strata revealed in the site investigation is shown in Fig. 13.13. Falling head tests in the silty sand below formation level gave results of the order of 10^{-5} m/s, but pumping tests at the site indicated the permeability was more probably of the order of 10^{-4} m/s, consistent with Hazen's rule applied to particle size tests on the sand. Due to the complexity of the interaction between pumping and recharge, a finite element numerical

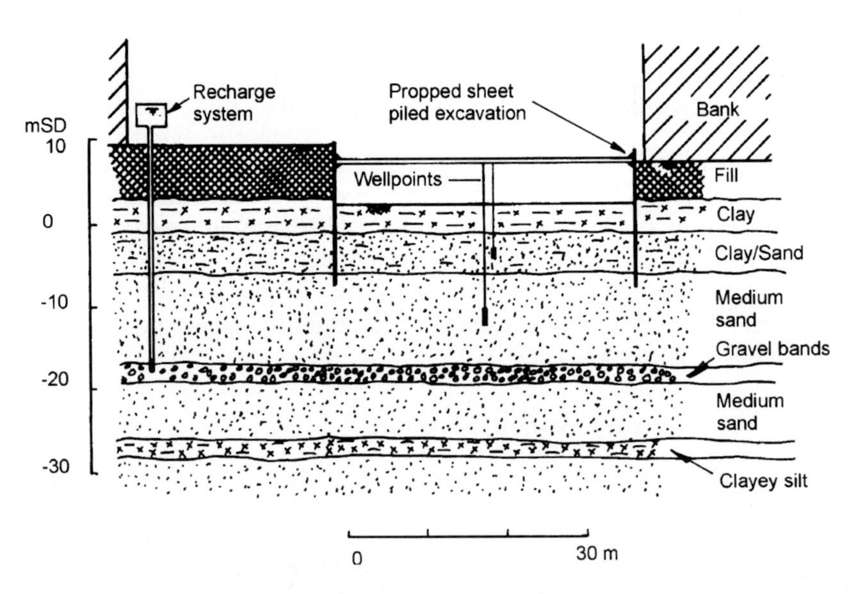

Figure 13.13 Bank Misr, Cairo: section through dewatering and recharge system (after Troughton 1987).

model was set up for the site. Based on the soil fabric in the alluvial soils, the horizontal permeability was assumed to be significantly greater than the vertical permeability.

The model was used to predict the pore water pressure reduction needed to safeguard the base of the excavation, and to limit the drawdown beneath adjacent structures to less than 1 m. Calculations indicated a 1 m drawdown would result in 12 mm settlement of the fill and alluvial soils, with differential movements within acceptable limits. Figure 13.14 shows a schematic plan view of the system, with an abstraction wellpoint system inside the cofferdam, and recharge wells between the excavation and critical buildings.

Abstraction wellpoints were installed to pump from strata at two levels at 13.5 and 21.5 m below ground level. The recharge wells were installed in three lines, with wells at 3 m spacing. The recharge wells (which consisted of 125 mm diameter screen and casing installed in a 250 mm diameter borehole) were designed to make use of the soil stratification at the site by

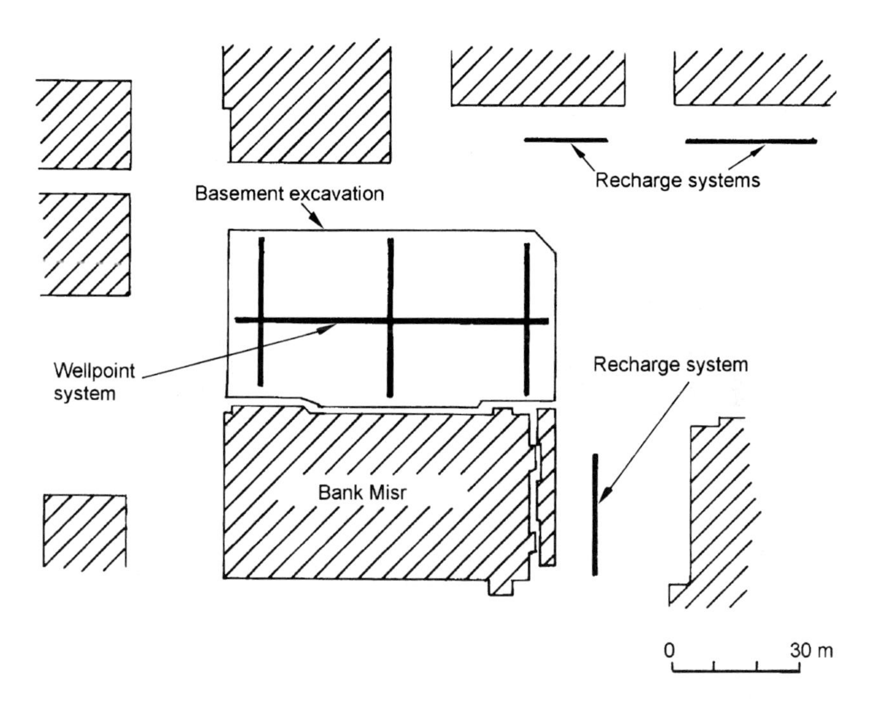

Figure 13.14 Bank Misr, Cairo: plan layout of dewatering and recharge system (after Troughton 1987).

Notes:

1 Recharge system consisted of recharge wells at 3 m centres.

2 Pumping system consisted of wellpoints at 3 m centres.

injecting water into the most permeable horizon present, a layer of gravel at 28 m below ground level. Water could be relatively easily injected into this permeable layer, from where it would feed to the other soil layers – analogous to the underdrainage dewatering effect (see Section 7.3) in reverse. The discharge water was fed to the recharge wells from a header tank 2 m above ground level; total flow rate was in the range 25–40 l/s. An array of pneumatic piezometers was installed to monitor pore water pressure reductions in the compressible alluvial clays. Conventional standpipe piezometers were used in the sands.

In operation the maximum drawdowns observed below structures were limited to 0.7 m, which resulted in settlements of 3 mm, around half the value predicted by calculation. As is often the case, the efficiency of the recharge wells reduced with time, and they had to be redeveloped by airlifting at monthly intervals to maintain performance. Two of the lines of recharge wells were located in a car park used during daytime business hours – these had to be re-developed at night.

As mentioned earlier, one of the aims for the recharge system was to dispose of the discharge water without overloading the antiquated sewer system. However, as reported by Cashman (1987) the dewatering field supervisor did not have a lot of faith in recharge and tapped into the sewer system with a hidden discharge pipe. So, in fact, much of the discharge water was actually going down the sewer. Unfortunately, over the Christmas period one of the main Cairo pumping stations broke down, and everything flooded back. The chairman of the main contractor received a telephone call from the Mayor of Cairo Municipality demanding his personal presence on site immediately. He was told that if this ever happened again, he, the chairman, would be immediately put in jail. This was a pretty convincing argument to ensure things were put right and the recharge system stayed in operation.

References

Brassington, F. C. (1986). The inter-relationship between changes in groundwater conditions and civil engineering construction. *Groundwater in Engineering Geology* (Cripps, J. C., Bell, F. G. and Culshaw, M. G., eds). Geological Society Engineering Geology Special Publication No. 3, London, pp 47–50.

Bauer, G. E., Scott, J. D., Shields, D. H. and Wilson, N. E. (1980). The hydraulic failure of a cofferdam. *Canadian Geotechnical Journal*, **17**, 574–583.

Burland, J. B. and Wroth, C. P. (1975). Settlement of buildings and associated damage. *Proceedings of the British Geotechnical Society Conference on Settlement of Structures*, Cambridge, pp 611–654.

Cashman, P. M. (1987). Discussion. *Groundwater Effects in Geotechnical Engineering* (Hanrahan, E. T., Orr, T. L. L. and Widdis, T. F., eds), Balkema, Rotterdam, p 1015.

Cliff, M. I. and Smart, P. C. (1998). The use of recharge trenches to maintain groundwater levels. *Quarterly Journal of Engineering Geology*, **31**, 137–145.

Driscoll, F. G. (1986). *Groundwater and Wells*. Johnson Division. Saint Paul, Minnesota.

Fetter, C. W. (1993). *Contaminant Hydrogeology*. Macmillan, New York.

Greenwood, D. A. (1984). Re-levelling a gas holder at Rhyl. *Quarterly Journal of Engineering Geology*, **17**, 319–326.

Harris, M. R., Herbert, S. M. and Smith, M. A. (1995). *Remedial Treatment for Contaminated Land*. Construction Industry Research and Information Association, CIRIA Special Publications 101–112 (12 Volumes), London.

Holden, J. M. W., Jones, M. A., Fernando, M-W. and White, C. (1996). *Hydraulic Measures for the Treatment and Control of Groundwater Pollution*. Construction Industry Research and Information Association, CIRIA Funder's Report FR/CP/26, London.

Lake, L. M., Rankin, W. J. and Hawley, J. (1996). *Prediction and Effects of Ground Movements Caused by Tunnelling in Soft Ground Beneath Urban Areas*. Construction Industry Research and Information Association, CIRIA Report PR30, London.

Merritt, A. (1994). *Wetlands, Industry & Wildlife: A Manual of Principles and Practices*. Wildfowl & Wetlands Trust, Slimbridge, Gloucester.

Nyer, E. K. (1992). *Groundwater Treatment Technology*, 2nd edition. Van Nostrand Reinhold, New York.

Palmer, R. C. and Lewis, M. A. (1998). Assessment of groundwater vulnerability in England and Wales. *Groundwater Pollution, Aquifer Recharge and Vulnerability* (Robins, N. S., ed.). Geological Society Special Publication 130, London, pp 191–198.

Powers, J. P. (1985). *Dewatering – Avoiding its Unwanted Side Effects*. American Society of Civil Engineers, New York.

Powers, J. P. (1992). *Construction Dewatering: New Methods and Applications*, 2nd edition. Wiley, New York.

Powrie, W. (1997). *Soil Mechanics: Concepts and Applications*. Spon, London.

Preene, M. (2000). Assessment of settlements caused by groundwater control. *Proceedings of the Institution of Civil Engineers, Geotechnical Engineering*, **143**, 177–190.

Preene, M., Roberts, T. O. L., Powrie, W. and Dyer, M. R. (2000). *Groundwater Control – Design and Practice*. Construction Industry Research and Information Association, CIRIA Report C515, London.

Privett, K. D., Matthews, S. C. and Hodges, R. A. (1996). *Barriers, Liners and Cover Systems for Containment and Control of Land Contamination*. Construction Industry Research and Information Association, CIRIA Special Publication 124, London.

Pyne, R. D. G. (1994). *Groundwater Recharge and Wells: A Guide to Aquifer Storage Recovery*. Lewis Publishers, Boca Raton.

Smethurst, G. (1988). *Basic Water Treatment*, 2nd edition. Thomas Telford, London.

Todd, D. K. (1980). *Groundwater Hydrology*, 2nd edition. Wiley, New York.

Troughton, V. M. (1987). Groundwater control by pressure relief and recharge. *Groundwater Effects in Geotechnical Engineering* (Hanrahan, E. T., Orr, T. L. L. and Widdis, T. F., eds), Balkema, Rotterdam, pp 259–264.

Chapter 14

Monitoring and maintenance of groundwater lowering systems

14.0 Introduction

Any groundwater lowering system will need, to some degree, monitoring and maintenance measures to ensure effective operation. This chapter describes the methods commonly used for monitoring of groundwater lowering works, including the use of dataloggers and automatic control systems. Maintenance issues are also discussed, as are methods of sealing boreholes on completion. The problems of corrosion, encrustation and biofouling, which sometimes result in the gradual deterioration of system performance, are described.

14.1 The need for monitoring

Once in operation, a groundwater lowering system is the end result of a lot of effort by a lot of people. It is a complex system dependent on a diverse range of hydrogeological, hydraulic, chemical, mechanical and human factors, but it will have a clear aim – to lower groundwater levels sufficiently to allow the construction works to proceed. It would be silly, having invested so much time and money in a system, not to monitor it to check that initially, and on a continuing basis, these aims are achieved. Yet, many systems are not adequately monitored, leading to poor performance, loss of time and money, not to mention a stressful time for those concerned.

Groundwater lowering systems should have specific targets – maximum allowable groundwater levels at particular locations. They achieve these targets by pumping groundwater. The two most important parameters to be monitored are water levels (see Section 14.2) and pumped flow rate or discharge (see Section 14.3). Other parameters may also need to be monitored (see Section 14.4).

Monitoring is important throughout the operational life of a system, but especially so soon after the start up of pumping. There have been many cases where carefully designed systems have not initially achieved the target water levels. This potentially embarrassing eventuality may result from some very simple problems with the mechanics of the system or the way it is operated. Occasionally, problems arise if ground and groundwater conditions differ significantly from those indicated by the site investigation – so called 'unforeseen ground conditions'. Monitoring during the early stages of pumping is vital to allow potential problems to be identified and solved.

In some cases, particularly on large projects or where site investigation information is limited, monitoring of system performance can be used as part of the observational method of design (see Section 7.1). The observational method involves developing an initial design based on the most probable conditions, plus contingency plans to allow modification of the system in light of conditions actually encountered. Adoption of the observational method can allow fine-tuning of the system in difficult ground conditions. There may be a temptation to install only the bare minimum pumping capacity or number of wells to achieve the drawdown. This temptation must be resisted, and allowance made for standby equipment (see Section 14.6), and long-term deterioration of the system (see Section 14.8).

Whatever monitoring is carried out, merely taking the readings is not enough. They need to be reviewed by personnel who can interpret them appropriately. Plotting of long-term trends of groundwater levels or discharge flow rates can aid the identification of problems or anomalies. On many projects this need not be done by a groundwater lowering specialist as such. The main contractor's site engineer could review them – provided they have been briefed by the system designer or installer as to what factors are important, and if there are any particular targets that must be achieved. Whoever reviews the data, specialist advice should be sought if there is any uncertainty about the effectiveness of system performance.

14.2 Monitoring of water levels

The most basic form of monitoring is the measurement of water levels in observation wells, to determine whether groundwater levels have been lowered to the target levels. The water level is most commonly measured using a 'dipmeter' or 'dipper' (Fig. 14.1); the process is known as 'taking dip readings'. The dipmeter consists of a graduated cable or tape with a probe at its tip. The probe is lowered down the well until it touches the water surface; the water completes a low-voltage circuit and a buzzer or light is activated on the dipmeter reel.

There is an art to obtaining reliable dip readings. It is best to just gently lower the probe to the water surface (when the signal will activate), then raise the probe a little (the signal should cease) and lower the probe back into the water to confirm the reading. The graduated tape is used to determine the depth to water. Commercially available dipmeters can be used in bores of 19 mm or greater internal diameter.

The most useful records of groundwater levels are taken from unpumped wells or observation wells, not from pumped wells. This is because a pumped well may show a significantly lower water level than in the surrounding aquifer, due to the effect of well losses (see Section 3.4). Observation wells constructed as standpipe piezometers respond to water levels in one stratum only (see Section 6.5) and tend to give more representative groundwater levels than unpumped dewatering wells, which may exhibit a hybrid or average water level influenced by more than one stratum. If taking dip readings in a pumped well, it is best to install a plastic dip tube of 19–50 mm diameter down which the dipmeter probe is lowered. This avoids the tape getting stuck around the pump, riser pipe or power cable.

While the measurement of groundwater levels by dipping is a very simple process, it is possible that misleading readings may be generated. Possible causes for rogue readings include:

(a) Clogged observation wells. If the well has a clogged screen the water level inside it will not be representative of the groundwater around it. Once a few days' readings are available, clogged wells are often easy to identify because readings will tend to be constant with time, and will not reflect changes in drawdown shown in other unclogged wells. Wells

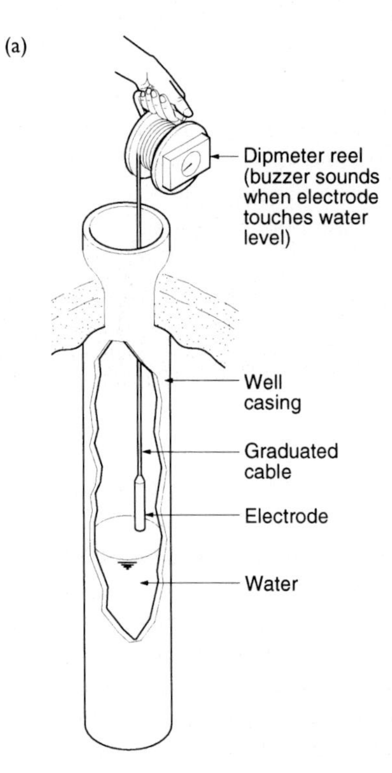

Figure 14.1 Dipmeter for measuring depth to water in a well or piezometer. (a) Schematic view of dipmeter in use (From Preene *et al.* 2000: reproduced by kind permission of CIRIA). (b) Dipmeter used to record water level in a monitoring well (courtesy of WJ Groundwater Limited).

(b)

Figure 14.1 Continued.

can sometimes be rehabilitated by flushing out with clean water or compressed air.

(b) Cascading of water into a well. If a well has a large depth of screen, and the water level inside the well is drawn down, water may cascade into the well from the upper parts of the screen above the true water level. When the dipmeter probe is lowered down the well, it may signal when it touches the cascading water, indicating a water level higher than the true level. Dipmeters with 'shrouded' probes are less prone to this problem, and should be used if at all possible.

(c) Saline water. The dipmeter relies on the groundwater acting as an electrolyte to conduct the low voltage current and trigger the signal. Saline or heavily mineralized waters are more conductive and can cause problems. The dipmeter may be triggered erroneously by small moisture droplets on the side of the well casing or may signal continuously once water gets onto the probe. Obtaining reliable readings in these conditions depends on the skill (and patience!) of the operator. It helps to dry the probe on a cloth between readings, and to repeat the reading at each well several times to be sure that a 'true' reading has been obtained. Some dipmeters have an internal sensitivity adjustment. If this is turned to a minimum setting the dipmeter may be more reliable in saline wells. Occasionally, poorly conducting water (with a low TDS) may be encountered, where contact with the water does not complete the circuit and activate the dipmeter. This has been solved by the simple expedient of adding a packet of salt to the observation well!

(d) Errors in recording readings. The person taking the readings may have made an error. The graduated tape of a well-worn dipmeter can be difficult to read when used in the field. It is not uncommon for the centimetre part of the reading to be correct, but the metre part to be in error by one metre. If the operator is recording all the readings on a sheet or notebook this error may be repeated for subsequent readings. This is because the operator may expect each new reading to be similar to the last, and will copy the metre portion of the reading from one record to the next. When checking the data, any sudden one metre changes in level may indicate this type of error.

(e) Malfunctioning dipmeters. Dipmeters sometimes malfunction. A dipmeter that does not work (perhaps because the battery is flat) is a practical problem, but does not generate false readings. A more subtle problem is when the dipmeter tape is damaged and the conductors are broken and exposed above the probe. The dipmeter will not signal when the probe touches water, but may do so when the exposed conductor reaches the water, indicating a lower water level than is actually present.

(f) Modified dipmeters. Dipmeters are sometimes repaired by shortening the graduated tape and re-jointing the probe. If the operator is not aware of this, the recorded water levels will be deeper than actual levels, with the difference being equivalent to the shortening of the tape. It is not good practice to shorten dipmeter tapes; they should be replaced with a new tape of the correct length.

It is not always that easy to spot rogue readings, but it becomes easier with experience. If in doubt, treat the readings with caution and investigate how they were taken. A lot of information can be gained by speaking directly to the person who took the readings. If the long-term trends of water levels are plotted and compared with any changes in pumping, rogue readings are often obvious.

In the field of hydrogeology for water supply water levels are often measured at weekly or even monthly intervals (see Brassington 1998), which is sufficient to identify long-term trends. For temporary works groundwater lowering applications, water level monitoring is normally carried out much more frequently. This is because short-term effects (such as fluctuation in drawdowns resulting from pump failures) are of greater interest. Typically, groundwater levels in all observation wells or piezometers should be recorded at least once per day, on every day of operation. On sites where the groundwater lowering system is critical to the stability of the works, groundwater levels may be recorded more frequently, perhaps two or three times per shift. If very frequent monitoring is necessary (perhaps on sites where water levels vary due to tidal effects) consideration should be given to the use of automatic datalogging equipment (see Section 14.5).

14.3 Monitoring of discharge flow rate

The discharge flow rate pumped by the system is another vital measure of performance. The way the flow rate is measured depends on the pumping method in use:

Wellpoint systems: total flow rate generally measured at common discharge point.
Deep well systems: flow rate may be measured at common discharge point, or measured for individual wells and then added together.
Ejector systems: flow rate may be measured at common discharge point, or measured for individual wells (calculated as the outflow from an ejector minus the inflow) and then added together.

Discharge flow rate is commonly measured by one of three methods.

(i) Flowmeters. Proprietary flowmeters may be of the totalizing type (which record total volumes of flow – flow rate is calculated from two readings at known time intervals), or the transient type (which measure flow rate directly). The most commonly used devices are propeller or turbine meters (where the flow rotates blades inside the meter) and magnetic flow meters (where the water flows through a magnetic field, inducing a voltage proportional to the flow rate). All meters should be installed in the pipework in accordance with the manufacturer's recommendations. Flowmeters should generally be located away from valves and with adequate lengths of straight pipe either side (typical requirements are a straight length of ten pipe diameters upstream and five downstream). Flowmeters may require periodic maintenance or recalibration.
(ii) Weir tanks. This method is rugged and reliable in the field, and involves passing the flow through a tank fitted with a V-notch (Fig. 14.2) or rectangular weir and measuring the depth of water over the weir. Calibration charts allow this measurement to be converted to flow rate. V-notch weirs are the most common type in use; calibration charts for this type of weir are given in Appendix 14A. The method is suitable for routine estimation of flow rate for monitoring purposes, but it can be

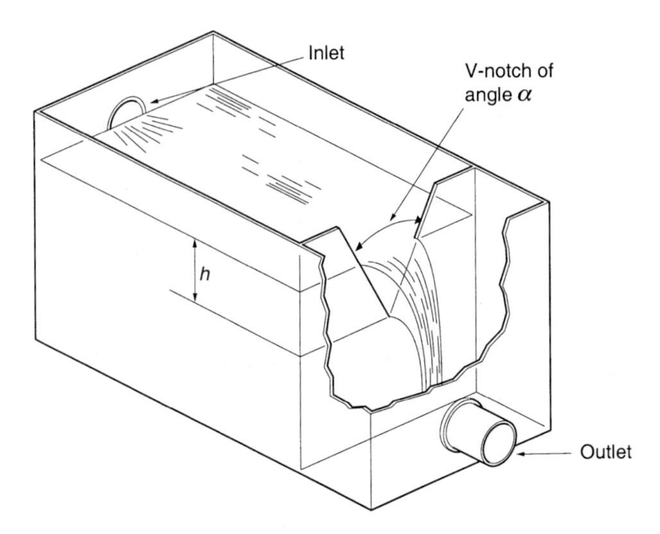

Figure 14.2 V-notch weir for measurement of discharge flow rate (From Preene *et al.* 2000 reproduced by kind permission of CIRIA). The depth of water h over the weir is measured above the base of the V-notch. The position of the measurement should be upstream from the weir plate by a distance of approximately 0.1–0.7 m, but not near a baffle or in the corner of a tank. Baffles may be required to smooth out any surges in the flow.

difficult to achieve high precision with weirs installed in small tanks (see BS3680: 1981).

(iii) Volumetric measurement. This simple method estimates flow rate by recording the time taken to fill a container of known volume (the method is sometimes called the container or bucket method). It is important that a sufficiently large container is used, to avoid too much water being lost by splashing, and to ensure the container fills slowly enough for accurate timing. For low flows (generally less than 5 l/s, but perhaps up to 10 l/s) this technique gives reasonable accuracy using equipment which is cheap, rugged and easily portable. Ideally, timing should be by stopwatch, and a relatively large container (40–200 l) should be used. The measurement should be repeated three times, and the average flow recorded.

If working in developing countries or in remote locations, the weir tank or volumetric methods are to be preferred for their simplicity. A jammed or broken flowmeter could be an embarrassment when located remote from the nearest service agent or replacement meter.

14.4 Other parameters that may be monitored

In addition to groundwater levels and flow rate, more complex projects may require other parameters to be recorded. Table 14.1 is a comprehensive list

Table 14.1 Observable parameters for dewatering systems (after Roberts and Preene 1994)

Parameter	*Method*	*Comment*
Soil stratification	Well borehole log Observation of jetting water returns	Should always be checked against site investigation
Drawdown	Dipmeter monitoring or datalogging of observation wells (standpipe or standpipe piezometer) Dipmeter monitoring or datalogging of unpumped well	Generally essential, should be monitored daily
Flow rate, system total	V-notch weir Flowmeter Volumetric measurement	Generally important, should be monitored daily
Flow rate, individual well	V-notch weir Flowmeter Volumetric measurement	Important for well and ejector systems only, monitoring intervals 1–6 months.
Water level in pumped well	Dipmeter monitoring of dip tube in well	Useful check on well performance
Mechanical performance	Vacuum (wellpoints) Supply pressure (ejectors) Discharge back pressure Engine speed (diesel pumps) Power supply alarms	Important for maintenance and/or monitoring continuous running, but data required only if flow rate or drawdown unsatisfactory
Water quality	On-site (pH, specific conductivity, etc.) Off-site (laboratory testing)	Necessary to assess clogging potential and environmental impact
Suspended solids	Condition of settlement tank or lagoon, turbidity or suspended solids content of discharge water	Always recommended to check for fines removal

Table 14.1 Continued

Parameter	*Method*	*Comment*
Settlement	Pre-construction building condition survey Level monitoring of selected locations Crack monitoring of selected locations	Sometimes necessary if risk of damaging settlements is significant
Tidal effects	Regular monitoring (at 15–60 minute intervals) of drawdown for a minimum period of 24 hours	Provides a useful check on data, detailed analysis of significance is complex
Rainfall	Rain-gauge on site Data from regional weather stations	Can be important for pumping test, or for assessing the impact of dewatering pumping on regional groundwater levels
Barometric pressure	Barometer Data from regional weather stations	Can be relevant for pumping test, or for assessing the impact of dewatering pumping on regional groundwater levels

of parameters that can be observed; it is unlikely that the full list would be monitored on a single project.

14.5 Datalogging systems

Collection of monitoring data is generally carried out by manually dipping water levels, reading flowmeters, etc. and recording the results in notebooks or record sheets. This can be avoided by using electronic datalogging equipment, connected to appropriate sensing devices to record the required data. The datalogger stores the data until the operator collects (or downloads) the data for inspection and interpretation. As each year passes, available datalogging equipment is becoming cheaper, smaller and more versatile. Equipment that, only a few years ago, might have been suitable for large-scale projects only is now cost-effective on even simple schemes.

The use of dataloggers can reduce the number of personnel needed for monitoring, especially if regular monitoring during day and night shifts is required. By downloading the data directly, there is no need for someone to laboriously key in manually recorded readings and the time and effort required to go from data gathering to plotting and analysis of data can be minimized. If e-mail communications are available, rapid dissemination of complete datasets is possible.

The most common application of datalogging equipment is to record groundwater levels. An electronic pressure transducer is installed below the water level in the well (Fig. 14.3). Such devices typically operate on the vibrating wire principle where the transducer contains a diaphragm in hydraulic connection with the groundwater. Changes in water pressure deflect the diaphragm and an associated tensioned steel wire (the vibrating wire). The deflection changes the frequency response of the vibrating wire when excited (via a connecting cable) by a readout unit or datalogger. Some of the most useful dataloggers are 'stand-alone', that is, they have sufficient battery power to operate for extended periods without an external power supply. Dataloggers can be multiplex, serving several transducers, or single station, serving one transducer only.

The pressure transducer is lowered down to a significant depth below the lowest anticipated drawdown level; typically the transducer is suspended from its own cable. The transducer records the pressure above its own level, so an on-site calibration is required to allow pressure readings to be converted to water levels. This may be done by accurately recording the depth at which the pressure transducer is suspended in the well. Alternatively, if the water level is manually measured at the same time as the datalogger takes a reading, the effective transducer depth can be estimated by adding the datalogged pressure head (measured by the transducer) to the observed depth to water. The datalogger should be installed and calibrated in accordance with the manufacturer's guidelines; some types of pressure transducers are affected

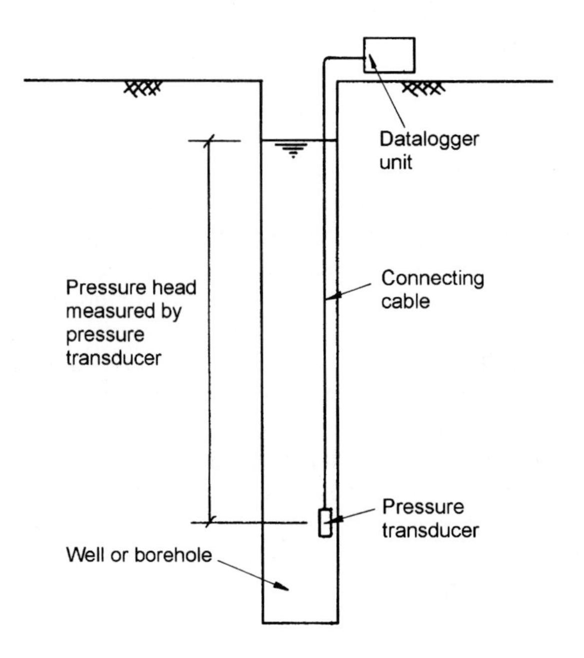

Figure 14.3 Schematic view of water level datalogger.

by barometric pressure changes, and appropriate corrections may need to be applied.

Flow rates can also be datalogged if the flow measuring device produces a suitable electronic output which can be fed to the datalogger. This is normally possible with magnetic flowmeters and certain types of propeller or turbine flowmeters.

As well as for operational monitoring, datalogging systems can be useful for gathering data during site investigation pumping tests (see Section 6.5 and Appendix 6C).

14.6 Mechanical factors and automation

The mechanical and electrical plant in the system will require periodic maintenance, depending on the type of equipment and the way it is being used. Diesel powered pumps or generators will require regular fuelling, and checking/replenishment of lubricant and coolant levels. Electrically powered equipment typically requires less maintenance, but should be tested regularly (see BS7671: 1992).

The provision of standby pumps or power supply should be considered for all groundwater lowering systems where relatively short interruptions in pumping will cause instability or flooding of the excavation. Wellpoint and ejector systems, which have relatively few running pumps, often have standby pumps provided – one standby for every two duty pumps is typical. Deep well systems, which may have many pumps, often do not have standby pumps provided. This is acceptable, provided that the system design has allowed for a few extra wells over and above the minimum number required. In this way the failure of one or two submersible pumps will not cause a problem while the pumps are replaced. Where systems are electrically powered (from mains or generator) a standby generator should be provided as a back up.

Standby pumps and generators should be tested regularly (at least weekly) by running on load. If possible it is good practice to use equipment in rotation, whereby one week a unit is run as the duty unit, and next week it acts as standby, and so on.

Electronic equipment is available to allow systems to be alarmed and, to some degree, automated. Sensors, similar to those used in datalogging, can be used to monitor groundwater levels, pump vacuum, ejector supply pressure, or the failure of the power supply or of individual pumps. The output from these sensors can be linked to monitoring equipment, which can send an alarm signal (which could be via a siren, flashing beacon, radio or telephone signal) to warn staff of problems and alert them to the need for remedial action. The remedial action might typically be to switch on the standby pumps or power supply and re-start the system. While this could be done manually by someone called out to site, the next logical step is to automate this process and allow the system to activate the standby plant in response to the alarm signal. This is the basis of automatic mains failure (AMF) systems. Such systems are used on only a small minority of groundwater lowering systems, but have been established practice on some of the larger water supply wells since the 1980s. Electronics and computer technology are advancing constantly; it is probable that in the near future automation and remote monitoring will be applied to a much greater proportion of systems than at present.

14.7 Backfilling and sealing of wells on completion

On completion of the groundwater lowering works, after pumping equipment has been removed, wells may need to be backfilled and capped off. This is required to:

i Remove the safety hazard of having open holes.
ii Prevent the well acting as a conduit for surface contamination to reach the aquifer (this is especially important for wells penetrating aquifers used for public supply, see Section 13.3).
iii Prevent uncontrolled flow of groundwater between strata penetrated by the well.

In some cases, particularly shallow wellpoint systems penetrating one aquifer only on uncontaminated sites, the wells may be left unsealed, with the wellpoint riser simply cut off below surface reinstatement level. In contrast, almost all deep wells and many ejector wells will need to be sealed.

Sealing of wells first involves removing the pumps and any headworks from the wells. The well casing and screen are normally left in place but are cut off just below ground level. The well bore is then backfilled with appropriate material. The materials used to backfill a well must be clean, inert and non-polluting. Suitable materials include clean sands and gravels, bentonite or cement grout (grouts should generally be placed via tremie pipes). It may be necessary to obtain agreement from the environmental regulatory authorities that the sealing methods proposed are adequate. In England and Wales the Environment Agency (2000) has produced guidelines on best practice.

14.8 Encrustation, bifouling and corrosion

Wells pumping groundwater for extended periods of time may reduce in performance (loss of yield or increase in drawdown in the well) gradually during operation. The loss of performance may result from clogging due to encrustation (precipitation of chemical compounds) or biofouling (bacterial growth and associated processes). Additionally, if the groundwater is saline or brackish, corrosion of metal components (such as pumps or pipework) may be of concern.

These effects have long been recognized in water supply wells, which have working lives of several decades. Such wells (and associated pumps and pipework) are typically rehabilitated at periodic intervals to ensure performance does not reduce to unacceptable levels (see Howsam *et al.* 1995). Gradual reduction in performance is less of a problem for temporary works groundwater lowering systems, largely because of the shorter pumping periods involved. Nevertheless, there have been a number of instances where systems operated for periods between several months and up to a few years have been affected. An understanding of the factors involved is useful when planning for the maintenance of systems.

The following sections mainly discuss clogging effects in wells, but problems can also occur with the encrustation or biofouling of pump internal components, pipework and flowmeters.

14.8.1 Chemical encrustation

Groundwater contains, to varying degrees, chemical compounds in solution (see Section 3.8). These compounds may precipitate in and around the well, being deposited as insoluble compounds to form deposits of scale on well screens and pumps. The water experiences a drop in pressure, and may be aerated by cascading, as it enters a well; these are ideal conditions for precipitation.

The principal indicators of the encrustation of potential of groundwater (from Wilkinson 1986) are:

i pH greater than 8.0
ii Total hardness greater than 330 mg/l
iii Total alkalinity greater than 300 mg/l
iv Iron content greater than 2 mg/l.

Common deposits include calcium and magnesium carbonates or iron or manganese oxides. For temporary works applications, chemical encrustation is rarely severe enough to affect operation. If groundwater analyses suggest it may be a problem, the wells should be designed to have as low a screen entrance velocity (see Section 10.2) as possible. If encrustation becomes a problem once the system is in operation, acidisation or chemical treatment of the wells, followed by airlifting or clearance pumping may help loosen and remove deposits (see Howsam *et al.* 1995).

14.8.2 Bacterial growth and biofouling

Groundwater naturally contains bacteria; problems occur because the wells and pipework forming a groundwater lowering system may offer an environment in which these bacteria can thrive. The most common, and hence most problematic, bacteria are the iron-related species such as *Gallionella* or *Crenothrix*. According to Howsam and Tyrrel (1990) these are sessile bacteria; this means that they attach themselves to surfaces. The action of the bacteria will cause a biofilm to develop on a surface. A biofilm consists not only of bacterial cells but also proportionately large volumes of extracellular slime. This can trap particulate matter and detritus from the water flowing by and provides an environment for the precipitation of iron and manganese oxides and oxyhydroxides. The most common form of biofilm (also known as biomass) is a thick red-brown gelatinous slime or paste which builds up in wells, pumps and pipework.

The practical result of this is that if conditions are favourable for bacterial growth a system may become biofouled. The biofilm can be surprisingly tenacious, and if not removed or controlled in some way may clog a system, dramatically reducing performance (Fig. 14.4). As biofouling deposits build up the discharge flow rate will decrease. If no action is taken, the drawn-down groundwater levels will rise, and may result in instability or flooding of the excavation. Regular monitoring of flow rate and water levels is essential for diagnosis of these problems. A programme of periodic rehabilitation of the system may be necessary to ensure continued satisfactory operation.

Figure 14.4 Biofouling of submersible pump due to iron-related bacteria. The submersible pump and riser pipe shown have been removed from the well that can be seen in the background after several weeks of operation. The upper part of the pump is coated with a thick red-brown biofilm slime.

Howsam and Tyrrel (1990) state that the following conditions may give an increased risk of biofouling: the use of iron or mild steel in the system (increasing iron availability); intermittent pumping, large well drawdowns, cascading of water into the well (all increasing oxygenation); and high flow velocities, such as through well screen slots or at valves (increasing nutrient uptake). Unfortunately, many of these conditions are almost unavoidable in groundwater lowering systems! Therefore biofouling should be considered as an operational risk for most systems.

Consider the conditions needed for the growth of iron-related bacteria (Howsam and Tyrrel 1990):

1 Nutrients. Bacteria in general need carbon, nitrogen, phosphorus and sulphur. Many natural groundwaters contain sufficient levels of these nutrients to sustain significant biofilm growth.
2 The presence of an aerobic/anaerobic interface. This is typically formed by a well, where anaerobic aquifer water can come into contact with oxygen. This interface is the point at which aerobic bacterial activity and chemical iron oxidation is initiated.

3 The presence of iron in the groundwater. The presence of dissolved iron is necessary because the oxidation of the iron provides energy for the bacteria's metabolism. This action precipitates the insoluble iron compounds, giving the characteristic red-brown colour to the biofilm.
4 Water flow. The flow of water transports nutrients to the biofilm. The faster the flow of water the more food is available for bacterial growth, and the faster the biofilm will grow. This is a key point, because it means that high flow wells may biofoul more quickly than low flow wells.

Conditions 1 and 2 will be present for most systems, so the likelihood of biofouling should be assessed from conditions 3 and 4, dissolved iron and rate of flow. Based on a number of case histories Powrie *et al.* (1990) produced Table 14.2 in terms of those two factors.

Table 14.2 shows that the risk of biofouling problems is also dependent on the type of system. Wellpoint systems (where much of the pipework is under vacuum) do not provide a good environment for the growth of aerobic bacteria, and are not especially prone to biofouling. Deep well systems using submersible pumps are prone to biomass building up on the outside of the pump and riser pipe, inside the pump chambers and inside the riser pipe. Because the riser pipe is normally of relatively large diameter, clogging of the riser pipe is only a problem in severe cases; biofouling inside the pump is more of a problem, often leading to pump damage if not removed by cleaning. Ejector systems are particularly susceptible to clogging from biofouling; this is because the smaller diameter pipework generally used is easily blocked by biomass. Artificial recharge systems are extremely vulnerable to biofouling (see Section 13.6).

If biofouling occurs, and significant loss of performance has been identified by regular monitoring of groundwater levels and discharge flow rates, the system may need to be rehabilitated. Possible methods include:

i Agitation, scrubbing or flushing out of the wells. Biomass may be removed from the wells by development techniques (see Section 10.6) that surge and agitate the well, thereby loosening the clogging material. Airlift surging and pumping, water jetting (using fresh water or a chlorine solution), and scrubbing (using a tight fitting wire brush hauled up and down inside the well) have all proved effective.
ii Cleaning of pumps and pipework. Submersible pumps, ejectors and risers can be removed from wells and cleaned at ground level by jet washing or scrubbing. Pumps and ejectors may need to be dis-assembled to allow cleaning or replacement of internal components. Riser, header and discharge pipes may be cleaned by jetting or the use of pipe cleaning moles.
iii Chemical treatment of wells and pipework. This might involve dosing the wells and rest of the system with a chlorine solution or other biocide

to kill the bacteria, and then using methods (i) or (ii) to remove the residue of biomass. This is a specialist technique, and should be carried out with care and due consideration for the handling and disposal of chemicals and effluents.

Systems operating for long periods may need rehabilitation several times during the their working lives.

14.8.3 Corrosion

Metal components in the system (such as pumps, pipework, flowmeters, steel well casings, etc.) may be subject to corrosion. Corrosion is a relatively

Table 14.2 Tentative trigger levels for susceptibility to *Gallionella* biofouling, modified after Powrie *et al.* (1990) (From Preene *et al.* 2000: reproduced by kind permission of CIRIA)

Pumping technique	*Susceptibility to biofouling*	*Concentration of iron in groundwater mg/l*	*Frequency of cleaning*
Wellpoints	Low	<10	Biofouling unlikely to present difficulties under normal operating conditions and times of less than 12 months
		>10	Biofouling may be a problem for long-term systems
Submersible pumps	Moderate	<5	6–12 months
		5–10	0.5–1 month
		>10	Weekly (system may not be viable)
Ejector wells (low flow rate; <10 l/min)	Moderate	<5	6–12 months
		5–10	Monthly
		10–15	Weekly (system may not be viable)
Ejector wells (high flow rate; >20 l/min)	High	<2	6–12 months
		2–5	Monthly
		5–10	Weekly (system may not be viable)
Recharge wells	Very high	Recharge wells are extremely prone to biofouling, which is likely to occur even if iron concentrations are below 0.5 mg/l To minimize biomass growth and encrustation extreme care should be taken to avoid aerating the recharge water It is not uncommon for recharge wells to require cleaning on a weekly or monthly basis Recharge wells may not be viable at high iron concentrations	

minor problem in water supply wells, but is rarely severe in the short term as the abstracted groundwater is usually relatively fresh (i.e. low in chlorides). Groundwater lowering systems, on the other hand, may sometimes pump water that is significantly brackish or saline (see Section 3.8) – this is likely in coastal areas or aquifers where saline intrusion has occurred. Systems pumping saline groundwater may be subject to very severe corrosion during even a few months of pumping. Fig. 14.5 shows corrosion of submersible pumps (made from grade 304 stainless steel) after between six and twelve months operation at a site with highly saline groundwater. This is an extreme example, but does show that corrosion may occasionally be a problem.

The indicators of the corrosion potential of groundwater (from Wilkinson 1986) are:

i pH less than 7.0
ii dissolved oxygen (which will accelerate corrosion even in slightly alkaline waters)
iii the presence of hydrogen sulphide
iv carbon dioxide exceeding 50 mg/l
v total dissolved solids (TDS) exceeding 1000 mg/l
vi chlorides exceeding 300 mg/l
vii high temperatures.

It should be recognized that these indicators are a guide only. Corrosion of groundwater abstraction systems is complex and may be influenced by other factors including:

(1) Electro-chemical corrosion. This is when corrosion is facilitated by the flow of an electric current. Two conditions are necessary: water containing enough dissolved solids to act as a conducting fluid (electrolyte); and a difference in electrical potential on metal surfaces. Differences in potential can occur on the same metal surface in heat or stress affected areas where the metal has been worked (such as threads or bolt holes) or at breaks in surface coatings such as paint. Such conditions allow both a cathode and anode to develop; metal is removed from the anode. Bimetallic corrosion occurs when two different metals are in contact and immersed in an electrolyte. This can affect pumps made from different metals; the more susceptible metal will corrode preferentially and may be severely affected.
(2) Microbially induced corrosion. The growth of a biofilm (formed by iron-related aerobic bacteria) on the surface of metal pumps and pipework can allow an anaerobic environment to form below the film. If sulphates are present in the groundwater the anaerobic condition can allow the growth of sulphate-reducing bacteria. These produce sulphides which are very corrosive to cast iron and stainless steel. Even though stainless steels are normally very resistant to corrosion, they become susceptible

(a)

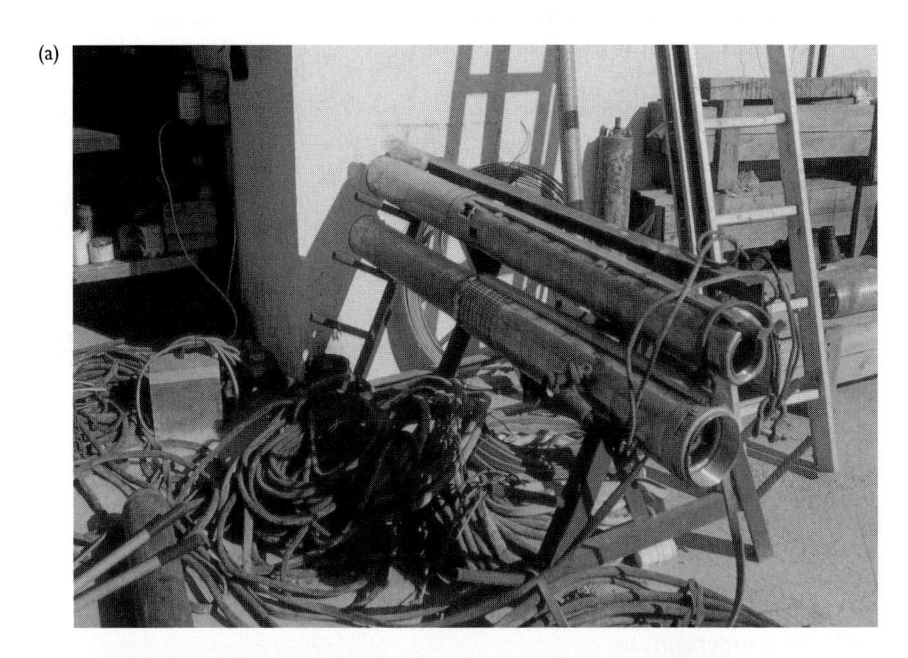

(b)

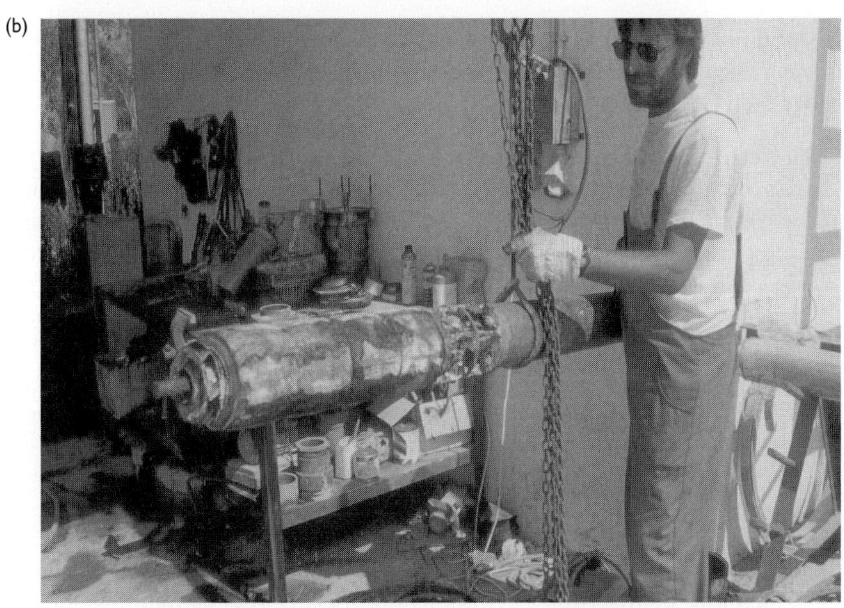

Figure 14.5 Extreme corrosion of submersible pumps in highly saline groundwater. (a) Pumps in pristine conditions prior to use. (b) Heavily corroded pumps. Note the previously shiny stainless steel pumps are now blackened and badly pitted.

to attack by chloride ions under such conditions. The passive oxide layer, which normally prevents corrosion, cannot re-form in the anaerobic zone beneath the biofilm.

If corrosive conditions are anticipated, the materials used in the system should be chosen with care. Mild steel (as is commonly used for pipework) is very susceptible to corrosion, stainless steel (from which many submersible pumps are made) is less so, and plastics are inert to corrosion. To reduce corrosion risk, as much pipework (above ground discharge pipes and below ground riser pipes) should be made from plastic. Consideration should be given to using pumps made from the more resistant grade 316 stainless steel, rather than the more common grade 304. If microbially induced corrosion is suspected, regular cleaning of the system to remove the biofilm will remove the environment for this form of corrosion.

14.9 Fault finding and problem solving

Like any complex system, there may be times when a groundwater lowering scheme does not adequately achieve its desired aims when switched on, or if it does work effectively at first, its performance may deteriorate with time. Identifying system problems and figuring out how to put them right is a vital skill for any practical groundwater engineer. Most problems with groundwater lowering systems will probably have a mechanical, hydrogeological or geotechnical cause, but it is worth thinking about all the factors which can have an effect:

i The ground. The nature of the ground will clearly affect the way a groundwater lowering system performs.
ii The pump. An inappropriate or poorly performing pump may cause problems.
iii The power supply. Every pump needs a power supply, generally diesel or electric. Unreliable or underpowered units may be the cause of difficulties.
iv The pipework. If the water cannot get to or from the pump, a system will perform poorly.
v The environment. External factors, such as changes in groundwater levels, bad weather, etc. may have an influence.
vi Human factors. Systems do not generally run themselves; the way they are operated will affect their performance.

As with any daunting problem, the secret is to identify the possible causes and eliminate the irrelevant ones until you are left with those directly linked to your woes. It requires a logical approach, and might even be enjoyable if it was not invariably carried out under pressure to solve the problem as soon as possible and allow the excavation to proceed.

Typical operational problems requiring correction include:

During initial period of running:
High flow problem. The flow rate is greater than the pump capacity, and the target drawdown is not achieved.
Low flow problem. The flow rate is less than the pump capacity and the target drawdown is not achieved.
Lack of dry conditions in excavation, when monitoring of groundwater levels indicates the target drawdown has been achieved.
After extended running:
Sudden loss of performance. The system operates satisfactorily, but after some time the flow rate or drawdown changes suddenly.
Gradual loss of performance. The system operates satisfactorily at first, but performance deteriorates gradually with time.

To diagnose any significant problem the raw material is monitoring data of the sort described in this chapter. If a system has not been adequately monitored, the first stage of any problem solving process will involve gathering data about the system performance. This will take time, probably taxing the patience of all concerned – neglecting regular monitoring of groundwater lowering systems is a false economy in terms of both time and money.

Appendix 14A: Estimation of flow rate using V-notch weirs

V-notch weirs are a common method for estimation of discharge flow rate in the field.

V-notch weirs are a form of thin plate weir, where the area of flow is a notch cut in the shape of a 'V' with an internal angle of α (see Fig. 14A-1), normally installed in a tank. The flow rate over the weir is a function of the head over the weir, the size and shape of the discharge area, and an experimentally determined discharge coefficient.

BS3680 (1981) provides formulae and discharge coefficients for V-notch weirs (and the less commonly used rectangular notch weirs). These formulae have been used to produce calibration charts for various standard V-notch angles; these charts are presented at the end of this appendix.

Knowing the angle of the V-notch, a measurement of the head of water above the base of the notch allows the flow rate to be estimated from the appropriate chart. The position of measurement should be upstream from the weir plate by a distance of approximately 1.1–0.7 m, but not near a baffle or in the corner of the tank. The tank should be positioned on a firm stable base, with timber packing used to ensure that the V-notch plate is vertical, and that the top of the notch is horizontal.

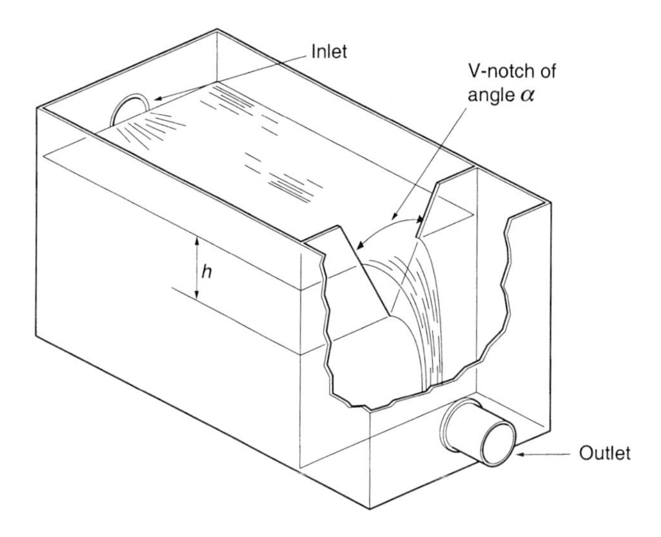

Figure 14A-1 V-notch weir for measurement of discharge flow rate (from Preene *et al.* 2000: reproduced by kind permission of CIRIA).

The calibration charts given below are based on generic formulae and discharge coefficients and are intended for use only to provide field estimates of flow rate from groundwater lowering systems. BS3680 (1981) gives guidance on the use of thin plate weirs for more accurate measurements. Calibration charts are provided for the following notch angles:

Name	*V-notch angle*
90° V-notch	90°
½ 90° V-notch	53° 8′
¼ 90° V-notch	28° 4′
60° V-notch	60°
45° V-notch	45°
30° V-notch	30°

The angles of ½ 90° and ¼ 90° V-notch weirs are not 45° and 22.5° as might be expected. These weirs are so called because they can pass one half and one quarter as much flow as a 90° V-notch weir.

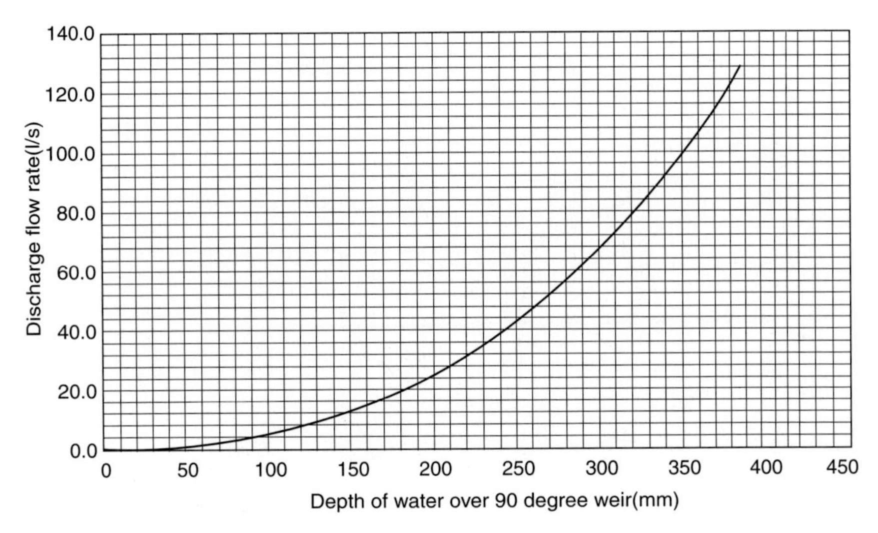

Figure 14A-2 Calibration chart for 90° V-notch.

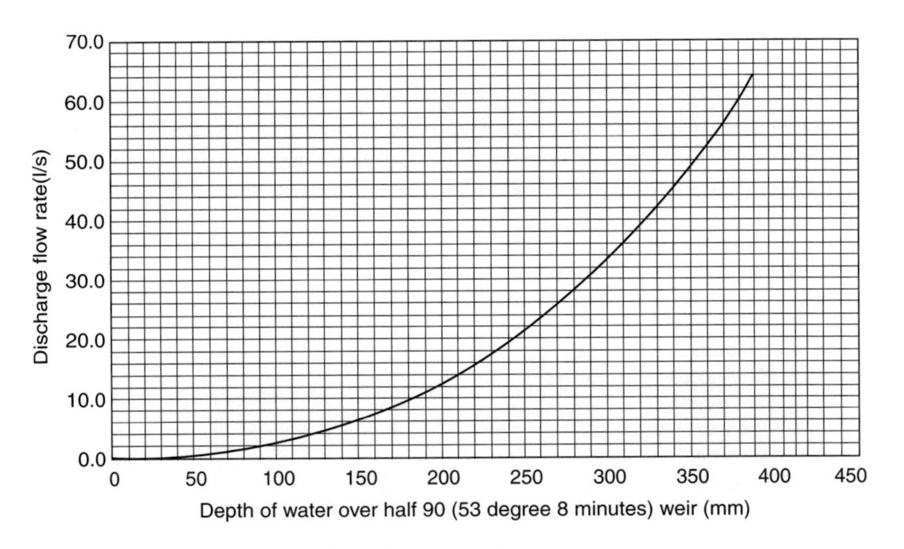

Figure 14A-3 Calibration chart for half-90° V-notch.

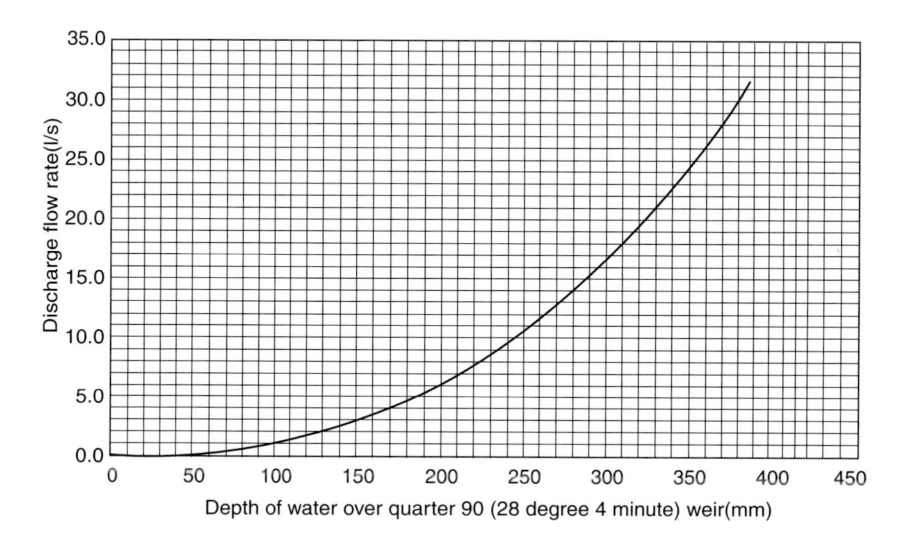

Figure 14A-4 Calibration chart for quarter-90° V-notch.

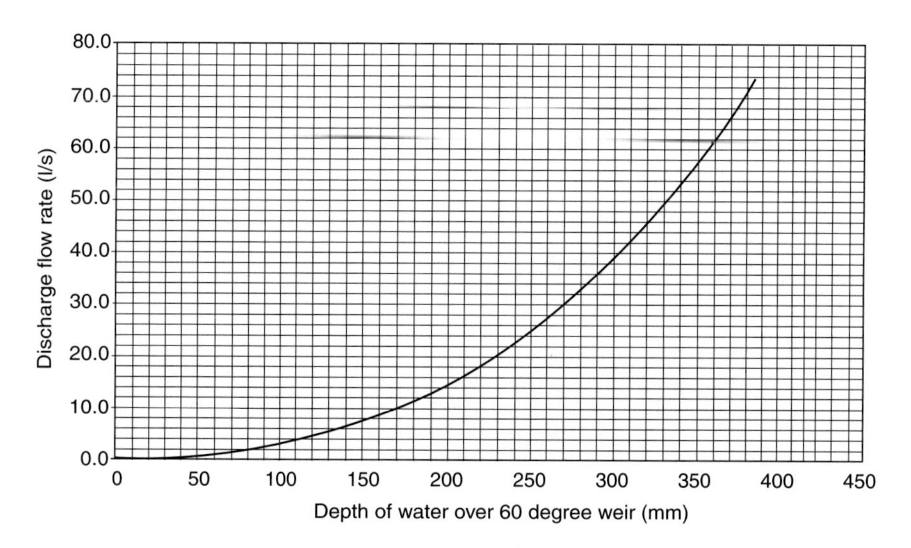

Figure 14A-5 Calibration chart for 60° V-notch.

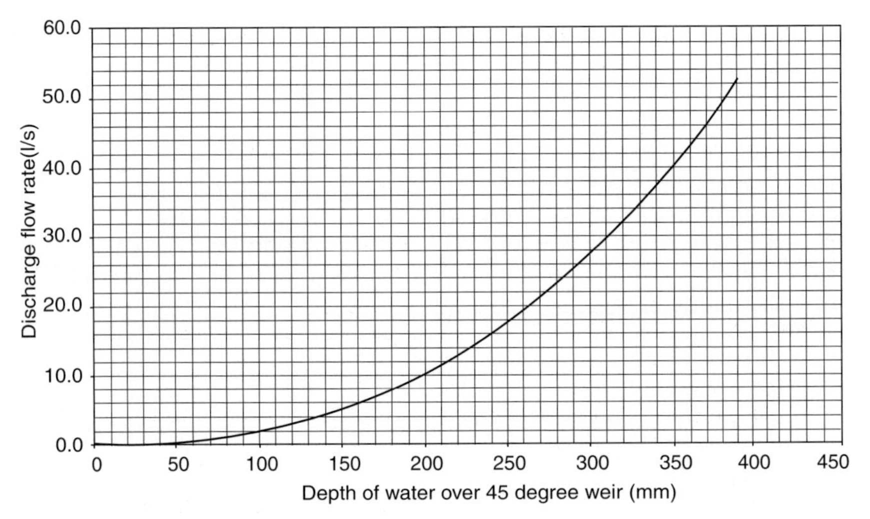

Figure 14A-6 Calibration chart for 45° V-notch.

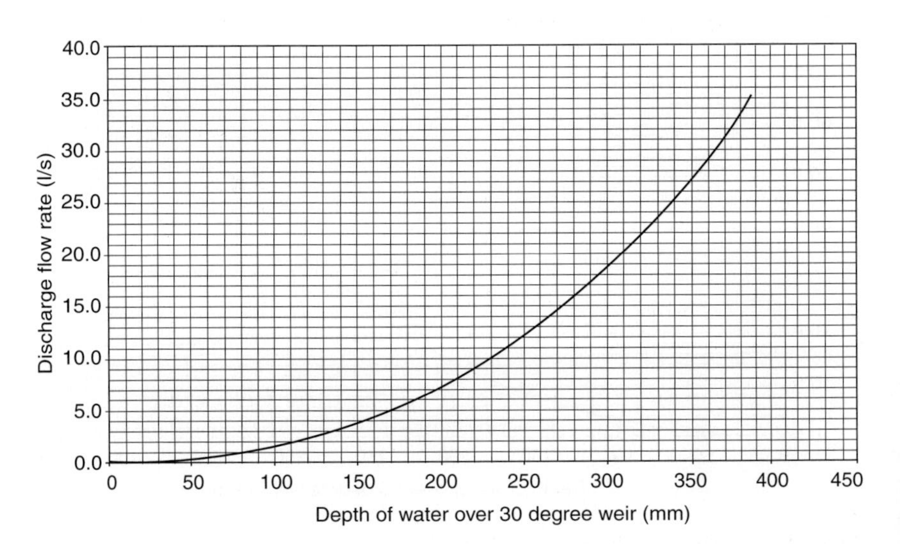

Figure 14A-7 Calibration chart for 30° V-notch.

References

BS3680 (1981). *Measurement of Liquid Flow in Open Channels: Part 4A Method Using Thin Plate Weirs*. British Standards Institution, London.

BS7671 (1992). Requirements for Electrical Installations: IEE Wiring Regulations, 16th Edition. British Standards Institution, London.

Brassington, R. (1998). *Field Hydrogeology*, 2nd edition. Wiley, Chichester.

Environment Agency (2000). *Decommissioning Redundant Boreholes and Wells*. National Groundwater and Contaminated Land Centre, Solihull.

Howsam, P., Misstear, B. and Jones, C. (1995). *Monitoring, Maintenance and Rehabilitation of Water Supply Boreholes*. Construction Industry Research and Information Association, CIRIA Report 137, London.

Howsam, P. and Tyrrel, S. F. (1990). Iron biofouling in groundwater abstraction systems: why and how? *Microbiology in Civil Engineering* (Howsam, P., ed.). Spon, London, pp 192–197.

Powrie, W., Roberts, T. O. L., and Jefferis, S. A. (1990). Biofouling of site dewatering systems. *Microbiology in Civil Engineering* (Howsam, P., ed.). Spon, London, pp 341–352.

Preene, M., Roberts, T. O. L., Powrie, W. and Dyer, M. R. (2000). *Groundwater Control – Design and Practice*. Construction Industry Research and Information Association, CIRIA Report C515, London.

Roberts, T. O. L. and Preene, M. (1994). The design of groundwater control systems using the observational method. *Géotechnique*, **44**(4), 727–734.

Wilkinson, W. B. (1986). Design of boreholes and wells. *Groundwater, Occurrence, Development and Protection* (Brandon, T. W., ed.). Institution of Water Engineers and Scientists, Water Practice Manual No. 5, London, pp 385–406.

Chapter 15

Safety, contracts and the environment

15.0 Introduction

Groundwater lowering is the means to an end – the excavation and construction of below ground works in stable and workable conditions. It will be one of many diverse activities carried out on a construction site during the life of a project. As such, dewatering must:

i Be carried out safely, and in such as way so as to minimize (as far as is practicable) the risk of harm to the workers or others.
ii Have an appropriate contractual arrangement, so that adequate information is available to allow design and construction, and to ensure that all parties are aware of their rights and responsibilities.
iii Be executed with a view to minimizing or controlling any adverse environmental impacts, and conforming to any specific legal restrictions imposed by environmental regulators.

This chapter presents brief details of the health and safety legislation in force in the United Kingdom, and outlines key issues relevant to dewatering works. Typical contractual arrangements used to procure and manage groundwater lowering works are outlined, including the traditional contractor sub-contractor relationship, as well as alternative forms of contract. Finally, the environmental law in relation to dewatering is discussed and the legal implications of dewatering works are outlined.

15.1 Health and safety

In the United Kingdom it is a legal requirement that all company and site management personnel ensure that the construction operations carried out by their company conform with safe working conditions and noise abatement regulations.

Since the beginning of the century there has been concern over accidents and working conditions on building works. In 1904 a Public Enquiry was held but it was not till 1926 that the first Building Regulations were passed into law. From time to time since then building and construction regulations have been augmented and strengthened by various specific laws. Currently the major legislation covering occupational health and safety of construction works are:

(a) *Health and Safety at Work Act 1974*
(b) *Management of Health and Safety at Work Regulations 1992*
(c) *Construction (Design and Management) Regulations 1994*
(d) *Construction (Health, Safety and Welfare) Regulations 1996.*

Legislation will change with time, as technology and working practices evolve. This chapter provides only a brief overview of some significant issues. It is essential that the reader checks to see what legislation is current, consults the original regulations and obtains specialist advice. This legislation applies to the United Kingdom. Naturally, this legislation is not applicable in other countries but the legal duties of the various parties defined in the *Health and Safety at Work Act 1974* are reflected in the legal systems of many other countries.

This section will deal specifically with health and safety issues particular to groundwater control works. Generic guidance on construction health and safety can be found in various publications (see for example, Health and Safety Executive 1996).

15.1.1 CDM regulations

The *Construction (Design and Management) Regulations 1994* (known as the CDM regulations) define the legal duties of the various parties to the construction process, including clients, their agents, designers and contractors. In simple terms the regulations require health and safety to be considered at every stage of a project from design through to construction, maintenance and demolition. This is a change in focus from the traditional approach to health and safety, which concentrated on the construction phase and the role of the contractor.

The CDM approach requires designers to consider whether it is reasonably practicable to exclude or reduce potential hazards. The risk to health and safety arising from a particular aspect of the design has to be weighed against the cost of excluding that feature by

- Designing to reduce foreseeable risks to health and safety.
- Tackling the causes of risk at source.
- Reducing and controlling risks by means that will protect anyone at work (rather than just the individual at work) and so yield the greatest benefit.

The cost is based not just on financial considerations but also fitness for purpose, environmental impact and 'buildability'. The purpose of the regulations is not only to reduce possible hazards and risks, but also to identify and record those which are unavoidable, as a warning to subsequent parties in the design/construction/maintenance/demolition process.

Detailed guidance on the application of the CDM regulations is given in the *Approved Code of Practice* by the Health and Safety Executive (1995) and in CIRIA Report 166 (Ove Arup & Partners 1997). The application of CDM to the design of groundwater lowering systems is discussed in Preene *et al.* (2000).

In essence, the CDM regulations formalize what a good dewatering designer should be doing as a matter of course. Firstly, considering the full range of options for a design and then, so far as is practicable, designing systems which will be effective but which can be installed without unnecessary risks to the workers or others. In other words to consider the 'buildability' of a system.

When considering options the designer must look at detail – for example if a line of wellpoints requires the jetting crew to work next to an open deep cofferdam, could the line of wellpoints not be relocated to avoid the risk? The designer must also, however, look at the bigger picture. If a sheet-pile cofferdam is dewatered, but relies on continued dewatering for stability is that too risky an option? What if the power supply fails? What if the standby generator fails too? Such risks will need to be assessed against the factors discussed earlier. In this case the options might be to avoid dewatering altogether and use a system based on groundwater exclusion. A multiple redundant pumping and power supply system may increase the cost, but may produce a significant reduction in the risk of catastrophic failure of the excavation. These are the kind of questions that should be part of the design process.

15.1.2 COSHH regulations

The *Control of Substances Hazardous to Health Regulations 1988* (known as the COSHH regulations) lay down the requirements for the assessment of risks from hazardous substances, and for the protection of people exposed to them.

The regulations cover almost all potentially harmful substances (although some substances such as lead and asbestos are covered in separate legislation). Suppliers of any chemical or hazardous substance must provide a data sheet to the purchaser, which must contain sufficient information to allow an assessment to be made of the risk to health from the use of the substance. Based on these assessments, employers have to ensure that, as far as is reasonably practicable, the exposure of employees to hazardous substances is prevented, or if this is not possible, then controlled. This may involve the use of personal protective equipment (gloves, eye protection, respirators, etc.) appropriate to the hazard.

15.1.3 Use of electricity on site

The use of electrical equipment on site is controlled by:

- *Electricity at Work Regulations 1989*
- *Electrical Equipment (Safety) Regulations 1994.*

Portable equipment used on construction sites is generally limited to 110 V supply, arranged so that no part of the installation is at more than 55 V (single phase) or 65 V (three phase) to earth. These voltage limits are intended to reduce the severity of injuries if electrocution occurs. Many dewatering systems are electrically powered (particularly deep well and ejector systems) and use 415 V three phase systems, in which each phase is at 240 V to earth. In the event of electrocution, injuries will be much more severe than with conventional equipment; unfortunately fatal incidents do still occur in the construction industry.

Due to the high voltages involved most dewatering electrical systems should be treated as permanent installations and should comply with the *Institution of Electrical Engineers Wiring Regulations* (which have been issued as BS7671 (1992)). Electrical switchgear must be installed, commissioned and regularly tested by an 'authorized person' defined under the regulations.

Many injuries caused by electricity arise from equipment being worked on by inexperienced or unauthorized personnel, or by poor communication between those working on the system. This can lead personnel to work inside panels or switchgear which they believe to be safe but which are, in fact, live. A significant number of accidents occur in this way. It is imperative that this situation is avoided by good communication, training and management.

15.1.4 Safety of excavations

A significant number of accidents occur in excavation works, particularly as a result of falls or resulting from collapses of poorly supported excavations. Legal requirements for safety in excavations are contained in the *Construction (Health, Safety and Welfare) Regulations 1996*. A key requirement is that a competent person is required to inspect the excavation at the start of each shift while persons are working in the excavation. Additional inspections are required after any event which may have affected the strength and stability of the excavation.

Practical guidance on methods of work in trenches up to 6 m depth is given in Irvine and Smith (1992).

15.1.5 Noise

Pumps and dewatering plant are normally operational continuously day and night, seven days a week. Any noise from the plant will be a potential health hazard to those working on site, and an annoyance to the public around the site.

The exposure of employees to noise is governed by the *Noise at Work Regulations 1989*, which sets various action levels in relation to daily exposure to peak noise levels; noise levels need to be assessed in the working area. Noise should be reduced at source, but if this is not practicable then appropriate hearing protection must be provided and worn.

The effect of noise on the public outside the site is covered by the *Control of Pollution Act 1974*, which gives local authorities statutory powers to set noise limits and allowable working hours.

Selection of dewatering plant has the greatest impact on ambient noise levels.

(a) The quietest systems are those electrically-driven from mains power. Deep well systems thus powered will be almost silent (since the pumps are below ground) and wellpoint or ejector systems will only generate a low 'hum' from the rotation of the pump and motor.
(b) If mains power is not available electrical plant should be run from super-silenced diesel-powered generators – modern equipment is very quiet.
(c) For wellpoint systems, diesel powered pumps are sometimes the only practicable option. Pumps with silenced or 'hushed' prime movers should be used whenever possible, not only to reduce ambient noise, but also to reduce exposure to operatives. If pumps with unsilenced prime movers are used, noise levels should be reduced by reducing the engine revs as far as possible.
(d) Acoustic screens can also be constructed around pumps, but these can cause problems with ventilation and overheating of pumps that have resulted, in extreme cases, in pumps catching fire. It is far better to start with suitably silenced equipment in the first place.

15.1.6 Working with wells and boreholes

Many groundwater lowering works will involve the construction, commissioning, maintenance and, ultimately, decommissioning of wells. The hazards associated with wells and boreholes can broadly be subdivided into three types:

i Hazards during drilling, relating to the drilling and construction works themselves. These include: crushing, penetration or impact injuries caused by equipment breakages or inappropriate handling of casing, etc; and health issues arising from ingestion or dermal contact with soil or groundwater on potentially contaminated sites.
ii Hazards during operation. During operation pumps will be installed in most wells, reducing the risk of falls into all but the largest diameter

wells. However, if any relatively large diameter wells are left without pumps during part of the construction period they should be temporarily covered or capped off.

iii Hazards following decommissioning. These only arise if the wells are not backfilled or capped off and secured at the end of the project as outlined in Section 14.7. Larger diameter wells, if left uncovered, clearly present a hazard. Even the smaller diameter wells (typically of a few hundred millimetres open bore) common on dewatering projects may present a hazard to small children if they can gain access to the site. It is essential that wells be decommissioned responsibly.

Guidance on safety during drilling works is given in publications by Association of Geotechnical and Geoenvironmental Specialists (1992a,b). Particular guidance on working with large diameter wells is given in *Safety in Wells and Boreholes* (Institution of Civil Engineers 1972).

15.2 Contracts for groundwater control works

Groundwater lowering is a specialist process which, when measured in cost or manpower terms, is often only a tiny part of the overall construction project. Despite this, groundwater (and its inadequate or inappropriate control) has historically been the cause of many construction disputes. Suitable contractual arrangements are an important part of managing dewatering works for positive results. This section discusses the background to contractual issues and outlines some arrangements used in practice.

15.2.1 The need for contracts

As has been described throughout this book, groundwater control is highly dependent on the ground conditions, over which the designer has no control. It follows that groundwater control must run a greater risk of poor performance than, say, reinforced concrete construction (where the designer can specify and control the materials used). The dewatering designer must gain his design information from the results of the site investigation (see Chapter 6). Sadly, it is still the case that not all investigations attain the standards that designers would aspire to – see *Inadequate Site Investigation* (Institution of Civil Engineers 1991) and *Without Investigation Ground is a Hazard* (Site Investigation Steering Group 1993).

Because groundwater control works are often carried out at the very start of a construction project (e.g. for construction of foundations) any problems or delays at that stage can have serious knock-on effects for the rest of the project. A review of cost overruns on groundwater lowering projects is given in Roberts and Deed (1994).

The specialized and perceived 'risky' nature of groundwater lowering works has led many clients and main contractors to view dewatering as a 'black art' best left to the cognoscenti. Apart from on the very largest projects, many contractors prefer to sub-contract dewatering works to specialist organizations who provide the expertise, experience and equipment to carry out such work. It is important that an appropriate contract exists between the various parties, and that the rights and responsibilities of each are clearly identified.

15.2.2 Traditional contract arrangements

In the United Kingdom, the traditional form of construction contract involved the client appointing a client's representative (called the Engineer under some forms of contract) to administer and supervise the works. Traditionally, the client's representative would also design the permanent works, but not the temporary works, which were the remit of the contractor. The client's representative would, via a bidding or negotiation process of some sort, arrange for a contractor to undertake the works. The contractor would be employed directly by the client under a form of contract such as one of the editions of the Institution of Civil Engineers conditions of contract (known as the ICE conditions).

Groundwater lowering works are almost always classed as temporary works and are the main contractor's responsibility. These works are commonly sub-contracted, and the contractor would employ a dewatering sub-contractor. The sub-contract between the dewatering company and the main contractor would typically be 'back to back', meaning that the rights and responsibilities of each party applies to the other. In essence this means that the dewatering sub-contractor is effectively acting on behalf of the contractor and takes over their responsibilities relevant to the dewatering. It also means that in the event of any changes or problems, the dewatering sub-contractor has the same rights as the contractor to apply for additional time or money via the clauses in the contract.

One of the most common type of serious disputes in traditional dewatering contracts occur when the groundwater lowering system does not achieve the target drawdown, and it is believed that the ground conditions may not be as represented in the site investigation data provided at tender stage. The ICE conditions contain clause 12 (commonly known as the 'unforeseen ground conditions' clause), which allows the sub-contractor and contractor to apply to the client's representative for additional time or payment. The 'claims' process is often protracted, whereby the contractors have to demonstrate that the physical conditions or artificial obstructions encountered could not reasonably have been foreseen by an experienced contractor. This process can sometimes distract from the real problem of trying finish the project and to deal with the conditions actually present in the ground. Some claims cannot be quickly resolved and may go to court and are finally resolved (one way or the other) several years after the end of construction.

15.2.3 Alternative forms of contract

In the United Kingdom there are moves away from traditional contracts, which are viewed as being too adversarial and leading to too many costly and time consuming 'claims'. When construction problems occur prompt and open sharing of information can be vital in developing solutions; this did not always happen under traditional contracts. Sometimes in the past such contracts were applied in such a way that rather more effort was spent trying to apportion blame than on solving the problem.

The nature of contracts is also changing as a result of increased use of so-called 'design and build' contracts. These involve the contractor designing the permanent works as well as temporary works, and change the nature of the relationships between client, client's representative and contractor.

Various non-adversarial forms of contract have been developed to try and avoid some of these problems. A number of different schemes are possible including:

(a) 'Partnering' – which implies the development of longer-term relationships between the various parties including client, permanent works designer, contractor and specialist sub-contractors.
(b) 'Open book' contracts – where information is shared and all parties are kept informed of what is going on, and are able to have input into relevant decisions.

These forms of contract can allow the 'risks' of unforeseen ground conditions (or other factors) to be shared between the various parties in an open and transparent way. Dispute resolution procedures exist within such contracts to allow problems to be quickly highlighted and examined without the need for claims or other confrontational procedures.

When these forms of contract are used the aim should be to control risks to the project. Any sub-contractor should be selected on the basis not merely of cost, but also in terms of quality, health and safety, and ability to meet the programme timescale. It is also important that, by involving all the parties, expertise and experience can be pooled to solve problems as quickly as possible. The need to overcome problems in a timely manner cannot be over-emphasized – on modern construction projects many cost overruns result mainly from time delays, rather than changes in methods. The control of risks in construction is discussed by Godfrey (1996).

15.2.4 Dewatering costs

Because of the varied nature of groundwater lowering works, and the wide variety of ground and groundwater conditions, the development of generic costs is not easy. It is not possible to estimate dewatering costs, even on an approximate basis, from the quantity of water pumped, the volume of soil dewatered or the depth of drawdown. Dewatering costs are more commonly broken down on an 'activity schedule' basis. Some of the activities will be costed on a unit basis (e.g. per well, per metre of header pipe) while the costs during the pumping period – pump hire, fuel, supervision, etc. – will be time-related charges (e.g. per day or per week).

15.3 Environmental regulation of groundwater control

In relation to construction works groundwater is viewed primarily as a problem, hence the need for groundwater control. In other contexts groundwater is a resource, used for public and private drinking water, and for industrial use. In many countries environmental regulations or laws exist to help safeguard groundwater resources. This section describes the environmental regulatory regime applicable in England and Wales and the implications for the planning of groundwater control works.

The person or organization responsible for the groundwater control has certain legal obligations to ensure that consents and permissions are obtained from the regulators. Under the normal forms of contract the party responsible is either the contractor or the client. The dewatering sub-contractor or pump hirer is not normally responsible for the consents, but they should satisfy themselves, before work commences, that the necessary consents and permissions have been obtained.

There are two main facets to the legal requirements. The first deals with pumping of groundwater (termed *abstraction*) and is intended to make sure the regulators can control groundwater abstraction to ensure that groundwater lowering systems do not cause nearby groundwater users to lose their supplies. The second deals with disposal of groundwater (termed *discharge*) and is intended to ensure that the pumped water does not itself cause pollution.

15.3.1 Groundwater protection

Groundwater protection is the collective term for the policy and practice of safeguarding the groundwater environment. This covers protection of both groundwater resources – the quantity of groundwater available for use – (to ensure the aquifers do not 'dry up') and of groundwater quality (to ensure that groundwater is not contaminated by man's activities).

The regulatory regime for groundwater protection in England and Wales is described in *Policy and Practice for the Protection of Groundwater* (Environment Agency 1998).

One way to protect groundwater resources is to require abstractors to be licensed. The licensing process allows the regulators (typically a governmental or quasi-governmental body) to scrutinize the applications and set limits on flow rates, or perhaps even prohibit abstraction in certain areas. Licensing of abstractions is a relatively modern concept, dating from the second half of the twentieth century. The history of the chalk aquifer beneath London, where unregulated abstraction in the late nineteenth and early twentieth centuries resulted in large and widespread lowering of the piezometric level, was described in Section 3.5. Such a situation would have been unlikely to occur had a system of regulation been in place at the time – modern regulators are, of course, left with the legacy of managing rising groundwater levels beneath London.

Over-abstraction of groundwater resources has occurred in several other British cities and in other countries. In Thailand, for example, literally thousands of wells were drilled beneath Bangkok between the 1950s and the 1990s. Abstraction lowered groundwater levels from almost ground level down to 50 m depth, depleting groundwater resources as well as causing considerable ground settlement (Eddleston 1996). This resulted in the Thai government passing laws, for the first time, to regulate groundwater abstractions. Similar regulations exist in many other countries.

The protection of groundwater quality is important because if groundwater becomes polluted it can be very difficult to rehabilitate. It is therefore better to avoid or reduce the risk of contamination than to deal with its consequences. Groundwater quality is protected by the imposition of controls on discharges to the water environment (groundwater and surface water). This approach sets allowable limits on the concentrations of many dissolved or suspended substances in discharge water.

15.3.2 Abstraction of groundwater

The abstraction of groundwater for beneficial use or supply in England and Wales is restricted by law. The relevant legislation is the *Water Resources Act 1991*, which replaced earlier legislation dating from 1945 and 1963. This means that an abstraction licence is required if groundwater is to be pumped for subsequent use, for example by a farmer or by a factory (although low volume abstractions for domestic use do not require a licence). The licence will set limits on the quantity to be pumped, and the requirement to obtain a licence ensures that the regulator can keep records of total abstractions from particular aquifers.

The regulator responsible for the protection of groundwater resources in England and Wales, the Environment Agency (EA), considers the merits of the application in relation to groundwater resources in the area affected. Applications which raise significant environmental concerns, or which are in areas where groundwater resources are over utilized may be rejected. Abstractions in Scotland and Northern Ireland are not controlled by a licensing system and common law rights apply. These do not impose restrictions on dewatering abstractions although they do not remove any existing liabilities for damages resulting from such pumping.

The water abstraction licensing system set out in the *Water Resources Act 1991* provides a small number of exemptions, which can allow groundwater to be abstracted without the need for a licence or consent. These exemptions apply if the total abstracted volume is very low, or if the water is being abstracted for one of the purposes exempt under the Act, such as domestic use.

The Act (under Section 29) states that abstractions for dewatering purposes are, in principle, exempt from legal restrictions if pumping is necessary to

> prevent interference with any mining, quarrying, engineering, building or other operations (whether underground or on the surface); or to prevent damage to works from any such operation.

The exemption from licensing does not generally extend to the use of the abstracted water for other purposes, such as for concrete production on site. A licence is still required for any beneficial use of the abstracted water.

This exemption might suggest that the EA does not have powers to object to proposed dewatering works, but that is not in fact the case. Section 30(1) of the Act requires that the EA be notified of any dewatering works prior to the installation of the first well or wellpoint. It is a *criminal offence* not to notify the authorities in advance of the works.

Once notified, the EA has an opportunity to serve a conservation notice under Section 30(2) of the Act. This is intended to avoid unregulated dewatering which could cause significant detrimental environmental impacts and may significantly derogate private water users or otherwise affect the wider groundwater or surface water environment. A conservation notice requires the party responsible for the groundwater control to take 'reasonable measures for conserving water'; these measures will be specified in the notice and are intended to monitor or mitigate some of the side effects of dewatering that are described in Chapter 13.

Another issue is that while dewatering abstractions are currently exempted from consents, site investigation well pumping tests and boreholes drilled specifically to obtain groundwater information are not. The *Water Resources Act 1991* requires prior consent for the construction of any well or other works with the intention of abstracting groundwater to investigate the presence, quantity or quality of groundwater. Hence, investigative pumping tests, whether for hydrogeological, geotechnical or environmental reasons will require a formal consent under Section 32 of the Act before a well is drilled. The application for the pumping test consent will also serve the useful purpose of alerting the EA to the likelihood of future dewatering works at the site.

The way that dewatering abstractions are regulated will change in the near future (Department of the Environment, Transport and the Regions 1999). Under the new regulations all abstractions above a small threshold quantity, for whatever purpose, will require prior authorization from the EA. Current exemptions based on the use of the abstracted water will be removed, and dewatering works will need prior authorization in the form of a permit or consent (Table 15.1).

15.3.3 Discharge of groundwater

Once abstracted from the ground, the pumped water must be disposed of (volumes generated are normally too great to be stored on site). The abstracted water from a groundwater control system is legally classified as *trade effluent*. In urban areas it may be possible to dispose of water into sewers or surface water drains – permission must be obtained from the water utility or

Table 15.1 Abstraction licences, consents and permits (based on Department of the Environment, Transport and the Regions, 1999)

Authorization type	*Notes*
	All abstractions above a threshold amount (20 m^3/day) will require one of the following forms of prior authorization from the EA. The only exceptions will be abstractions for firefighting, abstractions on boats, drainage of land and Ministry of Defence activities on training and operational duties.
Abstraction licence	Required for all groundwater abstractions where the water is put to beneficial use.
Abstraction consent	Required for medium-term to long-term dewatering abstractions (i.e. greater than one month), provided the water is discharged without beneficial use. If some of the water is used, an abstraction licence will be required for that portion of the abstraction.
Abstraction permit	Required for short-term dewatering abstractions, including 'emergency' abstractions, provided the water is discharged without beneficial use. The criterion for a permit is that the abstraction is short term (i.e. less than one month). Each Permit will be valid for not more than one month. The EA is likely to take steps to prevent abuse of the system by successive applications for permits on sites where dewatering is required for longer than originally intended. After one month such sites are likely to require a consent.

its agents before this can be done. Such permissions normally take the form of trade effluent licences, which may set limits on the quantity and quality of water that can be discharged. Charges (based on a cost per cubic metre) are generally levied for disposal of water in this way. For long duration discharge of substantial quantities these charges may add up to considerable sums.

However, in many cases the abstracted water must be disposed of to surface water including rivers, lakes and the sea or to groundwater (via recharge wells or trenches). All surface and groundwater are legally classified as *controlled waters* under the *Water Resources Act 1991*. In these cases, even if the volume of water to be disposed of is small and for a short duration, a discharge consent is still required under section 85 of the Act. Under this section it is a criminal offence to discharge poisonous, noxious or polluting material into any controlled waters, either deliberately or accidentally. Polluting materials include silt, cement, concrete, oil, petroleum spirit, sewage or other debris and waste materials. Measures to prevent the discharge of polluting materials are described in Section 13.4.

Even if the discharge flow is to be disposed of by artificial recharge (see Section 13.6) back into the same aquifer a discharge consent is still required. This means that the recharge water quality will have to be monitored to ensure it is within the quality limits set on the discharge consent.

15.3.4 Settlement resulting from the abstraction of groundwater

The lowering of groundwater levels resulting from the abstraction of groundwater will cause an increase in vertical effective stress in a soil. This will inevitably lead to some ground settlement, possibly over a wide area. Occasionally, the settlements may be large enough to cause distortion or damage of nearby structures or services. Engineering mitigation techniques are available to minimize the extent and effects of these settlements. Chapter 13 describes these problems and their potential solutions.

There has been considerable litigation over many years regarding settlement damage that has been alleged to result from groundwater abstractions for water supply or dewatering purposes. The law related to ground settlements arising from groundwater abstraction is found in common law rights, which are based on the precedents of judgements laid down in previous cases in the civil courts.

According to a number of reviews of the legal issues (such as Akroyd 1986; Powrie 1990) it is well established in common law that a landowner has no right of support from the water percolating in undefined channels beneath his land. Groundwater in almost all aquifers is effectively flowing in undefined channels and is therefore considered by common law to be a reservoir or source of supply, which is no one's property, but from which everyone has the right to abstract as much as they wish, in so far as it is physically possible to do so.

The courts have found that if groundwater abstraction causes settlement damage to a neighbour's property there is no right of action against the abstractor. The courts have also found that since there is no duty of care to avoid damage resulting from groundwater abstraction, there is also no cause of action under the law of nuisance or negligence.

Possible conditions where groundwater would not be considered to be in undefined channels include well-established subterranean solution channels known to exist in karst aquifers or water flowing in abandoned mine workings. From a legal point of view if water is believed to be flowing in a defined underground channel, the presence of the channel must be known. It is not necessary for it to be revealed by excavation or exposure, but its presence must be a reasonable inference from the available information.

This means that if dewatering abstractions pump clear groundwater (with no suspended solids) from most aquifers there is no legal liability if settlement damage occurs, even if mitigation measures could have been employed to prevent damage. This position is quite different from settlement damage arising from excavation or tunnelling, where the primary mechanism is the removal of the support from the soil and a legal liability exists.

However, for dewatering schemes if the water pumped is not perfectly clear and if it contains suspended silt or sand particles, the support of the ground is being removed. A landowner does have the right to the support of the ground beneath his land. If the pumping of silt or sand led to settlement damage, there would be a legal right of action against the abstractor. A consequence of this is that all wells and sumps must be equipped with effective filters to prevent the flow of groundwater removing fine particles from the soil. In practice this is most likely to be a problem with sump pumping systems, and these should be employed with caution if structures are present near the excavation.

Despite there being no legal liability for settlement damage resulting from the abstraction of groundwater, in the great majority of projects construction methods are designed to be sensitive to the risks of consequential damage. If there is believed to be a significant risk of settlement damage the project client or designer will generally require the use of settlement mitigation or avoidance measures (such as artificial recharge).

15.3.5 Environmental regulation of major projects

Very large projects are, in principle, subject to the same environmental constraints as smaller projects. However, the sheer scale and high profile nature of the projects may result in a slightly different approach being taken. This can be illustrated by the case history of the Medway Tunnel, constructed in southern England in the 1990s (Lunniss 2000; Thorn 1996).

The tunnel was of immersed tube design and required significant dewatering to allow construction of a casting basin and cut-and-cover tunnel sections on either side of the river. At planning stage the dewatering was anticipated to require extensive pumping from the chalk aquifer, which provided public water supply from a number of nearby wells. There was concern that the dewatering would derogate these sources, and that saline intrusion may be promoted by river dredging during construction.

As with many large infrastructure projects, construction of the tunnel was covered by a special Act of Parliament, in this case the *Medway Tunnel Act 1990*. Such acts require concerned bodies to petition for their 'rights' in the drawing up of the bill and request special protective provisions. The then National Rivers Authority (NRA) – the predecessors of the EA – obtained, via the act, protective measures as follows:

> The wardens (of the scheme) ... shall so design and construct the tunnel as to ensure by all reasonably practicable means that saline or other contaminating intrusion into water resources in underground strata does not occur by reason of such construction

The construction contract gave the contractor the responsibility of developing a dewatering and monitoring scheme acceptable to the NRA. Detailed liaison followed and a dewatering and monitoring strategy was developed; this strategy was robust enough to cope with an increase in anticipated pumping rates from around 250 l/s up to 400 l/s. The monitoring exercise was intensive. The data gathered in the initial stages of the project allowed a numerical model of the aquifer to be developed by the contractor to support their proposals for the later stages of dewatering.

While the extent of monitoring and discussion with the NRA is considerably more extensive than is routine, this project highlights that any major dewatering works are likely to require significant ongoing liaison with the environmental regulators. It is not unusual for the planned dewatering works to be varied in light of the regulator's requirements.

References

Akroyd, D. S. (1986). The law relating to groundwater in the United Kingdom. *Groundwater: Occurrence, Development and Protection*, Water Practice Manual No. 5, Institution of Water Engineers and Scientists. London, pp 591–607.

Association of Geotechnical and Geoenvironmental Specialists (1992a). *Safety Awareness on Investigation Sites*. Association of Geotechnical and Geoenvironmental Specialists, Beckenham, Kent.

Association of Geotechnical and Geoenvironmental Specialists (1992b). *Safety Manual for Investigation Sites*. Association of Geotechnical and Geoenvironmental Specialists, Beckenham, Kent.

BS7671 (1992). *Requirements for Electrical Installations: IEE Wiring Regulations 16th Edition*. British Standards Institution, London.

Department of the Environment, Transport and the Regions (1999). *Taking Water Responsibly*. HMSO, London.

Eddleston, M. (1996). Structural damage associated with land subsidence caused by deep well pumping in Bangkok, Thailand. *Quarterly Journal of Engineering Geology*, **29**, 1–4.

Environment Agency (1998). *Policy and Practice for the Protection of Groundwater*, 2nd edition. The Stationery Office, London.

Godfrey, P. S. (1996). *Control of Risk: A Guide to the Management of Risk from Construction*. Construction Industry Research and Information Association, CIRIA Special Publication 125, London.

Health and Safety Executive (1995). *Managing construction for Health and Safety: Construction (Design and Management Regulations 1994), Approved Code of Practice*. HMSO, London.

Health and Safety Executive (1996). *Health and Safety in Construction*. HMSO, London.

Institution of Civil Engineers (1972). *Safety in Boreholes and Wells*. Institution of Civil Engineers, London.

Institution of Civil Engineers (1991). *Inadequate Site Investigation*. Thomas Telford, London.

Irvine, D. J. and Smith, R. J. H. (1992). *Trenching Practice*. Construction Industry Research and Information Association, CIRIA Report 97, London.

Lunniss, R. C. (2000). Medway Tunnel – planning and contract administration. *Proceedings of the Institution of Civil Engineers, Transport*, **141**, February, 1–8.

Ove Arup & Partners (1997). *CDM Regulations – Work Sector Guidance for Designers*. Construction Industry Research and Information Association, CIRIA Report 166, London

Powrie, W. (1990). *Legal Aspects of Construction Site Dewatering for Temporary Works*. Unpublished MSc dissertation, Kings College, London.

Preene, M., Roberts, T. O. L., Powrie, W. and Dyer, M. R. (2000). *Groundwater Control – Design and Practice*. Construction Industry Research and Information Association, CIRIA Report C515, London.

Roberts, T. O. L. and Deed, M. E. R. (1994). Cost overruns in construction dewatering. *Risk and Reliability in Ground Engineering* (Skipp, B. O., ed.). Thomas Telford, London, pp 254–265.

Site Investigation Steering Group (1993). *Site Investigation in Construction, Volume 1: Without Investigation Ground is a Hazard*. Thomas Telford, London.

Thorn, B. (1996). Dewatering: Environmental Issues. *Temporary Works: Dewatering and Stability*, seminar notes, Institution of Civil Engineers, London, 2nd July.

Chapter 16

The future

by T. O. L. Roberts

16.0 Introduction

Imagine a world in which:

- All dewatering pumps, pipework and well screens are made of steel with virtually no plastic components or flexible pipes.
- Most dewatering projects involve steel wellpoints installed manually, deep wells with submersible borehole pumps are at the cutting edge, and ejectors are only found in textbooks.
- There is little or no regulation or environmental constraints on abstraction or discharge of groundwater.
- Monitoring of flow rates and drawdowns is an expensive luxury only appropriate for high-profile projects.
- Contracts and project management are deliberately confrontational; delayed payment and claims are commonplace.

This was the state of the dewatering industry in the United Kingdom in the 1970s. The earlier chapters have shown that much has changed. This chapter will look briefly at what further changes are likely in the first decades of the twenty first century.

16.1 Techniques

The basic laws of physics govern the mechanical performance of the principal groundwater lowering systems and it is difficult to see how there could be any fundamental improvement in these systems. However developments in materials, plant design and the application of information technology are bound to lead to some improvement in efficiency and reliability over time. Changes are likely to be incremental and should lead to a corresponding reduction in cost. These improvements are generally not driven by the needs of the construction dewatering industry. The components of dewatering systems are largely borrowed from other industries and it is their needs that are driving the changes: the water supply industry for borehole pumps, well screens and ejectors; the piling and grouting industry for drilling systems; and process engineering for centrifugal pumps, exhausters, valves and pipe systems.

Efficiency and reliability improvements are important since groundwater lowering techniques compete in the market place with other civil engineering processes used to control groundwater, such as grouting and artificial ground freezing. However, of greater interest are developments that offer the opportunity to widen the scope and application of dewatering systems. The main potential must lie in the control of pore water pressure in fine-grained low permeability soils. These soils also present a challenge to other techniques such as grouting. The use of ejector systems in the United Kingdom over the last ten years has extended the range of application of dewatering systems to fine-grained soils. There is some evidence from detailed monitoring of dewatering systems in fine-grained soils that there can be a significant variation in the yield from individual wells, installed in a similar manner in apparently similar ground conditions. It is not clear if this is due to some installation effect or undetected variations in the soil fabric local to the wells, or both. There would be significant cost and performance benefits if a system could be devised to ensure that all the wells in a system performed as efficiently as the best performing well. Any solution will probably involve modification to some or all of the current methods of well drilling, well installation, screen specification, filter-pack specification and placement and well development.

Another approach to control of pore water pressure in fine-grained soils is electro-osmosis (see Chapter 11). This was developed and used effectively, particularly in Canada, but is now rarely, if ever, used. Over the last few years there has been renewed interest in electro-osmosis for recovering heavy metal contaminants as a means of soil remediation. The technique is known as electrokinetic or electro-chemical remediation. A practical application of the technique is described by Trombly (1999). Studies have also been made into the use of conductive geotextiles in conjunction with electro-osmosis. Innovations such as these could herald a renaissance in electro-osmosis.

Opportunities for widening the application of groundwater lowering systems are also arising because of changes in the nature of the construction industry and changes in other construction processes. Recent examples include:

- Access constraints. Construction of new and improved infrastructure in cities must be carried out while minimizing disruption to existing services and infrastructure. The effect of this requirement is often to introduce significant access constraints; some can be overcome by installing inclined or horizontal wells or wells installed directly out of tunnels. Hartwell (2001) describes techniques for drilling wells from tunnels, but there remains significant room for improvement.

- Directional drilling and horizontal wells. In the last decade substantial improvements have been achieved in directional drilling techniques designed to allow services to be installed without trench works. These techniques can also be used to install horizontal wells, which could find an application in dewatering systems. Costs are significant, installation of filter packs is tricky or impossible, and development procedures are not established. Nevertheless, there are clear opportunities where surface access is constrained.
- Compressed air. Health and safety regulations mean that there are significant costs and risks associated with working in air pressures above 1 bar. In cases where pressures above 1 bar would normally be required, it has sometimes proved cost-effective to use a dewatering system to lower pore pressures so that working air pressures can be reduced to below 1 bar.
- Pressure relief on retaining walls. Controlling pore water pressures can reduce the need for temporary propping or anchoring of retaining walls. Controlling pore water pressures on the external 'active' side will reduce loading. Lowering groundwater levels on the internal 'passive' side to significantly below excavation level can improve the soil properties and increase the passive resistance.
- Soil remediation. An effective method of removing volatile contaminants from soils is by soil vapour extraction. A high water table prevents movement of vapour and curtails the effectiveness of this technique. Dewatering systems can be used to reduce the water level to improve the performance and reach of the remediation process.
- Landfill leachate. New engineered landfills have built-in leachate drainage systems. However, there are many older landfills where retrofitting of a leachate drainage system is needed. The combination of aggressive chemical conditions, substantial settlement and usually low permeability makes this a demanding environment for dewatering systems. Well systems are used with some successes but there is scope for improving well design and the reliability of the pumping systems.
- Pile construction. Bored piles are often installed in unstable soils below the water table using bentonite to provide temporary borehole support during pile construction. Reducing groundwater levels to below the toe of the piles can have a number of benefits. Less bentonite is used, the bentonite does not require de-sanding, the installation time is some 30 per cent faster, and the load capacity of the finished pile is greater. Whilst substantial drawdowns may be needed, for a large project the savings in cost and time can be significant.

There are probably many more examples of innovation and the ingenuity of engineers will find many more opportunities in the future.

16.2 Impact of information technology systems

The spectacular developments in computing and communications in the last thirty years are likely to continue and possibly accelerate. Despite these developments it remains the case today that the vast majority of small and medium-sized dewatering schemes, with a straightforward geological setting, are designed on the basis of experience, local knowledge and empirical 'rules of thumb'. Two-dimensional or radial steady-state analyses (see Chapter 7) may be undertaken to justify the design. Finite element numerical modelling may be used to look at the regional impact of a large scheme. Numerical modelling may also be used when the geometry of a scheme is not consistent with two-dimensional or radial-flow assumptions, as in the case of flow to a dewatering system under a partially penetrating cut-off. Some of these software models are immensely sophisticated and account for unsaturated flow and negative pore water pressures in two dimensions, three dimensions and four dimensions (i.e. with time-dependent effects). These programmes are moving beyond mere modelling towards 'virtual reality'. Clever interfaces and element building tools allow two-dimensional models to be put together quickly, but three-dimensional and four-dimensional modelling remains too time consuming for most projects. A further shortcoming of three-dimensional models is the inability to model usefully conditions in the immediate vicinity of a well. This is generally overcome by making simplifying assumptions, such as combining several wells into one sink and ignoring any seepage face at the wells. For the purposes of assessing total seepage flows to an excavation, these simplifications are reasonable and the resulting errors in flow estimates are generally small. However, such simplifications effectively prevent any direct assessment of the number, size and depth of dewatering wells required. Furthermore, the Institution of Civil Engineers (1991) note that site investigations are often woefully inadequate and it might be argued there is little to be gained from a sophisticated model with poor input data.

Modelling software has been exploited to some extent as a way of developing accessible design tables or charts for particular geometries or conditions. Clearly a computer model is much less costly than a fully instrumented field test. Good examples are the design charts given by Powrie and Preene (1992) which bridge the gap between radial analysis and two-dimensional plan analysis. A fruitful area of further study would be to establish the scale of the errors resulting from applying two-dimensional or radial analysis to problems that are strictly three-dimensional. This would put the relatively simple two-dimensional and radial analysis on a firmer footing and make it clearer where a three-dimensional model is needed.

Clearly, numerical modelling techniques are having an important impact on the design of dewatering systems. But their application is limited and considerable improvements in modelling software and site investigations will be needed for there to be any serious challenge to the central role of experience, local knowledge and empirical 'rules of thumb' in the design process. In the mean time modelling will continue to be used to support design assumptions (probably by sensitivity analysis), to examine environmental and settlement risks, and as a tool to back-analyse or interpolate monitoring data obtained during the works.

Information technology systems are also making a significant impact on monitoring. Effective monitoring involves linking site sensors (pressure transducers and flow meters) and dataloggers (to read and record the data) with good communications (to relay the data to the end user), data processing and presentation. With these systems in place performance data for a dewatering project can be widely accessible, probably in real time, extending opportunities for the application of the observational approach. The data sets obtained for individual projects may also increase our understanding of the flow regime in the vicinity of well arrays, which may lead to improvements in modelling software. Monitoring systems can also be readily applied to the mechanical performance of the plant (vacuum, headermain pressures, oil/bearing temperatures, power consumption, fuel levels, status of standby plant), which should improve reliability.

16.3 Regulation

The regulatory regime under which we operate in drives changes in working practices and innovation in construction techniques. For groundwater lowering systems environmental regulation and, to a lesser degree, health and safety regulations are of most significance. At present in the United Kingdom construction dewatering is exempt from the need for an abstraction licence or other formal permission. Preene (1999) indicates that this is likely to be changed in the foreseeable future as part of a general increase in the control and regulation of the environmental impact of construction works. It might be thought that this could represent a threat to the use of dewatering systems, but compared to alternative processes the environmental impact of groundwater lowering systems is relatively benign. The energy consumption for dewatering systems is modest, the impact on groundwater flows should be temporary, and only the well screens are left behind on completion (even these could probably be removed with a bit of ingenuity). The well screens can be sealed to avoid the risk of cross-links between aquifers. In comparison, artificial ground freezing has a heavy energy draw and uses freeze agents with questionable environmental credentials. Grouting involves pumping potentially large quantities of environmentally unfriendly material into the ground and leaving it for all time.

16.4 Conclusion

In the future society will continue to demand the construction of new and improved buildings and infrastructure. It is clear that groundwater control systems will continue to have an important role to play in this process. There have been substantial advancements in groundwater lowering operations in the last thirty years, and we may yet find ourselves surprised at the changes in the industry in the next thirty years.

References

Hartwell, D. J. (2001). Getting rid of the water. *Tunnels & Tunnelling International* 40–42.

Institution of Civil Engineers (1991). *Inadequate Site Investigation*. Thomas Telford, London.

Powrie, W. and Preene, M. (1992). Equivalent well analysis of construction dewatering systems. *Géotechnique*, **42**(4), 635–639.

Preene, M. (1999). EA tightens up dewatering rules. *Ground Engineering*, **32**(9), September, 25.

Trombly, J. (1994). Electrochemical remediation takes to the field. *Environmental Science and Technology*, **28**, 289–291.

Notation

A Area; Cross-sectional area of borehole casing
a Calibration coefficient; Length of well array
B Calibration coefficient; Partial penetration factor for radial flow
b Calibration coefficient; Width of well array
C Calibration coefficient
C_d Correction factor for effective stress
c_h Coefficient of consolidation of soil for vertical compression of soil under horizontal drainage
c_v Coefficient of consolidation of soil
D Aquifer thickness; Diameter of borehole test section; Thickness of soil layer
D_{10} Sieve aperture through which 10 per cent of a soil sample will pass
D_{15} Sieve aperture through which 15 per cent of a soil sample will pass
D_{40} Sieve aperture through which 40 per cent of a soil sample will pass
D_{50} Sieve aperture through which 50 per cent of a soil sample will pass
D_{60} Sieve aperture through which 60 per cent of a soil sample will pass
D_{70} Sieve aperture through which 70 per cent of a soil sample will pass
D_{85} Sieve aperture through which 85 per cent of a soil sample will pass
E_0' Stiffness of soil in one-dimensional compression
EC Specific conductivity of water
e Void ratio of soil
F Factor of safety; Shape factor for permeability test in borehole or observation well
f Angularity factor of soil grains
G Geometry shape factor for flow to rectangular equivalent wells in confined aquifers
H Excess head in test section; initial groundwater head
H_c Constant excess head during constant head permeability test
H_0 Excess head in test section at $t=0$
h Total hydraulic head; groundwater head; Maximum drainage path length for vertical drainage
h_w Groundwater head in pumped well or slot

dh	Difference in hydraulic head
$(H - h)$	Drawdown
$(H - h_w)$	Drawdown in a pumped well or slot
i	Hydraulic gradient
i_{crit}	Critical hydraulic gradient
i_{max}	Maximum hydraulic gradient at entry to a well
J	Empirical superposition factor
k	Permeability (also known as coefficient of permeability or hydraulic conductivity)
k_h	Permeability in the horizontal direction
k_v	Permeability in the vertical direction
k_e	Calibration factor between total dissolved solids and specific conductivity
L	Length of borehole test section
L_0	Distance of influence for plane flow
l	Length of flow path
l_w	Wetted screen length of a well
dl	Length of flow path
m_v	Coefficient of volume compressibility of soil
N	Standard penetration test blow count
n	Porosity of a soil or rock; Number of pumped wells
P	Depth of penetration into aquifer of partially penetrating well or slot
Q	Flow rate; Discharge flow rate from an equivalent well or slot
Q_{fp}	Discharge flow rate from a fully penetrating well
Q_{pp}	Discharge flow rate from a partially penetrating well
q	Flow rate; Discharge flow rate from a well
R	Average degree of consolidation of a soil layer
R_0	Radius of influence for radial flow
r	Distance from the test well to an observation well; Radial distance from a well; Radius of a well borehole
r_e	Equivalent radius of a well array
S	Storage coefficient of aquifer; Specific surface of soil grains
S_y	Specific yield of unconfined aquifer
s	Drawdown
s_w	Drawdown at a pumped well
Δs	Change in drawdown per log cycle
T	Transmissivity of aquifer; Basic time lag for permeability test in borehole or observation well
T_r	Time factor for radial groundwater flow
T_v	Time factor for vertical drainage
T_{50}	Time factor for 50 per cent dissipation of pore water pressure
TDS	Total dissolved solids
t	Time since pumping began; Time since permeability test started

t_o	Time at which a straight line through the observation well data intercepts the zero drawdown line in a Cooper–Jacob time-drawdown plot
t_{50}	Time for 50 per cent dissipation of pore water pressure
U	Degree of dissipation of pore water pressure; Uniformity coefficient
u	Pore water pressure; Argument of Theis well function
u_i	Pore water pressure at the start of a permeability test
u_o	Equilibrium pore water pressure
u_t	Pore water pressure at time t
Δu	Reduction in pore water pressure
$W(u)$	Theis well function
x	Linear distance from a slot; Length of a pumped slot; Fraction of the total mass of a soil sample
z	Depth below the water table in an unconfined aquifer
α	Internal angle of V-notch in thin plate weir
ϕ'	Angle of shearing resistance of soil
γ_s	Unit weight of soil
γ_w	Unit weight of water
λ	Partial penetration factor for confined slots
ρ_{corr}	Corrected compression of a soil layer
ρ_t	Effective compression of a soil layer at time t
ρ_{total}	Ground settlement
ρ_{ult}	Ultimate compression of a soil layer
σ	Total stress
σ'	Effective stress
τ	Shear strength of soil
τ_f	Shear strength of soil at failure

Glossary

Abstraction The pumping or removal of groundwater (such as from a well or sump) or surface water (such as from a lake or river).

Air lift A means of pumping water from a well using compressed air to aerate the water so that the air-water mixture rises to the surface. The method is capable of pumping silt and sand with the water, and is often used as a method of *well development*.

Anisotropic Having one or more physical properties that vary with the direction of measurement – the converse of *isotropic*.

Aquiclude Soil or rock forming a stratum, group of strata or part of stratum of very low permeability, which acts as an effective barrier to groundwater flow. This term is obsolete in American terminology, where confining bed is the equivalent term.

Aquifer Soil or rock forming a stratum, group of strata or part of stratum from which water in usable quantities can be *abstracted*. By definition an aquifer will be water-bearing (i.e. saturated and permeable).

Aquitard Soil or rock forming a stratum, group of strata or part of stratum of intermediate to low permeability which yields only very small groundwater flows, or allows groundwater to pass through very slowly.

Artificial ground freezing The process of reducing the temperature of the ground sufficiently to form a very low permeability freezewall of frozen soil. The freezewall can be used as a cut-off and support structure for below ground works.

Artificial recharge The deliberate re-injection of water (via pits, trenches or wells) into aquifers or aquitards. Sometimes used as a means of reducing drawdowns in the aquifer outside an excavation to minimize *consolidation settlements*.

Barrier boundary An aquifer boundary that is not a source of water.

Base heave Uplift or heave of the floor of an excavation as a result of unrelieved pore water pressures at depth.

Biofouling The clogging or deterioration of performance of wells, pumps and pipework as a result of bacterial growth.

Boil The turbulent jet of soil and water that rises to the surface around a placing tube when a wellpoint or well is installed by *jetting*. Also used to describe an uncontrolled upward seepage into an excavation that may wash *fines* out of the soil, leading to instability.

Borehole Strictly, a borehole is a hole drilled into the ground for any purpose, including site investigation boreholes. In groundwater terminology a borehole is often taken to mean a relatively small diameter *well*, particularly one used for water supply.

Capillary saturated zone The zone immediately above the *water table* in an unconfined aquifer where the soil remains *saturated* at negative (i.e. less than atmospheric) pore water pressures. The water in this zone is continuous with the pore water below the water table. The height of the capillary saturated zone is greater in finer-grained soils than in coarse-grained soils. The zone rises or falls with any variations in the water table.

Coefficient of permeability See *permeability*.

Cofferdam A temporary retaining wall structure used to exclude groundwater and surface water from an excavation.

Cone of depression A depression in the *water table* or *piezometric level* that, in theoretically idealized conditions, has the shape of an inverted cone and develops radially around a well from which water is being abstracted. It defines the *radius of influence* of a well.

Confined aquifer An aquifer in which the groundwater is isolated from the atmosphere by an overlying (or confining) aquiclude, and where the *piezometric level* is above the top of the aquifer. A confined aquifer is saturated throughout. Also known as an artesian or sub-artesian aquifer.

Consolidation settlements Ground settlements resulting from increases in vertical *effective stress* when groundwater levels are lowered or pore water pressures are reduced.

Construction dewatering See *dewatering*.

Controlled waters A definition used in environmental legislation in England and Wales to describe all surface water, lakes, seas and all groundwater.

Darcian velocity The mean flow velocity across a unit area as defined by *Darcy's law*. When divided by the *porosity*, the mean groundwater flow rate through the pores may be calculated.

Darcy's law The expression, developed by Henri Darcy in 1856, relating the flow rate through a porous medium to the *hydraulic gradient*, cross-sectional area of flow, and the *permeability* of the medium.

Deep well A groundwater abstraction well for the purpose of groundwater lowering (as opposed to a well for the purpose of water supply, which is termed a *borehole*).

Derogation The reduction in yield or other adverse effect on a water supply *borehole* or *spring* as a result of groundwater pumping by others.

Desk study The review of all available information (such as maps, aerial photographs, historical records) as part of the *site investigation* process.

Development See *well development.*

Dewatering A colloquialism for *groundwater lowering* that originated in America and is now widely used in Europe.

Dipmeter A portable device for measuring the depth to water in a *well, borehole, standpipe* or *piezometer.* Also known as a dipper or water level indicator.

Discharge The flow of water from a pump or a groundwater lowering system.

Drawdown The lowering of the *water table* (in unconfined aquifers) or the *piezometric level* (in confined aquifers) as a result of the abstraction of groundwater. It is measured as the vertical distance between the original water table (or piezometric level) and the current level during pumping.

Drift Geologically recent superficial strata of sand, gravel silt or clay (see also *soil*). Excavations below the groundwater level in drift deposits may be unstable without adequate *groundwater control.*

Eductor See *ejector.*

Effective stress The difference between the total stress (due to self weight and any external loading) and the *pore water pressure* at a point in a soil mass. The effective stress is a vital controlling factor in the strength and compressibility behaviour of soils.

Ejector A water jet pump which pumps water (and air) via a nozzle and venturi arrangement. When used in low yielding wells a vacuum can be developed if the top of the well casing is sealed.

Ejector well A small diameter groundwater lowering well pumped by an *ejector.*

Electro-osmosis A rarely used method of *pore water pressure control* applicable to very low permeability soils such as silts and clays. An electric potential is applied across the area of soil to induce groundwater flow from the anode to the cathode.

Equivalent well A conceptual large well, used to represent a groundwater lowering system that consists of a ring of many wells. By considering the real system as an equivalent well, simpler and more accessible methods of analysis can be used.

Filter pack Sand or gravel placed around a *well screen* to act as a filter. This allows water to freely enter the well, while preventing movement of *fines* toward the well.

Fines The smaller particles in a soil stratum or analysis. Generally taken to refer to clay, silt and fine sand-size particles. See *loss of fines.*

Fissures Natural cracks, fractures, joints or discontinuities in *rock* that allow groundwater to pass. The extent and characteristics of fissures greatly influences the permeability of most rocks.

Flow, steady-state The flow regime when the magnitude and direction of the groundwater flow rate are constant.

Flow, transient-state The flow regime when the magnitude or direction of the groundwater flow rate is changing with time.

Flowing artesian conditions A special case of *confined aquifers*, where the piezometric level of the aquifer is above ground level. A well drilled through the confining bed into the aquifer will overflow at ground level without the need for pumping.

Formation level The final dig level of an excavation, after all digging and filling, but before any concreting or construction.

Fully penetrating The case where a well penetrates the full thickness of an aquifer – the converse of *partially penetrating*.

Ground investigation The physical investigation of a site as part of the *site investigation* process. Ground investigation includes drilling and probing, in-site tests and laboratory testing.

Groundwater Water contained in and flowing through the pores and fissures of soil and rock in the *saturated zone*. In this zone the water body is essentially continuous, except for an occasional bubble of air. In legal documents groundwater is sometimes described as water in underground strata.

Groundwater control Methods and techniques to control groundwater to allow stable excavations to be formed for construction purposes. The two main approaches are *groundwater exclusion* and *groundwater lowering*.

Groundwater exclusion The construction of artificial barriers to prevent groundwater from flowing into an excavation. The barriers can be formed from physical cut-off walls, or by reducing the permeability of the in-situ soil and rock. See also *cofferdam*.

Groundwater lowering The temporary reduction of groundwater levels and pore water pressures around and below an excavation by some form of pumping, sufficient enough to enable excavation and foundation construction to be carried out safely and expeditiously in water-bearing ground.

Heterogeneous Non-uniform in structure or composition or composed of diverse elements – the converse of *homogeneous*.

Homogeneous Uniform in structure or composition – the converse of *heterogeneous*.

Hydraulic conductivity The hydrogeological term for *permeability*. In general, engineering literature uses permeability, while hydrogeological references use hydraulic conductivity.

Hydraulic gradient The change in *total hydraulic head* between two points, divided by the length of flow path between the points.

Hydrogeology The study of groundwater or underground waters. Also known as groundwater hydrology.

Hydrological cycle The interlinked processes by which water is circulated from the oceans to the atmosphere to the ground surface and thence returned to the oceans as surface water and groundwater flow.

Hydrology The study of water. In a civil engineering context, generally taken to be the study of surface waters.

Isotropic Having the same physical properties in all directions – the converse of *anisotropic*.

Jetting A method used to install wellpoints and, less commonly, deep wells or ejectors. A jet of high pressure water is used to allow the penetration of a steel placing tube into the ground, within which the wellpoint or well is installed, following which the placing tube is removed. The jetting water returns to the surface as a *boil*, bringing with it the displaced soil.

Laminar flow A flow regime where flow is steady and continuous without any turbulence. It is characteristic of most groundwater flow through porous media and finely fissured rock formations – the converse of *turbulent flow*.

Leaky aquifer An aquifer confined by a low permeability *aquitard*. When pore water pressures in the aquifer are lowered by pumping, water will flow (or leak) from the aquitard and recharge the aquifer. Also known as a semi-confined aquifer.

Loss of fines The uncontrolled movement of *fines* carried by groundwater flow toward a well, sump or soil face where filters are inadequate or absent. Also used to describe the washing of fines from samples recovered from water filled boreholes during site investigation drilling.

Numerical model A method of analysing a groundwater conceptual model using computer software. The elements of the conceptual model are expressed as the geometry, boundary conditions and numerical values of the problem. The resulting equations are solved numerically (often by iteration) using the software, generally run on a desktop computer.

Observation well An instrument (a *standpipe, piezometer* or unpumped *well*) installed into the ground in a selected location for the purpose of measuring the level of the *water table, piezometric level* or *pore water pressure*.

Open area The proportion of the surface area of a *well screen* that is perforated or slotted, and allows water to pass. Open area is normally expressed as a percentage.

Overbleed A commonly used colloquial term to describe residual groundwater seepages into an excavation, as might occur when a *perched water table* is penetrated. Overbleed normally describes low rates of seepage that are a nuisance, rather than causing significant flooding. Nevertheless, even at small rates of seepage, if *fines* are being continuously transported by the seepage, this should be counteracted immediately. The guiding principle is: do not try to stop the flow of seepage water as this will cause a build up of pore pressures but do prevent continuous *loss of fines*, perhaps by placing a suitable filter.

Partially penetrating The case where a well does not penetrate the full thickness of an aquifer – the converse of *fully penetrating*.

Particle size distribution The relative percentages by dry weight of the different particle sizes of a soil sample, determined in the laboratory by mechanical analysis (e.g. by sieving). Also known as PSD, soil grading, sieve analysis.

Perched water table Water trapped (or 'perched') in an isolated saturated zone above the water table. Perched water tables may exist naturally or may be caused when groundwater levels are lowered by dewatering; very low permeability silt or clay layers will inhibit downward seepage of groundwater, trapping water above them. When an excavation is dug through a perched water table, water will enter the excavation as residual seepage or *overbleed*.

Permeability A measure of the ease with which water can flow through soil or rock. Also known as coefficient of permeability or hydraulic conductivity.

Phreatic level The level at which pore water pressure is zero (i.e. equal to atmospheric pressure), and thus is at the base of the *capillary saturated zone*.

Piezometer An instrument, installed into the ground to act as an *observation well* to measure the pore water pressure or piezometric level in a specific stratum, layer or elevation. This is achieved by having only the relevant section of the borehole exposed to the ground, the remainder of the borehole being sealed with grout.

Piezometric level The level representing the *total hydraulic head* of groundwater in a *confined aquifer*. Also known as potentiometric surface.

Plane flow A two-dimensional flow regime in which flow occurs in a series of parallel planes. This type of flow occurs toward a pumped slot or long line of closely spaced wellpoints, when the direction of flow is perpendicular to the slot or line of wellpoints.

Pore water pressure The pressure of groundwater in the pores of a soil, measured relative to atmospheric pressure. Positive and negative pore water pressures are greater and less than atmospheric respectively.

Pore water pressure control The application of *groundwater lowering* to low permeability soils. The fine-grained nature of such soils means that relatively little water drains from the pores, and desaturation does not occur. Nevertheless, pumping from vacuum-assisted wells can reduce *pore water pressures*, thereby controlling effective stress levels and preventing instability.

Porosity The ratio of the volume of voids (or pore space) to the total volume of a soil sample or mass.

Pre-drainage The methods which lower groundwater levels in advance of excavation. Pre-drainage methods include *wellpoints*, *deep wells* and *ejectors*. The converse of pre-drainage is *sump pumping*, where water is allowed to enter the excavation, from whence it is removed.

Pumping test A method of *in situ* permeability testing involving controlled pumping from a well while the *discharge* flow rate and *drawdown* in the aquifer are recorded.

Radial flow A two-dimensional flow regime in which the flow occurs in planes, which converge on an axis of radial symmetry. This type of flow occurs toward an individual well or sump, when flow lines converge as they approach the well.

Radius of influence The radial distance outward from the centre of the pumping well to the point where there is no lowering of the water table or reduction of the piezometric level – the edge of the *cone of depression*. Also known as the distance of influence.

Recharge Water which flows into an aquifer to increase or maintain the quantity of groundwater stored therein. Recharge may be from several sources: infiltration from surface waters; precipitation; seepage from other aquifers or aquitards; or *artificial recharge*.

Relief well A well installed within an excavation to act as a preferential pathway for flow to relieve any *pore water pressures* trapped beneath the excavation. As the excavation is deepened the well will overflow, relieving pressures. Also known as a bleed well.

Rock A geological deposit formed from mineral grains or crystals cemented together – this is distinct from uncemented *soil* or *drift*.

Running sand A colloquial term for the conditions when saturated granular soils are so unstable that they are unable to support a cut face or slope and become an almost liquid slurry. Running sand is not a type or property of the material, it is a condition in which a granular soil can exist under unfavourable seepage conditions. Effective *groundwater lowering* can change running sand into a stable and workable material.

Saline intrusion The movement of saline water into a fresh water aquifer as a result of hydraulic gradients. Problems of saline intrusion can be created or exacerbated as a result of pumping from wells.

Saturated The condition when all the pores and fissures of a soil or rock are completely filled with water – the converse of *unsaturated*.

Saturated zone The zone of *saturated* soil or rock, all parts of which are in hydraulic connection with each other. In a confined aquifer the saturated zone comprises the entire aquifer. In an unconfined aquifer the saturated zone is the area below the water table, plus the *capillary saturated zone* above the water table.

Settlement lagoon A large lagoon or pit through which the pumped discharge is passed to settle out sand and silt-sized particles prior to disposal.

Settlement tank A small self-contained tank (often containing baffles) through which the pumped discharge is passed to settle out sand-sized particles prior to disposal. Such relatively small tanks are ineffective at settling out silt-sized particles, due to the short retention time.

Shoestring A colloquial term used to describe a narrow, but perhaps extensive, lens of soil markedly different from the stratum of soil within which it is encompassed. The term is usually applied to a lens of permeability significantly greater than the surrounding soil.

Site investigation The overall process of obtaining all relevant information about a site, to allow the design and construction of a civil engineering or building project. Elements of site investigation include the *desk study* and the *ground investigation*.

Slimes A mining term used to describe the residual waste products remaining after washing mined mineral extracts. Sometimes referred to as washings or tailings.

Soil The term used in civil engineering to describe uncemented deposits of mineral (and occasionally organic) particles such as gravel, sand, silt and clay – this is distinct from cemented *rock*. Excavations below the groundwater level in soils may be unstable with adequate *groundwater control*. See also *drift*.

Specific yield The storage coefficient in an unconfined aquifer.

Spring A natural outflow of groundwater at the ground surface. In rural areas springs may be used as water supplies.

Standpipe An instrument, typically consisting of a perforated casing or pipe, installed into the ground to act as an *observation well*. A standpipe is open to water inflows from all strata that it penetrates, and may give a water level reading that is a hybrid of levels in different aquifers. A *piezometer* should be used to determine water levels or pore water pressures in a specific stratum.

Storage coefficient The volume of water released by gravity drainage from a volume of aquifer. Normally expressed as a dimensionless ratio or percentage. Also known as storativity.

Storage release The release of water from storage in an aquifer. During the early stages of pumping (the first few hours or days) storage release can significantly increase the discharge flow rate above the steady-state value. Storage release is most significant in unconfined aquifers of high permeability.

Stratum A geological term for a layer of any deposited soils or rock. The plural of stratum is strata.

Sub-drain A perforated or open-jointed pipe sometimes found immediately beneath sewers or pipelines that were laid in trenches. During construction the sub-drain is laid in the trench just ahead of the main pipe, to carry water away from the working area.

Submersible pump A pump designed to operate wholly or partly submerged in water. The most common forms of submersible pumps are electrically powered. Special slimline submersible pumps are suitable for use in *wells* or *boreholes*.

Sump A pit, usually located within an excavation, in which water collects prior to being pumped away.

Sump pump A robust pump, capable of handling solids-laden water, used to pump groundwater and surface water from sumps.

Sump pumping A method of *groundwater lowering* that allows water to enter the excavation, from whence it is pumped away. In some circumstances the flow of water into the excavation may have a destabilizing effect, and the use of *pre-drainage* methods (in preference to sump pumping) may reduce the risk of instability.

Surface water Water contained in rivers, streams, lakes and ponds.

Total hydraulic head The potential energy of water due to its height above a given level. The total head controls the height to which water will rise in a piezometer. The total head at a given point is the sum of the elevation head (i.e. the height of the point above an arbitrary datum) and the pressure head (i.e. the height of water above the point recorded in a standpipe piezometer). Also known as total head or total hydraulic potential.

Transmissivity A term used in *hydrogeology* to describe the ease with which water can flow through the saturated thickness of the aquifer. Transmissivity is equal to the product of *permeability* and saturated aquifer thickness.

Turbulent flow A flow regime where the flow lines are confused and heterogeneously mixed. It is typical of flow in some surface water bodies – the converse of *laminar flow*.

Unconfined aquifer An aquifer whose upper surface is not confined by an overlying impermeable stratum and whose upper surface is directly exposed to atmospheric pressure. Also known as a water table aquifer.

Underdrainage A method of *groundwater lowering* that uses soil layering to advantage by pumping directly from a deeper, more permeable stratum, in preference to the overlying lower permeability layers.

Pumping from the deeper layer promotes vertical downward drainage from the lower permeability layer into the stratum being pumped.

Uniformly-graded Containing a narrow range of particle sizes, so that most of the particles are of similar size – the converse of *well-graded*. Also known as poorly-graded.

Unsaturated The condition when the pores and fissures of a soil or rock are not completely filled with water, and may contain some air or other gases – the converse of *saturated*.

Unsaturated zone The zone of *unsaturated* soil or rock. In an unconfined aquifer the saturated zone is the area above the *capillary saturated zone*. Within the unsaturated zone the forces of surface tension render the water pressure negative (i.e. lower than atmospheric).

Vadose zone See *unsaturated zone*.

Void ratio The ratio of the volume of voids to the volume of solids in a soil mass.

V-notch weir A thin plate *weir* where a V-shaped notch of a specified angle is cut. Calibration charts allow the flow rate to be estimated from measurements of the height of water flowing over the weir.

Water table The level in an unconfined aquifer at which the pore water pressure is zero (i.e. atmospheric). See also *phreatic level*.

Weir A structure used to control the flow of water, often so it can be measured. See also *V-notch weir*.

Well A hole sunk into the ground for the purposes of abstracting water. Wells for groundwater lowering purposes are generally categorized by their method of pumping as *deep wells, ejector wells*, or *wellpoints*. In water supply terminology, a well is often taken to mean a large diameter shaft, as may be dug by hand in developing countries. A smaller diameter well, constructed by a drilling rig, is termed a *borehole*.

Well casing The unperforated section of the *well liner*, installed at depths where any groundwater present is to be excluded from the well (e.g. if several aquifers exist, but drawdown is not required in all of them). Also known as plain casing or solid casing.

Well development Increasing or maximizing the yield of a well by removing drilling residue and fine particles from the well and the aquifer immediately outside the well. Normally carried out following well construction, but prior to the installation of pumping equipment.

Well liner A generic term for *well casing* and *well screen*.

Well loss The head loss (or additional drawdown inside the well) that occurs when water flows from the aquifer, through the *well screen* and *filter pack* and into the well itself.

Well screen The perforated or slotted section of the well liner installed to allow water to enter a well where it penetrates an *aquifer*. Also known as perforated casing or slotted casing.

Well-graded Containing a wide range of particle sizes – the converse of *uniformly-graded*. A soil formed from a mixture of various proportions of gravel, sand, silt and clay-sized particles would be described as well-graded.

Wellpoint A small diameter shallow well for groundwater lowering purposes. Normally installed by jetting, and pumped by a suction pump of some kind.

Yield The discharge flow rate of a well. The yield is generally taken to be controlled by the well, rather than the pump – a high yielding well would produce a low flow rate if an under-sized pump were installed. Also known as well yield.

Conversion factors

Length

In the metric system: 1 km = 1000 m; 1 m = 100 cm; 1 m = 1000 mm.

Conversion from metric to imperial			*Conversion from imperial to metric*		
from	*to*	*multiply by*	*from*	*to*	*multiply by*
km	mile	0.622	mile	km	1.609
m	yd	1.094	yd	m	0.914
m	ft	3.281	ft	m	0.305
cm	in	0.394	in	cm	2.54
mm	in	0.0394	in	mm	25.4

Volume

In the metric system: $1\,m^3 = 1000$ litres; $1\,m^3 = 10^6\,cm^3$; $1\,cm^3 = 1$ ml.

Conversion from metric to imperial			*Conversion from imperial to metric*		
from	*to*	*multiply by*	*from*	*to*	*multiply by*
m^3	cu yd	1.309	cu yd	m^3	0.764
m^3	cu ft	35.32	cu ft	m^3	0.028
m^3	gal (UK)	220	gal (UK)	m^3	0.0045
m^3	gal (US)	264	gal (US)	m^3	0.0038
litre	gal (UK)	0.220	gal (UK)	litre	4.546
litre	gal (US)	0.264	gal (US)	litre	3.785
litre	pint (UK)	1.76	pint (UK)	litre	0.568
litre	pint (US)	2.11	pint (US)	litre	0.473

Flow rate

In the metric system: 1 m^3/s = 1000 l/s; 1 m^3/hr = 3.6 l/s; 1 m^3/day = 0.0115 l/s.

Conversion from metric to imperial			*Conversion from imperial to metric*		
from	*to*	*multiply by*	*from*	*to*	*multiply by*
m^3/day	gal/day (UK)	220	gal/day (UK)	m^3/day	0.0045
m^3/day	gal/day (US)	264	gal/day (US)	m^3/day	0.0038
m^3/hr	gal/hr (UK)	220	gal/hr (UK)	m^3/hr	0.0045
m^3/hr	gal/hr (US)	264	gal/hr (US)	m^3/hr	0.0038
l/s	gal/min (UK)	13.2	gal/min (UK)	l/s	0.076
l/s	gal/min (US)	15.85	gal/min (US)	l/s	0.063

Permeability

In the metric system: 1 m/s = 8.64×10^4 m/day; 1 m/s = 3.16×10^7 m/year.

Conversion from metric to imperial			*Conversion from imperial to metric*		
from	*to*	*multiply by*	*from*	*to*	*multiply by*
m/s	ft/day	2.83×10^5	ft/day	m/s	3.53×10^{-6}
m/day	ft/day	3.281	ft/day	m/day	0.305
m/year	ft/year	3.281	ft/year	m/year	0.305

Pressure and head

In the metric system: 1 m H_2O = 9.789 kPa; 1 bar = 100 kPa; 1 bar = 750 mm Hg.

Conversion from metric to imperial			*Conversion from imperial to metric*		
from	*to*	*multiply by*	*from*	*to*	*multiply by*
mm Hg	in Hg	0.0394	in Hg	mm Hg	25.4
m H_2O	in Hg	2.896	in Hg	m H_2O	0.345
m H_2O	ft H_2O	3.281	ft H_2O	m H_2O	0.305
m H_2O	psi	1.42	psi	m H_2O	0.703
kPa	psi	0.1451	psi	kPa	6.884
bar	in Hg	29.551	in Hg	bar	0.0339
bar	ft H_2O	33.48	ft H_2O	bar	0.0299
bar	psi	14.51	psi	bar	0.0689

Abbreviations used

km	kilometre
m	metre
cm	centimetre
mm	millimetre
l	litre
ml	millilitre
h	hour
s	second
kPa	kilopascal
mm Hg	millimetres of mercury
m H_2O	metre head of water
yd	yard
ft	foot
in	inch
gal	gallon
in Hg	inches of mercury
ft H_2O	foot head of water
psi	pounds per square inch

Index